Einstein, Bohr and the Quantum Dilemma
From Quantum Theory to Quantum Information
Second Edition

Quantum theory, the most successful physical theory of all time, provoked intense debate between the twentieth century's two greatest physicists, Niels Bohr and Albert Einstein. Quantum information theory has emerged from intensive study of the structure and interpretation of quantum theory to become one of the fastest growing areas of twenty-first-century science. This study was stimulated by the seminal analysis of John Bell in the 1960s, but behind Bell lay the intensive debate between Niels Bohr and Albert Einstein which raged in the 1920s and 1930s. The debate concerned the nature of quantum theory, and the major contradictions and conceptual problems at its heart.

This second edition contains sympathetic accounts of the views of both Bohr and Einstein, and a thorough study of the argument between them. It includes non-technical and non-mathematical accounts of the development of quantum theory and relativity, and also the work of David Bohm and John Bell that restored interest in Einstein's views. It has been extensively revised and updated to cover recent developments, and the account of ongoing work has been brought up to date. A new chapter is devoted to describing the whole area of quantum information theory, from the work of Richard Feynman and David Deutsch that initiated the study of quantum computation to the theoretical and experimental approach to quantum cryptography.

This book provides a fascinating account of the development of quantum theory, which will appeal to anyone with an interest in the fundamental questions of physics, its philosophy and its history.

ANDREW WHITAKER is Professor of Physics at the Queen's University of Belfast, where he has worked since 1988. His main research has focussed on the foundations of quantum theory, particularly interpretations of quantum theory and the quantum Zeno effect. He also has a keen interest in the history and philosophy of quantum theory, and in particular the life and works of John Bell, on whom he has written several articles. He is co-editor with Mark McCartney of the book *Physicists of Ireland: Passion and Precision*.

Einstein, Bohr and the Quantum Dilemma

From Quantum Theory to Quantum Information

Second Edition

ANDREW WHITAKER

Department of Pure and Applied Physics
Queen's University Belfast

CAMBRIDGE
UNIVERSITY PRESS

CAMBRIDGE
UNIVERSITY PRESS

University Printing House, Cambridge CB2 8BS, United Kingdom

One Liberty Plaza, 20th Floor, New York, NY 10006, USA

477 Williamstown Road, Port Melbourne, VIC 3207, Australia

4843/24, 2nd Floor, Ansari Road, Daryaganj, Delhi - 110002, India

79 Anson Road, #06-04/06, Singapore 079906

Cambridge University Press is part of the University of Cambridge.

It furthers the University's mission by disseminating knowledge in the pursuit of education, learning and research at the highest international levels of excellence.

www.cambridge.org
Information on this title: www.cambridge.org/9780521671026

© A. Whitaker 1996, 2006

First published 1996; second edition 2006

A catalogue record for this publication is available from the British Library

ISBN 978-0-521-67102-6 Paperback

for Joan

Contents

Preface

At the beginning of the twenty-first century, quantum information theory is an exceptionally lively area of scientific research. Quantum computation and quantum cryptography promise powerful and efficient methods way beyond the scope of classical physics for solving practical problems. Quantum teleportation suggests the even more exciting possibility of achievements that would not even be dreamed of classically, and as minds adapt to new ways of thinking we must expect many other examples.

Quantum information theory emerged from a rich brew of scientific ideas that developed particularly in the 1980s and early 1990s. The ideas centred around the idea of quantum entanglement – the fact that, when two quantum particles have once interacted, even after separation their properties will be mutually dependent in a totally non-classical way. Crucially important were the theoretical analysis and experimental testing of the so-called Bell inequalities, the brain-child of John Bell, a physicist from Ireland.

Bell's main work and the initial experiments were performed in the 1960s and 1970s, respectively, and behind him was the famous Bohr–Einstein debate of the 1920s and 1930s. The Bohr–Einstein debate occurred at a critical point in the intellectual history of the twentieth century. By 1926, the 'new' quantum theory of Heisenberg and Schrödinger promised to provide the exact theoretical basis for the physics of atoms, but important questions remained about fundamental aspects of the theory.

Niels Bohr's approach implied the abandonment of *determinism*; in absolute contrast to Newtonian physics, complete knowledge of the present may provide only statistical information about the future. Also abandoned would be *realism*, at least in the form of *naive realism*, according to which any physical quantity – position, speed and so on – had a precise value at all times. Einstein was not prepared to go along with these revolutionary changes, so the great debate ensued during the next decade.

The debate was not so much about the contents of any theory, as about what a scientific theory ought to be; it was not *just* about the nature of the Universe, but about what kind of description of the Universe should be regarded as meaningful. It may not be over-dramatic to call it a battle for the *soul* of physics.

The practically universal view at the time that Bohr won the debate played a large part in Einstein, perhaps the most famous scientist ever, becoming practically a scientific recluse for the last quarter of a century of his life. Yet 50 years after Einstein's death, and thanks largely to the work of David Bohm and Bell himself, Einstein's views are being taken very seriously. Questions that seemed to have been settled 70 or 80 years ago are back under very serious consideration, and experimental techniques that Bohr would never have dreamed of have been perfected, many working with individual atomic particles. The Bohr–Einstein debate is very much alive!

In this book, after a brief discussion of Einstein and Bohr in Chapter 1, I give rather a full account of classical physics – physics before relativity and quantum theory – in Chapter 2; practically every part of classical physics plays a role in understanding quantum theory. Chapter 3 contains a brief account of relativity, which has a two-fold importance in the rest of the book. Relativity is at the heart of the crucial experiments suggested by Bell to test important aspects of quantum theory. More fundamentally, though, I argue that the mathematical sophistication of Einstein's theory of general relativity, coupled with its enormous importance, led to Einstein developing his view that quantum theory itself could progress only as a special case of a new highly mathematical and complex theory, a theory he was never able to produce.

Chapter 4 gives an account of the development of quantum theory, together with some idea of its basic structure and its many successes. It also introduces the initial conceptual difficulties brought up by the theories of Heisenberg and Schrödinger.

Chapters 5 and 6 describe Bohr's solution to these difficulties – his framework of *complementarity*, and Einstein's arguments against Bohr in the debate itself. They include the famous Einstein–Podolsky–Rosen argument. These chapters form the centre-piece of the book. Chapter 7 discusses the work of Bohm and Bell, which re-invigorated and extended the original debate, and played a major part in the creation of quantum information theory. Chapter 8 gives a brief account of some recent developments, theoretical and experimental, in the fundamental aspects of quantum theory, paying special attention to their relevance for the Bohr–Einstein debate.

Throughout the earlier part of the book I stress the manner in which the intellectual struggles over quantum theory led on to quantum information theory, and in Chapter 9 I reach the theory itself. Following a broad overview of the classical information theory and theory of computation, I study the work centred on Richard Feynman that introduced many of the ideas and developments required for quantum information theory, but did not quite manage to put them together in a coherent way. In particular I explain the idea of the reversible computer.

I then reach the work of David Deutsch, whom I consider to be the real father of quantum computation, the person who realised that the classical foundations of computer theory given by Alan Turing had to be rewritten from scratch to produce the new field of quantum computing. I sketch some of the algorithms that show clearly that quantum computation can achieve more than classical computation, the Deutsch and Grover algorithms, and in particular the all-important Shor algorithm.

Many ways of implementing the quantum computer have been studied; I briefly describe two of the most important, nuclear magnetic resonance and the ion trap. I then sketch the two most important areas of quantum information theory apart from quantum computation, namely quantum cryptography (or quantum key distribution) and quantum teleportation.

Lastly, in Chapter 10 I give a brief assessment of the work of Bohr and Einstein, particularly studying to what extent their particular perspectives encouraged Bell's own developments and the emergence of quantum information theory.

In the early chapters, references are given mainly to books, and wherever possible they are chosen to be non-technical. For the later chapters, though, and in particular Chapters 7 and 8 where current developments are being discussed, inevitably the references are nearly all to recent papers, often fairly technical, in the scientific literature. I might mention that I have perhaps over-represented papers by myself and collaborators. I make no apology for this since it may help to draw the attention of readers to accounts that may be more technical than those in the book, but are similar in approach, and so may serve as useful bridges to the more general literature.

The style of the book is as non-mathematical as possible; there are few actual equations, but there is a little reasonably simple mathematics at some points in the text; the only alternative would have been to paraphrase it, which would have been contrived, and probably harder to get to grips with. I have made every effort to avoid including too many technicalities, and the book should be accessible to anybody who is, or wishes to become, interested in quantum theory and its interpretation.

The expert may complain about *technical* simplification – about almost total restriction to the Schrödinger formalism, leading to discussion of a 'spin wavefunction', and occasional use of the term 'state' in a non-technical manner.

However, I feel that there has been no requirement for watering-down the *conceptual* arguments. The reader should need no expertise in mathematics or previous knowledge of physics to obtain an understanding not only of the main conceptual factors involved in discussing quantum theory and quantum information theory, but also of the disagreements which still exist.

As is almost inevitable, much of the content of the book is stated in historical terms, and it is impossible to avoid certain philosophical allusions, but readers should be warned or promised that the book is fundamentally on physics, and in the end it is the *physical* arguments which I hope they come to appreciate. I have tried to give the main arguments as much as possible in the actual words of the various participants; I would like to thank all those who have given permission for their writings to be quoted and, in particular, those who have allowed reproduction of diagrams, and in some cases provided suitable copies.

For the benefit of those who have studied the previous edition of this book, *Einstein, Bohr and the Quantum Dilemma*, published in 1996, I will explain briefly the changes introduced in the current book. Foremost, of course, is the entire new chapter on quantum information theory, Chapter 9, though I have also tried throughout the previous chapters to indicate the sources for quantum information theory in the earlier work. In 1996 quantum information theory was less a lusty infant than a babe in arms, and it achieved only the briefest of mentions in the previous edition.

The other main area of progress in the last decade has been the large amount of theoretical and experimental work on the Bell inequalities, and a full account of this work has been included in Chapter 7.

All the various sections of Chapter 8 on modern developments have also been updated. Some of the main additions are the following: the work principally of David Deutsch and David Wallace on modern views of many-worlds interpretations; Kurt Gottfried's reply to Bell's criticisms concerning his approach to measurement; the move from the knowledge interpretation of quantum theory due principally to Rudolf Peierls to the information interpretation of Anton Zeilinger; the contributions of Guido Bacciagaluppi and others to the theory of stochastic interpretations; the passage of the quantum Zeno effect from an abstract and almost despised curiosity to a practically universal component of any description of change of quantum state; and the progress of Tony Leggett towards macroscopic quantum mechanics.

Other parts of book have been updated and revised where appropriate, in particular the discussion of the famous EPR argument in Chapter 6. Overall, of the 567 references in this second edition, about 340 are new to this edition, and of these 230 have been published since the publication of the first edition and 130 date from the year 2000 or later.

I would again like to give my thanks to those with whom I have performed and published work on quantum theory in the past – Dipankar Home, Ishwar Singh, Euan Squires, Lucien Hardy and John Dennison. I have learned much from them. I would like to thank Adrian Kent and Tchavdar Todorov for discussions on the consistent histories and stochastic interpretations of quantum theory, respectively; Martin Lamb for discussions on the Mach–Zehnder interferometer; and Reinhold Bertlmann and Anton Zeilinger for hospitality in Vienna during discussions on the life and work of John Bell, and attendance at the conference of reference [215]. I would like to thank Dipankar Home again, and the S. N. Bose Centre and the Bose Institute, both in Calcutta, for hospitality during the final stages of completing this edition.

I would like to thank Simon Capelin, Vince Higgs, Steve Holt, Wendy Phillips, and others at Cambridge University Press, for their helpfulness and efficiency in the production of this book.

Lastly I would like to renew my thanks to my immediate family – Joan, John and Peter, for their support and tolerance during the preparation of this second edition of the book.

Acknowledgements

I am very grateful to the following publishers for granting me permission to reprint copyright material. Elsevier Science B. V., Amsterdam, the Netherlands and The Boulevard, Langford Lane, Kidlington OX5 1GB, UK (*Studies in the History and Philosophy of Science* (reference [47]) © 1990, *The Framework of Complementarity* (H. J. Folse) [55] © 1985, *Niels Bohr: His Life and Work As Seen by Friends and Colleagues* (S. Rozental, ed.) [74] © 1968, *Niels Bohr - Collected Works*, Vol. 6 (J. Kalchar, ed.) [66] © 1985, *The Lesson of Quantum Theory* (J. de Boer, E. Dal and O. Ulfbeck, eds.) [77] © 1986). Macmillan Press Ltd (*The Born–Einstein Letters* (M. Born, ed.) [5]. Macmillan Magazines Ltd (*Nature*, N. Bohr, Supplement, 14 April 1928, pages 148–58) © 1928). Kluwer Academic Publishers (*The Physicist's Conception of Nature* (J. Mehra, ed., 1973), *Quantum Paradoxes and Physical Reality* (F. Selleri) [140]). Oxford University Press (*Subtle Is the Lord* (A. Pais) [1], J. S. Bell in *Quantum Gravity 2* (C. Isham, R. Penrose and D. Sciama, eds., 1981, pp. 611–37)). Les Editions de Physique (J. S. Bell in *Journal de Physique*, Colloque C2, Supplément au numéro 3, Tôme 42, 1981, pages 41–61). Plenum Publishing Corporation (J. S. Bell in *Sixty-Two Years of Uncertainty* (A. I. Miller, ed., 1990). *Journal of Statistical Physics* [295]). The Nobel Foundation (J. S. Bell in *Proceedings of the Nobel Symposium 65: Possible Worlds in Arts and Science* (S. Allen, ed.) © 1986). John Wiley and Sons Inc. (*The Philosophy of Quantum Mechanics* (M. Jammer) [67] © 1974, *The Conceptual Development of Quantum Mechanics* (M. Jammer) [39] © 1966, *Atomic Physics and Human Knowledge* (N. Bohr) [68] © 1958). McGraw-Hill, Inc., (*Quantum Mechanics* (L. Schiff) [54] © 1955). Faber and Faber (*The Whitsun Weddings* (P. Larkin, 1964), *Larkin at Sixty* (A. Thwaite, ed.) [85]). Farrar, Straus and Giroux (*The Whitsun Weddings* (P. Larkin) (US rights)). Princeton University Press (*Quantum Profiles* (J. Bernstein) [78], *Mathematical Foundations of Quantum Mechanics* (J. von Neumann) [87]). University of Pittsburgh Press (*Frontiers of Science and Philosophy* (R. G. Colodny, ed.) [64] © 1962). University of Chicago Press (*The Shaky Game* (A. Fine) [94] © 1986 by the University of Chicago). American Philosophical Society (*Proceedings of the American*

Philosophical Society **124**, 1980, 323–45 (translation by J. D. Trimmer) [100]). Open Court Publishing Company, Chicago and LaSalle, Illinois (*Library of the Living Philosophers*, Vol. VII, *Albert Einstein, Philosopher-Scientist* (P. A. Schilpp, ed.) pp. 85, 89, 668, 674, 681 [73]). International Thomson Publishing Services Ltd (*Quantum Implications* (B. J. Hiley and D. Peat, eds.) [127]). World Scientific Publishing Co. Pte. Ltd (*Symposium on the Foundations of Modern Physics* 1985 (P. Lahti and P. Mittelstaedt, eds.) [104], *Symposia on the Foundations of Modern Physics* 1993 (K. V. Laurikainen and C. Montonen, eds.) [271], *Stochastic Evolution of Quantum States in Open Systems and Measurement Processes* (L. Diósi and B. Lukács, eds.) [304]). Società Italiana di Fisica (*Il Nuovo Cimento* [227]). American Institute of Physics (B. de Witt in *Physics Today*, September 1970; also *Physics Today* [166, 293]). Institute of Physics (*Journal of Physics A* [314, 315]). University of Bristol (*Observation and Interpretation in the Philosophy of Physics* (S. Körner, ed.) [52]). Niels Bohr archive (F. Aaserud, Director) (published writings of Niels Bohr).

1

Bohr and Einstein: Einstein and Bohr

Albert Einstein and relativity

If I were to ask a number of people in the street what they think was the most important new theory in physics in the twentieth century, and who has been the greatest physicist, I am fairly sure that – of those able to express an opinion at all – a substantial majority would say that relativity has been the greatest theory, and Einstein the greatest physicist.

Indeed Albert Einstein has probably achieved the remarkable feat of not just becoming the best-known practitioner of any branch of science among the general public, but retaining that position for 85 years, 50 of those years since his death in 1955.

It was in 1905, when he was 26, that Einstein astonished the scientific community by producing four pieces of work of the very highest quality. These included his first paper on relativity, in which he introduced what is now called the *special theory of relativity* (a term I shall explain in Chapter 3). The three other papers will be referred to in due course. What was astonishing was not *just* the quality of the work, but that the author was not an academic of note, or even of promise, but was working as a patent inspector after rather a mediocre student career. (A recent pleasantly written biography of Einstein is that of Abraham Pais [1], himself a well-known physicist; Pais gives references to many other accounts of Einstein's life. Another, more recent, biography is that by Fölsing [2].

After this explosion of 1905, Einstein was reasonably soon settled into a series of university positions of rapidly increasing prestige in Switzerland, Austria and Germany, and recognised as highly exceptional by the scientific community, though scarcely by the general public. His major return to relativity theory came just before and during the First World War, when he spent several years developing his *general theory of relativity*, essentially a theory of gravitation. (See Chapter 3.) This work required him to undertake a lengthy study of the rather complicated mathematics of the tensor calculus, followed by numerous attempts to apply it to the problem of gravity, but by 1916 the work was complete.

A crucial test – to check the bending of light from various stars as it passed the Sun – was undertaken by a team under Arthur Eddington on the island of Principe during the solar eclipse of 1919. The news that not only did the bending take place, but also the amount agreed quite well with Einstein's predictions, was bound to make his position at the head of the community of physicists unassailable. Also, and much more surprisingly, it made him famous among ordinary people across the world.

It is fascinating to ponder why, at the end of a war of unspeakable ferocity, and facing, in many cases, a totally uncertain future, the nations of the world should have elected as unofficial hero a man of massive intellect, with rather weak ties to any particular country, whose work was felt to be not only abstruse beyond belief, but of no application whatsoever to real life, for good or for evil.

There were early well-known newspaper reports that only a handful of people understood relativity. These were always nonsense; probably the bulk of specialists came to understand and appreciate both special and general relativity rather quickly considering their revolutionary nature. For the non-scientist, though, perhaps the attribution of crucial significance to such apparently meaningless phrases as 'non-simultaneity of time' and 'curvature of spacetime' gave an awareness of a realm of pure thought transcending the horrors of the past and the struggles of the present and immediate future.

For others, the mistaken feeling that the real meaning of the theory was covered in the simple slogan 'everything is relative' may have created a (false) sense of security, the hope that, while the details of the theory might be complicated, its basis was comprehensible, indeed something they had taken for granted all along.

Whatever the reasons, with the populace in general, relativity was undoubtedly a hit, and so, even more so, was Einstein. To the initial interest in his ideas was added delight in his character. For a world tired of boastful politicians and warmongering statesmen, his charm, his courtesy, his sense of fun and his modesty were supremely attractive. Add a sensitive smile and breathtakingly liquid eyes, and it is scarcely surprising that Einstein's face became one of the best-known worldwide in the 1920s, rivalled only perhaps by that of Charlie Chaplin.

In the political turbulence of the inter-war years, such a relatively carefree existence could not survive long. As the Nazis rose to power in Germany, Einstein – a Jew, a liberal, an intellectual – typified everything they detested. He was forced to emigrate to America in 1933, and he spent the rest of his life there, at the Institute for Advanced Study at Princeton. The best-known pictures of him date from this period and suggest rapid ageing. Though Einstein remained physically strong, his face became gaunt and lined, his hair and moustache turned white, and his eyes lost all sparkle and seemed to give visual proof of his fears for humanity.

He retained his modesty and simplicity. He played music with a few friends and made time for the neighbourhood children but kept himself aloof from close companionship. Of his scientific work at this stage of his life – he continued hard at it right up till his death – much more will be said later; here I shall just mention that his attitude to quantum theory removed him from influence over, and significant interaction with, the great majority of his fellow physicists. He devoted considerable effort during this period to political matters, appeals for world government and international control of nuclear weapons, and calls for scientists to work for peace.

But it is, of course, for relativity that he is chiefly remembered, and I would like to stress that he stands virtually alone as its sole creator. It is only fair to mention that the Dutch physicist Hendrik Lorentz and the Frenchman, more mathematician than physicist, Henri Poincaré produced, independently of Einstein (and each other), most of the equations of special relativity. However, the work of these scientists, and brilliant scientists they both certainly were, was rather plodding and piecemeal, and not particularly convincing. To compare it with the approach of Einstein, where, from a very few postulates, easily stated though conceptually challenging, all the results flowed in a straightforward and meaningful way, serves to increase, rather than decrease, one's impression of Einstein's genius. Lorentz and Poincaré produced new equations; Einstein gave us a new physics.

For general relativity, the only name other than that of Einstein needing to be considered is that of David Hilbert, the great German mathematician. Hilbert produced equations similar enough to those of Einstein to make Einstein briefly suspicious that Hilbert had cynically taken advantage of his own mathematical struggles. Hilbert himself, however, recognised Einstein as discoverer of the general theory, and this has never been questioned.

Relativity and quantum theory

But I now want to go right back to the beginning of the book and the question raised there, because, if it were physicists rather than laypeople being questioned on the greatest physical theory of the twentieth century, I don't believe they would pick relativity at all. Instead, they would go for the other major development in physics of this period – quantum theory. This would especially be the case if their field of interest concerned the behaviour of substances or systems existing naturally, or fairly easily manufactured, on Earth – semi-conducting solids, lasers and the like. Those more interested in astrophysics or cosmic speculation *might* stick to relativity.

To explain the reasons for this choice, I first need to explain under what circumstances relativity is required, and then do the same thing for quantum theory. I shall start from the point that the pre-1900 physics, the mechanics of Isaac Newton and the electromagnetism of James Clerk Maxwell in particular, gave correct answers for the areas of experience available for study at that time. (This physics is often called classical physics, and I shall outline its principal features in the following chapter.) Specifically classical physics worked well when the speeds of the particles were much lower than that of light (which is always referred to as c), and provided that their dimensions were much greater than those of atoms.

(In this book, I shall use scientific notation. If a little off-putting at first sight for those who don't understand it, it is not difficult to learn, and then it really does make things much easier. In this notation, a thousand, 1 followed by three noughts, is written as 10^3, and a million, 1 followed by six noughts, as 10^6. A one-thousandth, 0.001, is written as 10^{-3}, and a one-millionth, 0.000 001, as 10^{-6}, and so on. According to these rules, 1 may be written, if we wish, in the rather unlikely form of 10^0. So c, the speed of light, which is 300 000 000 metres per second, is best written as 3×10^8 metres per second, or, conveniently applying the same argument to the units, $3 \times 10^8 \, \mathrm{m\,s^{-1}}$. Atomic dimensions may be thought of, roughly, as between 10^{-10} m and 10^{-9} m, that is, between a ten-thousand-millionth and a thousand-millionth of a metre.)

If we start from the motion of a football, or that of the Moon in orbit round the Earth, speeds are certainly much less that that of light, c, and dimensions much greater than those of atoms. So Newton's laws apply extremely accurately to these processes; if they had not done so, Newton would never have become particularly famous! For these speeds, the predictions of relativity are (almost exactly) the same as those of Newton.

But when speeds approach that of light, relativity theory tells us that Newtonian theory is only an approximation, a good one if the speed is $c/10$, say, but rather bad by the time it reaches, say, $c/3$. On the other hand, the equations of relativity give the correct answers, that is to say, the results that are found in any experiment.

According to relativity, by the way, bodies can never travel as fast as light. This is true, at least, for all the objects we are familiar with, which have masses greater than zero. (If you are not familiar with the term 'mass', it just means what in everyday speech we call 'weight'; we shall meet it a little more formally in Chapter 2.) Rather peculiar particles with mass zero *have* turned up in physics, and we shall meet them in Chapter 4; they *must* have speed equal to c at all times. (I should mention that, comparatively recently, the existence of particles called *tachyons* has been suggested; such particles would *always* have speed *greater* than c. Many people think that the existence of such particles would cause conceptual paradoxes. Be that as it may, it can definitely be said that those looking for tachyons have not so far found any.)

Similarly, as dimensions are decreased, Newtonian ideas become unsatisfactory; their predictions become a deteriorating approximation to the truth. This is when quantum theory is required. Often the terms 'microscopic' and 'macroscopic' are used to denote, roughly, objects the size of atoms and objects large enough to be dealt with directly by our senses. So quantum theory and Newton agree for the macroscopic case, but for the microscopic case they disagree, and it is quantum theory that is correct.

At first sight, then, it might be suspected that *neither* relativity *nor* quantum theory should be very important for our everyday life on Earth. After all, we don't travel at speeds of 10^8 m s^{-1}, and the objects we interact with directly are macroscopic, as I just defined it. But at least for the case of quantum theory, this impression would be most misleading.

It *is* true that, when one drives a car, the dynamical properties, the relation of the acceleration to the power of the engine and the gradient and slipperiness of the road, are covered in a totally satisfactory way by Newton's mechanics. But one could scarcely even attempt to understand in any fundamental way the strength of the metal of the car body, or the energy provided by the petrol, without the use of quantum theory.

Even more so, the electrical properties of the car radio or the on-board computer would be incomprehensible without a good knowledge of quantum theory; indeed, the materials and electronic circuitry that constitute these devices could not have been designed without such knowledge. The individual atoms, their configuration and their electronic properties are crucial in the tasks of the (macroscopic) system.

Quantum theory has enabled us to understand the properties of atoms, molecules and nuclei, and, to some extent, the *constituents* of nuclei. (For readers not quite sure of the meaning of some of these terms, explanation will come in the following chapters.) For instance it tells us why two oxygen atoms combine to form a molecule, and why only certain nuclei are stable. In this paragraph I am unashamedly using microscopic terms – atom, molecule and so on. Of course human beings cannot sense or react to individual atoms, but our bodies certainly require oxygen molecules, and would respond badly to too much radioactivity caused by decay of unstable nuclei. Thus quantum theory is decidedly relevant to our life as human beings on Earth. In contrast, for most of the questions I've mentioned here, relativity provides only a very small correction factor compared with the Newtonian case, so it does not really play a role.

When we move from planet Earth to larger-scale physics like cosmology, or the origin of the Universe, relativity theory does play an important part. But so does quantum theory; so indeed do such areas of physics as quantum field theory and the study of elementary particles (which will be touched on in Chapter 4). Overall, then, I believe it should be quantum theory, not relativity, that should receive the laurels from the physicist.

Albert Einstein: universal genius

Does this mean then, I hope the reader is now asking, that Einstein must be dethroned as leading physicist of the century? My answer would be, emphatically, no, for three good reasons. The first is that, while, unlike relativity, quantum theory was not the creation of one scientist, but of many over a period of a quarter of a century, Einstein was one of its main creators. Especially if one includes the achievements of physicists he influenced and encouraged as well as his own direct contributions, his role will be seen as immense and often crucial. (See Chapter 4 in which the development of the theory is sketched.)

It is well known that his 1921 Nobel Prize was awarded, not for relativity, but for his theory of the photoelectric effect, an important step in the development of quantum theory (and the second of the 1905 papers mentioned in the first section of this chapter). What happened is that the Royal Swedish Academy of Sciences was under great pressure to award Einstein the Prize for Physics, but did not quite have the courage, or the necessary advice, to award it for relativity. In any case, Einstein ignored the details of the citation, and gave his Nobel Prize address on relativity. This last remark should not, though, be taken as an admission that the work on the photoelectric effect was *not* good enough for the Nobel Prize; it definitely was!

After making such massive contributions to the development of the quantum theory, it must seem strange that Einstein rejected its final form, or, at least, some elements of what most physicists of the time thought had to be its final form. I shall say no more on that subject now, as it is one of the main topics of the rest of the book and Chapter 6 in particular.

The second reason for retaining Einstein at the top of one's list is that, even apart from relativity and quantum theory, he did exceptionally important work in other areas of physics. Of his remaining two major papers of 1905, the first was a seminal study of the diffusion of solid particles through liquids. The second was an analysis of Brownian motion, the means by which molecules were (indirectly) observed for the first time; particles large enough to be visible, and suspended in a liquid, appear to move erratically under bombardment from the (invisible) molecules of the liquid.

Only by the standards of relativity and the photoelectric effect do these papers rate as less than amazing triumphs. Indeed the Nobel Prize committee considered awarding Einstein the Prize for his work on Brownian motion. Its reason for refraining from doing so was not that the work was too weak for a Nobel Prize, just that it would look most strange to award the prize for this, when the papers on relativity and the quantum theory were obviously more important. Happy indeed should be the theoretical physicist whose third-rate work (third-best paper in a given year) is good enough for a Nobel Prize!

My third reason is just that – almost, seemingly, independently of his actual achievements – it was recognised by every other physicist that he stood supreme. His loftiness of approach, his intellectual superiority, his independence of mind, were enough to make him their acknowledged leader, even though his attitude to quantum theory meant that, for most of them, and for almost half his career, he seemed a lost leader.

So there are undoubtedly good reasons for expecting a physicist to choose Einstein ahead of all others – but I expect that any who didn't would choose instead Niels Bohr.

Niels Bohr and quantum theory

If one were to suggest that what Einstein was to relativity, Bohr was to quantum theory, the remark would be most misleading unless highly qualified. While Einstein, as I've already said, built relativity practically single-handed, very many scientists, including, of course, Einstein himself, contributed to the development of quantum theory, a process that began in 1900.

Bohr himself took one major step, the introduction of the so-called Bohr atom, in 1913. Important as this was in stimulating the discoveries that led to the final form of the theory, which was reached around 1925, the Bohr atom does not actually appear in this final form, indeed is rather at odds with it. It is fair to add, though, that certain components of Bohr's theory *were* lasting and important – the energy-level diagram and the concept of the transition between different levels and its corresponding frequency. For all this, see Chapter 4. (For further information on all aspects of Bohr, Pais [3] has written a readable and detailed biography, and another account of Bohr's life and work has been written by Blaedel [4].)

Bohr was equally or more important in two other ways. From 1921, the date of the foundation of his Institute of Theoretical Physics in Copenhagen when

Bohr was 35, right up to his death in 1962, the institute was *the* world centre for research in theoretical physics (though with competition from Munich and, in particular, Göttingen in the early years, while, in the later period, the pendulum had swung in the American direction to a considerable extent). Many of those chiefly involved in the development of quantum theory spent long periods at Copenhagen, learning from Bohr, arguing with him, often assimilating much of his approach to physics. Among the major figures were Werner Heisenberg, Wolfgang Pauli and Hendrik Kramers, all of whom will figure substantially in later chapters of this book. Paul Dirac was another very important contributor to quantum theory who spent long periods in Copenhagen, though he retained very much his own approach. There were scores of visitors of somewhat less significance.

It is also interesting to note the main exceptions. There were those whose main inspiration came from Einstein – Louis de Broglie and Erwin Schrödinger. Max Born had a strong base in Göttingen, at least till the rise of Hitler, and Arnold Sommerfeld reigned supreme at Munich.

(Mention of Born reminds me to point out that a number of the physicists we are discussing have unfortunately similar names: Bohr, Born, and shortly we shall meet Bohm, who is not, as you may be inclined to suspect, a misprint for Bohr, but a very important quantum physicist in his own right. Please don't get confused!).

Perhaps Bohr's greatest lasting significance, though, lay in the interpretation, not the creation, of quantum theory. In order to explain this remark, it may be necessary to say why such a thing as an 'interpretation' is required, and again a comparison with relativity may be useful.

Everybody will agree that the ideas of relativity are difficult to come to terms with. The way of looking at things is just very different from what one is used to under Newton. Nevertheless, once one has taken the required mental steps, the new set of concepts is perfectly well defined. The equations of the general theory may be difficult to solve for important cases, and the more arcane cosmological aspects – black holes, the birth of the Universe and so on – are certainly thought-provoking, but I don't think practioners of relativity lie awake at night worrying at least about the *bases* of their subject.

Such is not the case with quantum theory – at least not necessarily. It is certainly possible to concentrate on the immensely successful and unproblematic calculational aspects of the theory, and thus sleep perfectly well at night. Bohr, however, was not content with this approach. He realised that, until the mathematical parameters and processes were given coherent physical meaning, any deductions from them were illegitimate. It was, however, by no means obvious that such 'interpretation' was possible, or, if so, how it should be carried out.

Bohr's solution to the problem was his fairly general framework of what he called 'complementarity', and the specific application to quantum theory became known as its 'Copenhagen interpretation'. The parameters kept their conventional meanings, but their simultaneous application was restricted. (This admittedly brief and obscure statement will be extended and, I hope, clarified in Chapter 5.) Such was Bohr's prestige that this set of ideas became very generally 'accepted wisdom'. Even most of those who had not studied them at all carefully had no hesitation is giving lip service to the Copenhagen virtues.

The leading exception was, of course, Einstein. He and Bohr has several well-publicised debates in the late 1920s and into the 1930s. (The orthodox view was that Bohr won them hands down.) Of those physicists noted above as being principally inspired by Einstein, Schrödinger remained reasonably close to Einstein, agreeing with him to a large extent on the defects of the Copenhagen interpretation, if not necessarily on what should replace it. De Broglie initially put forward his own interpretation, but was persuaded by Pauli to reject it and become a rather reluctant convert to Copenhagen. (In the 1950s, David Bohm was to renew interest in de Broglie's ideas and overcome Pauli's objections. This development will be an important part of Chapter 7.)

There were other towering names who were hostile towards Copenhagen – Max Planck and Max von Laue, but their important contributions, which will be described in Chapters 2 and 4, had been made in previous decades, and the overwhelming majority of quantum physicists had no hesitation in considering them, like Einstein, to be giants of the past, who had lost the mental flexibility to adapt to novel conceptual developments.

As Einstein's direct influence fell off from the 1930s, Bohr's, if anything, increased. Rather differently from Einstein, he had a strong attachment to family and nation. His base was in Copenhagen throughout his life, and he was recognised as Denmark's leading citizen for many years. But he had an important year in England in 1913, during which his great contribution to quantum theory was made, and he was forced to leave Denmark for three years of the Second World War which he spent in the USA. As an international man of science, he travelled widely.

During the 1930s, he did important work himself in nuclear physics, inventing the so-called compound nucleus model which was important for describing nuclear reactions, and playing a central part in the elucidation of nuclear fission, the process at the heart of the atomic bomb.

After 1945, Bohr did little science himself, though he continued to interact with the world's leading physicists. He spent much of his time preaching disarmament and international control of atomic weapons, a self-imposed task that had been begun during the war itself, when Winston Churchill came close to accusing him of treachery. Through the years of the cold war up to the time of his death he really had little success.

Such work obviously linked him with Einstein, though Bohr's constant travelling from country to country in search of their common goals contrasted with Einstein, who stayed at home signing letters and making occasional broadcasts, and this political unity, and the respect they always felt for each other as individuals, added extra poignancy to their continued scientific differences.

Bohr still maintained total scientific, as well as personal, respect for Einstein. Right up to Einstein's death, Bohr's work seemed motivated towards converting him to complementarity, and even when Einstein had died he was still Bohr's unseen opponent, at whom his argument were aimed.

On the other hand, Einstein's attitude to Bohr's scientific views may *possibly*, I feel, have lapsed into irritation at times. There is little direct evidence for this, but perhaps some indirect suggestions. For much of their lives, Einstein and Max Born exchanged letters [5]. A good deal of the content of these was concerned with world politics following the outbreak of fascism in Europe, which sent both men into exile, Einstein in America and Born in Scotland, and continuing into the cold-war period. This portion is moving and stimulating. However, there were also some fairly sharp exchanges on the interpretation of quantum theory. (Born was a strong advocate of Copenhagen.) Einstein occasionally showed clear signs of extreme impatience.

Another hint comes from the comments of Schrödinger – who, as I've already said, was reasonably close in spirit to Einstein – on Bohr, which are reported in Moore's biography of Schrödinger [6]. Schrödinger denounced complementarity as a 'thoughtless slogan', and says that, if he were not convinced that Bohr was honest, he would describe its use as 'intellectually wicked'. It would not surprise me greatly if such ideas were reasonably close to those of Einstein.

The debate continues

Bohr himself died in 1962. Since then the practically monolithic subservience to his views on quantum interpretation has fragmented somewhat. The leading spirit in the process of re-evaluation has been a physicist from Ireland, John Bell, who was stimulated both by the views of Einstein and by Bohm's work mentioned already. His work is discussed with that of Bohm in Chapter 7.

Many other physicists have joined in the discussion of these ideas, analysing the ingenious difficulties for the Copenhagen interpretation thought up by Einstein, Schrödinger, Bell and others, and putting forward interpretations of their own. A few of these ideas are discussed in Chapter 8. Some of these writers have been very critical of Bohr. Murray Gell-Mann [7], himself a winner of the Nobel Prize for Physics, for example, has accused Bohr of 'brain-washing' the physics community into thinking that the problems had been solved.

It is easy to understand why many of these writers have felt that it helps their cause to claim support from Einstein for their own ideas. Bell, for instance, claimed to be a follower of Einstein, and was quite critical of some, though by no means all, of Bohr's opinions. To this extent, then, the Bohr–Einstein debate may be said to have outlived the scientists themselves. It is quite common these days, in strong contrast to almost universal opinion 50 years ago, to hear the view that Bohr's ideas have been vanquished by one or other of contemporary theory or experiment, or both, and that Einstein's position has been vindicated. Others, of course, deny this equally vehemently! My own views will gradually be made clear through the second half of this book.

Bell's work brought the rather esoteric aspects of the Bohr–Einstein debate into touch with experimental reality. Then, very surprisingly, indeed, from the perspective of the 1930s, 1940s or 1950s, totally amazingly, many of the techniques, theoretical and experimental, of those studying the ideas of Bohr, Einstein and, in particular, Bell have recently become central to the new, extremely interesting and, at least in principle, very practical subject of quantum information theory, for which see Chapter 9.

The Bohr–Einstein debate has been one of the most important controversies in scientific history, not least because the contemporary verdict on it must have played a large part in driving the most-lauded scientist for 300 years practically into scientific exile. The debate is not so much about the content of particular scientific theories as about what a scientific theory ought to be. It is not *just* about the nature of the Universe, but about what kind of description of the Universe we should regard as meaningful. In terms of both personalities and ideas, I feel it has been a fascinating controversy and I think it is well worth trying to analyse the source of the disagreement, and forming an opinion as to who really got the better of the argument. This is the main purpose of the book.

2

The peace before the quantum

Classical physics

Quantum theory was produced roughly between 1900 and 1925; it changed our view of the Universe in a revolutionary way. Such is the central preoccupation of this book. Relativity was another revolutionary theory produced between 1905 and 1916; it is less central here. The clear implication of these statements is that, prior to 1900, there was an established body of theory that was widely successful and was thought to provide final answers to fundamental physical questions. This body of theory is known as *classical physics*.

The above is a simplified account of the development of physics over the last few hundred years. There is a good deal of truth in it. By 1900, Newton's Laws of mechanics had been established for over 200 years; they had been overwhelmingly successful in describing a huge range of phenomena, both terrestrial and in the solar system. They were held to be among the greatest human achievements, indeed the very greatest strictly scientific achievement, and practically a direct revelation of divine intent, and by 1900 it seemed unthinkable that they could be challenged.

The same could not quite be said of electromagnetism, an area of physics that encompasses electricity, magnetism and, as will be seen later in this chapter, optics. By 1900, what we now take to be the complete and final theory of *classical electromagnetism*, that of James Clerk Maxwell, was over 30 years old, and the discovery of *radio waves* by Heinrich Hertz which was acknowledged to confirm the theory, more than ten years old. There was still, though, a fair amount of misunderstanding, which surfaced in the attempt to establish the nature of *X-rays*, following their discovery in 1895. From a modern perspective X-rays have a very specific place in Maxwell's theory, but this did not become clear for over

15 years. Nevertheless, certain features of electromagnetism did seem beyond challenge – in particular, that light was an (electromagnetic) wave.

The revolutionary nature of quantum theory and relativity is clearly shown in that many seemingly unchallengeable aspects of classical physics *were* refuted (or, to put things more meaningfully, shown to be approximations, certainly very good ones in the regions of physics where they had been successful prior to 1900, but not very useful in areas studied increasingly in the new century).

So the straightforward view of a well-established set of ideas (or *paradigm*, to use Thomas Kuhn's term from his famous book, *The Structure of Scientific Revolutions* [8]), overturned by a revolution, is not too far from reality. It cannot be the whole truth, though, as shown by the development of ideas on heat. The thoroughly classical discipline of *thermodynamics* had largely been completed in the nineteenth century. While outstandingly successful as a predictor of the results of experiments and well able to account for the limitations in the practical uses of engines, it did little to *explain* the nature of the scientific processes involved, certainly not in terms of the individual particles of the participating substances.

For this, the alternative discipline of *statistical mechanics* (or *statistical thermodynamics*) is required. This subject applies the laws of physics to a statistical assembly of the individual particles of the substances. Either classical or quantum laws may be used, and two separate disciplines of classical and quantum statistical mechanics are produced; each may be appropriate in a given range of circumstances.

During the nineteenth century, there were a great many very important contributions to classical statistical mechanics, particularly those of Maxwell (again) and Ludwig Boltzmann. But by 1900 there were still important disagreements and uncertainties in the foundations of the subject. In the first decade or so of the century, the discipline of classical statistical mechanics was still being completed – Einstein was an important contributor.

The situation was complicated by the fact that quantum theory even of individual particles is itself statistical in nature (or I would prefer to say probabilistic, as will emerge later, but for the moment I shall not stress any difference between the terms). Also, some of the very earliest problems requiring quantum ideas were statistical in the sense of involving many particles. So three inter-connected disciplines, each using statistical ideas, namely quantum theory itself, and classical and quantum statistical mechanics, were being developed simultaneously. It was not surprising that progress was slow and confusion rampant.

I talked about statistical mechanics dealing with the 'individual particles' of the substances, and of course I am referring to atoms. *Atomic theory* constitutes the other important area in which progress could be described as ongoing around the year 1900. Of course quantum theory and atomic theory are closely linked; quantum theory may be spoken of as the physics of atoms, and since 1900 they have often developed together.

But the idea of atoms has a very long history. One is expected to mention here the name of Democritus (around 400 BC), but more important is the fact that, from the seventeenth century onwards, and particularly since the work of Robert Boyle and Newton himself, many physical phenomena were analysed in terms of very large numbers of minute indivisible particles ('corpuscles' or 'atoms') moving in all directions. In the early nineteenth century, John Dalton explained many features of chemical combination and behaviour by the use of different atoms for the different chemical elements – hydrogen, oxygen and so on, and during the rest of the century, chemical explanation became increasingly, and eventually virtually wholly, atomic in nature.

It was not till the end of this century and the beginning of the twentieth, though, that experiments which impinged directly on individual atoms and their constituents were carried out. These included the discovery of the *electron* and *radioactivity*. A decade later came the discovery of the *atomic nucleus*, and the unambiguous ordering of elements provided by the *atomic number*. The *neutron* was not discovered for another 20 years still. So even the most basic information about what may loosely be called the classical theory of atomic structure mostly emerged after the initial steps in the development of quantum theory.

In the remainder of this chapter I shall sketch the state of physics at the time of the birth of the quantum theory, not just the stable areas of mechanics and electromagnetism, but also the still problematic areas of statistical mechanics and atomic theory.

Virtually every aspect of classical physics will be important at some point in the later chapters. The reader may wish to study the account of classical physics given in this chapter straight through, or alternatively to use it as a source of reference when reading the rest of the book.

This may be a suitable place to state clearly what type of book this is. I am particularly referring to this chapter and the next two, where many statements that appear to be historical in nature are made – scientist X did particular experiments, or developed a new theory, in year N. When making such statements, I shall do my best to avoid outright error. Indeed, I shall take the liberty of pointing out sometimes where many accounts of these ideas – textbook and semi-popular – do mislead historically (though it is almost inevitable I shall myself offend on occasion).

Nevertheless, this is not, and could not be, a book on the history of science, a discipline that requires not just dates and events, but also an account of the historical context in which events take place, an attempt to view scientific work in this context, rather than just as a stepping-stone to our present knowledge, and a willingness to see growth in scientific understanding as just one aspect of a wider intellectual history. This is unashamedly a book about physics. I shall be more than pleased if the reader takes away a reasonable understanding of the various theories and interpretations of theories discussed. A few references to genuinely historical works will be given.

Newton's inheritance

Classical mechanics was to so great an extent the achievement of one man, Isaac Newton, in the second half of the seventeenth century, that it is very often just called Newtonian mechanics. It is certainly true, though, that Newton had the benefit of the work of several great thinkers of the previous century or so [9].

Ancient Greek thought largely had distinct approaches to terrestrial physics and to cosmology. Formally, *terrestrial* processes *could* be discussed in terms of objects finding a natural place in the overall *heavenly* structure, fire moving up, stone down and so on. In practice, though, the two areas presented different types of problems, and were analysed in totally different ways. It was the great achievement, we may call it the achievement that initiated modern science, first of Galileo, but more conclusively and triumphantly of Newton, to unite these apparently unconnected areas of discourse.

On the cosmological side, the simple and beautiful system used, though not invented, by Aristotle in the fourth century BC visualised the Earth at the centre of the Universe, surrounded by a series of concentric transparent spheres on which the Sun, Moon and planets travelled. This explained elegantly many of the central features of the Universe, but failed in many other important ways, in particular in the analysis of planetary orbits. For this reason it was embellished, over many years, but particularly by Ptolemy in the second century AD. A whole host of mathematical complications was used to distort the original idea of steady motion along circular paths centred on the Earth. Any conceivable physical meaning was lost, mathematical agreement between calculation and observation becoming all-important.

The step taken by Nicolaus Copernicus in the mid sixteenth century has justly become famous. It was, of course, the proposition that the Sun rather than the Earth was the centre of planetary motions, the Earth being relegated to being just one of the moving planets, being circled itself by the Moon, and itself spinning daily about its axis This was, or at least was the start of, *The Copernican Revolution* [9], to use the title of Kuhn's informative and readable book.

Copernicus certainly meant this to be a genuine physical change, though the *detailed* handling of planetary motion still required most of the mathematical

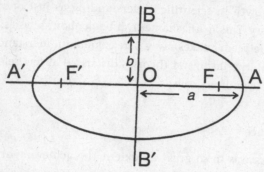

Fig. 2.1. An ellipse. The distances OA and OB are the semi-major and semi-minor axes, a and b. The eccentricity, e, or amount of 'squashing', is equal to the square root of the difference between the squares of a and b, divided by a. F and F′ are the foci of the ellipse; OF and OF′ are each equal to a times e.

complications of Ptolemy. He did, however, explain many of the general features in a more natural way. For example, Mercury and Venus always appear close to the Sun because they are nearer to the Sun than is the Earth. Retrograde motion, the apparent doubling-back on their paths exhibited by all planets from time to time, is easily explained, as is the variation of the time taken for a particular planet to complete an orbit.

For Copernicus, as for Aristotle, an unquestioned constraint on planetary orbits was that they should be constructed from circles and that motion round the circles should be at constant speed. Indeed, his main complaint against Ptolemy was the smuggling in of elements breaking this rule. But after many years of intense struggle reconciling mathematical model with observation, Johann Kepler, who had access to the particularly accurate planetary data laboriously obtained by Tycho Brahe in the last decades of the seventeenth century, was forced to drop these restrictions. It was indeed the very accuracy of the data that led him to the conclusion that planetary orbits were not circles but ellipses. This explained, of course, why both Ptolemy and Copernicus had had so much difficulty constructing orbits from a combination of circular motions.

Figure 2.1 shows an ellipse, together with its two *foci*, and its *semi-major* and *semi-minor axes*, a and b. A mathematical quantity, the eccentricity of the ellipse, e, tells us how 'squashed' the ellipse is. It is defined as the square root of the difference between the squares of a and b, divided by a. Also the distance from the centre of the ellipse to either focus is equal to the product of a and e.

The circle may be regarded as a special case of the ellipse in which a and b are equal, so the eccentricity, e, is zero (no 'squashing'), and both foci are at the centre. The orbits of most planets are quite close to circles; the eccentricity of the Earth's orbit is only 0.017, and all other planets have eccentricities less than 0.1 except for Mercury (0.206) and Pluto (0.249).

Kepler's three famous laws were stated in 1609 and 1619. The first announced the ellipse, and the others gave information about the speed of the planet round its orbit. The Second Law stated that the planet did not travel at constant speed, but moved so that a line joining the planet to the Sun swept through equal areas of the ellipse in equal times. The Third Law related the average distance between the planet and the Sun to the period of the orbit; for each planet the ratio of the *cube* of the average distance to the *square* of the period was the same. Kepler's Laws were to provide a detailed testing-ground for Newton later in the century.

While Kepler's work was primarily mathematical, his contemporary, Galileo, provided strong backing for Copernicus both from physical and from mathematical points of view. He was the first to use the telescope, which had been invented by a Dutch lens grinder, for astronomy, demonstrating that the Moon and Sun were irregular 'imperfect bodies', contrary to Aristotle; that Jupiter and its moons behaved like the Copernican solar system; and that Venus displayed phases analogous to those of the Moon. Such evidence was directly physical; Galileo, in any case, was not prepared to hide behind any form of words implying that the motion of the Earth was merely a convenient assumption – an honesty for which he paid dearly at the hands of the church [10, 11].

He also studied to great effect motion on Earth. Aristotle had propounded the very sensible view that the 'natural' motion of an isolated body that possesses weight is towards the centre of the Earth. An 'unnatural' motion, then, required an agent acting throughout the course of the motion, but many before Galileo had understood that this idea could not easily be applied to projectiles, objects hurled in sport or war from one point to another. A javelin retains its motion parallel to the Earth's surface, although there is no agent propelling it once it has left the hand of the thrower. This suggests that, air resistance in this case apart, and again concentrating on Aristotle's 'unnatural motion' parallel to the Earth's surface, a free body must have constant speed, not, as Aristotle said, zero speed.

Galileo required this type of argument for his Copernican proposition that the Earth was moving round the Sun and spinning about its axis. In these circumstances, Aristotle would say that a body dropped vertically downwards should be left behind as the Earth moves and hit the ground away from your feet, contrary, of course, to experience. Galileo's analysis above, though, says that the body retains its motion around the Sun as it falls, and comes to rest at your feet. Galileo described very coherent analogies, using people playing deck-games on (smoothly sailing) ships, and being unaware of the motion. This analysis of

Galileo was very important for Newton, and it was also the first example of relativistic thinking, as I shall show in the next chapter.

The famous story of Galileo dropping spheres of different materials off the Tower of Pisa, and finding that, contrary to another of Aristotle's beliefs, the heavier did *not* reach the ground first, is only legend. He did, though, reach an understanding of the important facts about falling bodies – if one can ignore air resistance, the speed reached is proportional to the time of fall and the distance to the square of the time, and the rate of fall is the same for all bodies. Again, this was to be of much benefit to Newton.

'Let Newton be'

Newton inherited, then, the idea of a Universe where planets moved on well-defined orbits, so that it was reasonable to hope to explain their motions physically, together with a robust terrestrial mechanics with many of the basic concepts fairly clearly understood. It was his genius to produce not just a coherent and universal system of laws that described in detail the motion of bodies on and close to the Earth, but also a mathematical analysis of gravitation. This united those aspects of terrestrial mechanics concerned with falling to Earth with the behaviour of planets in their orbits, and was able to explain, at least in outline, a considerable range of astronomical phenomena [12, 13].

His system was based on his three famous Laws of Motion. In explaining the first two, I shall assume that we have a good instinctive idea of two concepts. The first – *mass* – is, as I mentioned in Chapter 1, just what is usually referred to as 'weight', so the mass of a person or thing is measured in pounds or kilograms. (Note, though, that whereas weight varies slightly at different points on the Earth's surface, and is reduced by a factor of about seven on the Moon, mass is unchanged – it is a property of the body, irrespective of location.) The second term is *force*, by which I mean, at least to start with, a push or a pull. Later there will be fairly obvious extensions to gravitational, electrical or magnetic forces, which have the same kind of effect.

Using this term, *Newton's First Law* states simply that, if no force is applied to a body, its state of motion remains unchanged, that is to say it will continue to move at the same speed in the same direction. This is clearly very much the Galilean perspective as explained above, and very much opposed to Aristotle, for whom 'no force' would imply 'no (unnatural) motion'.

The word *acceleration* just means 'rate at which speed changes', and so the First Law can equally well be stated as 'no force' implies 'no acceleration'. The *Second Law* then tells us what happens if a force *is* applied to the object. It produces an acceleration that is proportional to the strength of the force and inversely proportional to the mass of the object. (Or just acceleration equals force divided by mass, or force equals mass times acceleration.)

To take an example, consider a car that needs pushing from A to B (along a flat road; we won't bring gravity in yet). To start it moving (to achieve an acceleration), we will certainly need to provide a force (a push). The harder the push we, and as many as possible of our friends, are able to provide, the faster we will bring the car up to a reasonable speed. But the more massive the car is, the harder we will struggle to do so, agreeing with the Second Law that acceleration is inversely proportional to mass, for a given applied force.

When we reach B, however, assuming that we have built up a fair speed, we must go through the opposite procedure; we will move swiftly to the front of the car and provide a force in the opposite direction, to cause a deceleration, a reduction in speed, in order to prevent the car rushing forwards and demolishing B.

This all agrees with Newton, but we now confront an apparent snag. On our way from A to B, we will almost certainly find that we must keep pushing to keep the car moving at the same speed. This seems to *disagree* with Galileo and Newton, who argue for persistence of motion, and so, as the sweat trickles down our backs, we may feel a renewed respect for Aristotle. The missing ingredient is, of course, *friction*. When we attempt to move an object, there is always a frictional force to overcome, and it always opposes motion. Applied force must be greater than frictional force to move the object at all; once it does move, acceleration will be related to the net force, the amount by which applied force exceeds frictional force.

Friction between two solid objects sliding over each other may arise from roughness at the macroscopic level, but, even when the surfaces involved are exceptionally smooth, we must remember that the surface plane of each object consists of a two-dimensional array of atoms; we may think of it, loosely but for this purpose usefully, as rather like a rack of billiard balls. Naturally, pushing two such racks over each other will require a force, though its magnitude will depend on the nature of the atomic structures. As rather a different example of friction, I may mention air resistance, which means that, in practice, the motion of a projectile parallel to the surface of the Earth does decrease somewhat.

I have discussed the Second Law in terms of an agent or experimenter applying a force to an object, itself assumed inanimate. Thus we have not inquired about any effect back on the agent. Now I want to change the focus to a collision, or interaction, between two bodies of similar type. They may be billiard balls; remembering that Newton was very keen on atomic notions, we may like them to be atoms colliding. For the billiard balls, there will naturally be no force between them except during the short period of contact, the impact. (This is ignoring gravitational forces, which will be exceptionally weak.) During the impact, *Newton's Third Law* comes into play – it tells us that, in any interaction between two bodies, the forces on the two are equal in magnitude and in opposite directions. In the atomic case, it may be that there are electrical or magnetic forces; if so, the two bodies will interact even when they are well separated, but again and at all times the forces must obey Newton's Third Law.

To widen the area of discussion, let us think of sitting on a chair. The chair experiences a force downwards; it applies an (equal and opposite) force upwards on you, which may make you a little sore. A soft chair is more comfortable than a hard one because the distortion of the chair surface allows the force to be spread over a wider area of your person. Another example occurs in firing a rifle. Because a high acceleration is given to the bullet, quite a large force is required, even though the mass of the bullet is small. An equal and opposite force is exerted on the firer's shoulder, and the result may be painful if a good technique is not employed. (For the last example in particular, Newton's own expression of the Third Law is extremely appropriate – 'Action and reaction are equal and opposite', the action being the force forwards on the bullet, the reaction the force back on the firer.)

Gravity on Earth

To complete an account of Newton's synthesis, I must go on to his discussion of gravity as an idea that unified physics in the terrestrial and heavenly regions. To keep to the terrestrial side first, it seemed obvious to Aristotle that the 'natural' fall of a heavy object towards Earth was an entirely different type of phenomenon from the motion of an object subjected to a push or pull. Actually Einstein would agree with this. (See Chapter 3.) Yet for a period of about 250 years, it seemed that Newton had been overwhelmingly successful in explaining gravity as 'just' another force.

For let us postulate a force on any body in a downwards direction and of magnitude equal to its mass times a quantity that I shall not discuss for a moment but just call g. By virtue of Newton's Second Law, the object will accelerate downwards, the magnitude of the acceleration being equal to the force divided by the mass. The masses cancel out, so it turns out that the magnitude of the acceleration is just g. This agrees precisely with what Galileo said. All objects fall with the same acceleration, and g may just be called the *acceleration due to gravity*. (Here again I have neglected air resistance, which will be a very good approximation for metal spheres, though very bad for pieces of paper or a parachute.)

(One important point in the above is that 'mass' comes in twice, once in the force due to gravity and once in Newton's Second Law. Why does the same quantity fulfil two different roles? Even if this question seemed interesting at first, everybody soon became so used to it that it appeared, if not obvious, at least so reasonable that one hardly noticed it. It was Einstein who reminded us that it was far from obvious; he made it a major plank in his *principle of equivalence*, itself a central component in his general theory of relativity.)

The approach to gravity above is a part of Newton's vision, but far too parochial to encompass it. Newton's genius, inspired perhaps by the famous apple, was to recognise that the same force that brought the apple to Earth caused the planets to orbit the Sun and determined their paths. To discuss this, I need to generalise a few concepts.

More about motion

So far I have written about motions, like pushing a car, in a straight line, where the force is either in the direction of the motion, giving acceleration, or opposed to it, giving deceleration.

For planetary orbits we need to work in two dimensions. (More generally, of course, we need three, but two will be sufficient here.) In the study of two-dimensional motion, the force may make any angle with the (instantaneous) direction of motion. Because force is related to the *rate of change* of the motion, not the motion itself, things may be quite complicated, which can be a source of difficulty in analysis, but also of richness in possible outcomes. (In contrast, Aristotle's ideas, which effectively relate the direction of force to that of the motion, could produce a planetary orbit only if the planet were physically pushed round. This is, indeed, as far as Kepler was able to go.)

To progress much further, I shall have to define some of our quantities more precisely, particularly in their directional aspects. (Everything will come in useful later in the book.) I shall introduce a distinction between two types of quantity – *scalars* and *vectors*. A scalar is a quantity totally specified by giving its size or magnitude, such as a mass or a time. For a vector, though, as well as a magnitude, a direction must be specified. A good example is a force; clearly while the magnitude of a push or a pull is important, so is its direction – is it a life-saving grab from under the wheels of a lorry, or a push under them? Direction may be specified in any convenient way – 'up', 'to the right', 'in the direction of motion', or, more mathematically, when we have defined an x, y, z set of axes, 'along the y-axis', or whatever. Newton's force of gravity, for example, is always 'downwards'.

Turning to the motion of a body, clearly direction is important. It has been traditional in science, though, to use the word 'speed' *without* taking direction into account. (So speed is a scalar.) Thus we may say an object's speed is

5 m s^{-1} whether it is up, down, or round a circular path. In contrast, we use the word 'velocity' to represent a vector, with *both* magnitude *and* direction. So a velocity may be 5 m s^{-1} to the right, or 5 m s^{-1} along the *x*-axis, for example. The important conclusion is that the velocity of an object changes, *either* if its speed changes, *or*, even if its speed is constant, if the direction of its motion changes.

Just a quick mention of notation – I shall represent a vector in bold type, so a velocity is written as v; the *magnitude* of this vector is respresented by the same letter in italic type, so *v* is the magnitude of the velocity, but this, of course, is just the speed.

Finally, let's think about acceleration. We must define it as the rate of change of a *velocity* (not just of speed) so it too must be a vector, specified by magnitude *and* direction. For linear motion, things are easy; if the body is speeding up, the direction of the acceleration is the same as that of the velocity, whereas when it's slowing down, the directions are opposed. (We can drop the word 'deceleration' now, because the increase or decrease of speed is taken care of in the statement of the direction of the acceleration.) In two dimensions, though, a body has non-zero acceleration if its direction of motion changes, even if its speed is constant.

In general, with perhaps both velocity and acceleration varying in magnitude and direction with time, motions can be very complicated, but the simplest and most important example is motion at constant speed around a circle (which is close to the motions of most planets, though, as Kepler showed, not quite correct). Since the direction of motion is changing constantly, so is the *velocity* of the object, even though the *speed* is constant.

In turn, since its velocity is changing, as I have explained, the acceleration of the object is not zero. What direction must the acceleration take? Certainly it cannot be along the tangent to the circle, for, if it were, the speed of the object would continuously increase or decrease. Rather it must be along the radius, at right angles to the direction of the motion, and towards the centre of the circle. (If you are uneasy about this, please wait till I discuss the force; I hope this may make things a little clearer.) We don't actually need to know the *magnitude* of the acceleration at the moment, but a little analysis shows that it is given by the square of the speed divided by the radius.

Let us now come back to Newton's Second Law and a slight, though fairly obvious, extension of our previous statement. We have said that both force and acceleration are vector quantities; they have magnitude *and* direction. So it is natural that when Newton's Second Law equates acceleration to force divided by mass, it is a relationship between directions as well as magnitudes; the direction of the acceleration is the same as that of the force. (Mass, of course, has no direction.) This is no surprise; if you push someone in a particular direction, you naturally expect them to be accelerated in that direction.

So to come back to the object moving round a circle at constant speed, there must be a force on the object directed along the radius vector towards the centre. It may be easier to look at this the other way round. It is this force that makes the object travel in the circular path in the first place; in the absence of the force, Newton's First Law would say that it suffers no acceleration, and it should travel in a straight line, not a circle.

This brings us right to everyday experience. A young child may tie a string to a stone and whirl it round. The stone travels in a circle; it does so because the string is taut, and the tension in the string acts as a force on the stone inwards along the radius. Should the string snap, this force no longer acts, and the stone flies off in a straight line.

While it is travelling in the circle, we may calculate the force on it, as the product of its mass and its acceleration, and I wrote an expression for the latter a few paragraphs back; the force is thus equal to the mass of the stone multiplied by the square of the speed and divided by the radius. All this makes good sense. The more massive the stone, the higher its speed, the harder the string must pull; if the circle is small, the stone is orbiting rapidly, and again it is natural that a large force is required.

As another example of circular motion, consider a car on a circular race-track, or just turning a corner sharply. Again there must be a force towards the centre; in this case it is provided by friction at the tyres. On an icy day, there might not be enough friction and the car may skid. In any case, if we go too fast, the force may turn the car over.

In the following section, I use these results for circular motion to help discuss planetary orbits. Then, in Chapter 4, it will emerge that Niels Bohr's great contribution to the development of quantum theory, the *Bohr atom*, used the very same ideas, in this case the body performing a circular orbit being the electron, a sub-atomic particle, and the required force being an electrical attraction between the electron and another sub-atomic particle, the proton, which is at the centre of the circle.

Planetary motions and gravity

I now wish to consider a planet orbiting the Sun, using, for the moment, the approximation that the orbit is a circle traversed at constant speed. (This is a very good approximation for nearly all the planets, as noted above.) It seems that there must be a force on the planet towards the Sun. Newton's Third Law reminds us that there must be an equal and opposite force on the Sun towards the planet. The force must be one of *attraction*; the planet is attracted towards the Sun, the Sun towards the planet.

Though the forces on Sun and planet are equal in magnitude, the accelerations are very different. To calculate the acceleration we divide the force by the mass. Since the mass of the Sun is so much bigger than that of any of the planets, its acceleration must be very much smaller that that of the planet. Thus to an extremely good approximation the planet circles the Sun rather than vice versa. (Copernicus was right!) To a slightly better approximation, they both circle a point on the line between planet and Sun, but this point is *so* close to the centre of the Sun that this refinement makes no real difference.

What we have obtained so far is important, but only partial information. We need to know how the force depends on the masses of Sun and planet, and the distance between them. *Any* such force-law can give a circular orbit (though not every case will be stable; in others, a slight perturbation to the orbit, such as that caused by a passing comet, would cause the planet to spiral away from its orbit).

However, the circle is always a special case, and a given force-law allows a more general orbit. Knowledge that an orbit is elliptical still does not specify the force-law; if one were to suppose that force is proportional to distance apart, for example, one would be led directly to elliptical orbits. It was the great good fortune of Newton that he had the full information of Kepler's Laws available and it was Kepler's Third Law in particular that led to the unique conclusion that the force between two masses is inversely proportional to the square of the distance apart. This is *Newton's Inverse-Square Law of Gravitation*, certainly one of the very great achievements of science, and for two and a half centuries triumphant in explaining so much of physics and astronomy, yet, ironically, shown by Einstein's general theory of relativity to be not only an approximation to, but conceptually incompatible with, the newer and better theory.

I will spell this law out – if any two particles of masses m_1 and m_2 are separated by distance r, there is an attractive force on each, proportional to m_1 and m_2, and inversely proportional to the square of r. The constant of proportionality is always written as G, Newton's gravitational constant, which is related to, but very different from, g, the acceleration due to gravity introduced above.

I have used the word 'particles'; this is to indicate that we are thinking of them as point masses, with dimension either strictly zero, or infinitesimally small. Only thus can one sensibly talk of a unique distance between them. Some of the most important sub-atomic particles, such as the electron, are thought to be point particles, whereas others, such as protons and neutrons, have dimensions that are not strictly zero, but may be approximated to it. I should also mention that the law has m_1 and m_2 to the first power; I have not yet justified this, but shall do so shortly.

When I return now to consideration of the gravitational attraction between a planet and the Sun, it is quite clear that neither can be treated as a single particle; each contains many particles and occupies a considerable volume, and we should really add together the gravitational interactions between *each* particle in the Sun and *each* particle in the planet. For each interaction, the relevant distance and the direction of the force will differ somewhat. Nevertheless, because the radii of Sun and planet are very much smaller than the distance between their centres, it will certainly be at least a very good approximation to lump together the total mass of each body at its centre, and use the simple equation above with r the distance between the centres. We do this almost automatically.

However, things are very different when we look at the fall under gravity of an object close to the Earth. On the positive side, it is clear that Newton's analysis of gravitation *does* explain this fall. The object experiences a gravitational attraction towards the Earth. (The Earth, of course, experiences an equal and opposite force towards the object. While the *forces* are equal, the *accelerations* are very much not; since acceleration is equal to force divided by mass, the acceleration of the Earth towards the falling object is enormously less than that of the object towards the Earth.)

It is more difficult to obtain a simple expression for the magnitude of the acceleration. Whether or not the dimensions of the object itself are small enough to be ignored, it is clear that those of the Earth are not. At first sight it would appear that an enormous calculation is required. The closest parts of the Earth may be only a metre or so from the object, others further away than the diameter of the Earth.

It is a remarkable fact, and it was one of Newton's hardest tasks to prove it, only true for a sphere, and only true because the law is inverse-square, that the gravitational attraction of a spherical mass of any size is identical to that of a mass equal to the mass of the sphere placed at its centre. (People are often told this at an early age, and come to believe that it is obvious, and should apply to a mass of any shape. It isn't, and it doesn't!)

Thus for an object close enough to the Earth that its distance from the centre of the Earth is not appreciably different from the Earth's radius, R_E, the magnitude of the force of gravity on the object must be equal to G, the gravitational constant, multiplied by the product of the masses of the object and the Earth, m and M_E, and divided by the square of R_E. Since R_E, M_E and G are constants, the force has the form of m times a constant; this is precisely in line with our very first remarks on gravitational fall, and this constant must be equal to g, the acceleration due to gravity. This analysis, then, concludes the demonstration of how Newton's mathematical study of gravity unified the terrestrial and heavenly regions.

I must briefly mention two points. First, I have assumed that the Earth is a perfect sphere. As is well known, it is, in fact, slightly flattened at the poles, as a result of which g varies a little over the surface of the Earth. Again, both the mass and the radius of the Moon are less that those of the Earth; the net result is that the acceleration due to gravity on the moon is smaller than g by a factor of about seven, and we have all seen pictures of astronauts demonstrating this fact.

Secondly, the fact that the force on an object falling under gravity was proportional to its mass was a result of the masses appearing to first order in Newton's Law of Gravitation, and so justifies that feature of the law, which I left unexplained when I introduced it.

I have made a point of describing Newton's work as a *mathematical account* rather than a *theory* of gravitation, and he emphasised this himself. There was no explanation of what the gravitational interaction actually was, how it was carried from one body to another. Newton and his contemporaries were profoundly suspicious of 'action at a distance', which seemed a reversion to Aristotle, and they continued to look for a mechanical basis for gravity. However, as the outstanding mathematical success of his ideas became more and more evident, such worries were initially relegated to second place, and then forgotten, and gravity became accepted as an intrinsic property of the interacting particles.

The gravitational attraction could be described mathematically as taking place via a *gravitational field*, a mathematical quantity that took non-zero values in the region between the particles, but this field was not thought of as having real physical existence. Like much else, this opinion has been revised in the present century, and it is now supposed that the gravitational interaction is transported by particles, the 'field particles' of the gravitational field, called *gravitons*. The disturbance caused by individual gravitons is so small that it seems unlikely they could ever be detected.

For confirmation of the idea of gravitational attraction, I would like to mention the work of Henry Cavendish [14] at the end of the eighteenth century. In exquisitely controlled experiments, he was able to demonstrate and measure the gravitational attraction between quite small lead spheres, by using this force to twist a wire of known stiffness.

But to conclude this section I want to return to planetary orbits. Till now I have assumed that the planets experience the gravitational attraction only of the Sun. Because the Sun is so much more massive than any planet, this is an exceptionally good approximation, but not quite correct. When the influence of other planets is taken into account, it is found that the elliptical orbits do not remain quite stationary, as Kepler had suggested, but rotate or 'precess' extremely slowly in a fixed plane. Considerable success was achieved in comparing calculations with observations.

However, in the middle of the nineteenth century it became clear that there was precession in the orbit of Uranus over and above the theoretical prediction. To explain this, the existence of an as yet unknown planet was postulated at a specific position in the solar system. The planet was found in the predicted region and called Neptune, and this was another great triumph for Newton.

Yet the same type of analysis on Mercury ended in failure [15]. This planet had a total precession of around 5600 seconds of arc per century (where 3600 seconds of arc is one degree), and this number was known to better than one part in 10 000. Calculation using perturbations from all known planets gave a figure of 5558 seconds of arc per century, accurate this time to one part in 25 000. There is a discrepancy of over 40 seconds of arc per century, a tiny amount, but clearly well beyond experimental or calculational uncertainty. Similar manoeuvres to those which located Neptune were tried, but all failed, and the difficulty remained a continuing sore for the rest of the century, though few would have expected it to be insoluble by Newtonian means given enough ingenuity. In the end it was explained only by the general theory of relativity, a theory it played no part in generating; the extra precession was an unexpected and triumphant result.

The Newtonian Universe

I have sketched the broad outline of the Newtonian system, and I now want to consider what it tells us about the nature of the Universe. The first point to stress is the predictability of the system. At any particular moment, it would be supposed that every particle in the Universe has a given position and a given velocity (speed *and* direction). (As will emerge, this seemingly innocuous statement is flatly contradicted by any orthodox interpretation of quantum theory.) The forces on each particle are then fully determined by the laws of classical physics. In this book we will not discuss all these in detail, particularly the electrical and magnetic forces, but they are, of course, well understood.

Newton's Laws then determine the acceleration of each particle, that is to say, how the system will (instantaneously) develop in time. I say 'instantaneously' because everything – positions, velocities, forces, accelerations of each particle will vary continuously. Some very simple cases may be analysed mathematically and exactly; most certainly cannot. But in all cases the behaviour of the system is *deterministic*; the state of the system at a given time determines its state at all later times.

To make this idea a little more concrete, the *Laplace demon* was dreamed up. Through a long career from the 1770s to the 1820s, Pierre Laplace applied Newton's Laws in great detail to the heavenly region. He now imagined a calculator of vastly greater power even than himself, able to take in detailed information about every particle in the Universe, and announce the entire future of everything.

Such a suggestion is worrying, of course, to most people – at least if applied to themselves. If we contemplate a great number of atoms shut up in a box and colliding with each other and the walls – very much the kind of model used for kinetic theory later in this chapter – there is nothing very unpleasant about the idea that the behaviour could be predicted in advance. But what about human beings? Do we not have free will? Can our actions justifiably be termed good or bad if they are all implied in the state of the Universe at its very beginning? Is there not something in us – our humanity – over and above the immediate and physical, that is not merely a facet of Newton's Laws? (If so, is it there in diluted form in a dog, and considerably more diluted in an earthworm?)

In this book I cannot, of course, trace these arguments, so important in the interaction of science, philosophy and theology, through the centuries. I will, though, just mention a book by Squires [16], in which he argues that, far from determinism being incompatible with free will, it is a requirement of it. What I will stress, though, is that the arguments lay outside physics proper. Though physicists might take part in them, the arguments played no part in their day-to-day work.

In the quantum era, these arguments have become, if not centre-stage, at least lurking in the wings. Quantum theory fairly conclusively sweeps away determinism. More dramatically, a special place for human 'observers' and a role for consciousness are possibilities taken seriously in order to solve definite physical problems, though they remain controversial.

They form part of what is called the debate over *realism*. While I have noted Newton's caution about the gravitational interaction itself, for him and his successors there could have been little doubt about the 'reality' of the objects they investigated and their properties. To scientists it would have seemed idle to question whether the objects they studied, through a telescope or on Earth, actually existed, or whether their properties, such as position and velocity, existed independently of observation. Newton's mathematics appeared to describe, and where required predict, a real physical Universe, which it exactly mirrored.

These beliefs *were* questioned, of course, but by philosophers. It could not be ruled out that the stimuli reaching one's senses were artificial, deductions from them were meaningless, and there was no 'real' Universe outside one's own brain. It could never be proved that objects existed, or had particular properties, in the absence of observation. But scientists would have rejected such speculation as ridiculous.

To their great surprise and, in some cases, horror, such possibilities now have to be faced up to. Quantum theory does seem to suggest that, if not the objects themselves, at least their properties are created by observation. Not everybody agrees with this, of course, and arguments for and against play a large part in the second half of this book.

To conclude this section, I shall discuss a fairly recent scientific development which is related to determinism, the theory of *chaos* [17, 18]. To explain what is involved, let us return to the Laplace demon, and imagine it performing its initial measurements, and recording its initial data. It would be freely admitted that there must be *some* uncertainty in the data; even by limiting the number of decimal places written down, *some* error must be introduced. Laplace would have accepted this for his own planetary data. He would have felt, though, and so would everybody else till quite recently, that small errors in input would have led to small errors in output, errors that would have increased as the time between input and output was extended, but increased slowly and steadily. Laplace could justifiably have claimed that the success of his celestial calculations justified this view.

Chaos theory probably had to wait to be discovered until computer simulations became routine. They showed that there are other types of system – often quite simple, governed by simple mathematics – which behave very differently. Small inevitable errors in input grow speedily and in ever-increasing fashion, so that the output rapidly becomes totally meaningless. A pendulum surrounded by a suitable array of magnets can behave in this way. So can the population of a particular species in successive generations. So too can weather systems, leading to an apparent limitation in principle of the length of time ahead for which forecasting could be possible, at least for some initial configurations.

What effect does this have on the earlier discussion of determinism? I think the answer is not much. The Laplace demon may have its nose put out of joint by chaotic systems, but that does not affect the fact that, even for such systems, determinism *does* hold, provided, of course, that everything remains classical. The state of the system at one time still uniquely determines its state at a later time, even though in practice it would scarcely be possible to calculate it. There is a great difference between the lack of calculability manifest in (classical) chaotic systems, and the inherent lack of determinism of quantum systems.

A sketch of classical mechanics

I want in this section, which is a little more technical than much of the book, to give a few more details of how Newtonian mechanics actually works,

Fig. 2.2. Each diagram shows vector addition.

and to introduce some further physical quantities required in discussion of any mechanical system, classical or quantum.

So far I have used velocity, force and acceleration. Now I introduce a quantity called *momentum*, which is very important for an alternative statement of Newton's Laws. The momentum of an object is just defined as the product of its mass and its velocity. It is important to note that, since velocity is a vector, with a magnitude *and* a direction, the same applies to momentum. (Mass, of course, has no direction.)

An important question for us now is that of how to find the total momentum of two or more bodies, with velocities not necessarily in the same direction. This is just an example of a general technique – the addition of a number of vectors. The general rule is shown in Fig. 2.2(a), the so-called 'triangle rule'. Note that, except in some simple cases, the magnitude of c, defined as the vector sum of vectors a and b, is not equal to the sum of those of a and b, and the direction of c is different from those of both a and b. Simple examples are shown in Figs. 2.2(b) and (c). In Fig. 2.2(b), a and b point along the same direction; c is in the same direction, and its magnitude is the sum of those of a and b. In Fig. 2.2(c), a and b are in opposite directions; c points in the direction of the one

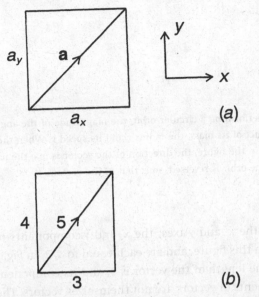

Fig. 2.3. Each diagram gives an example of the components of a vector.

with the larger magnitude, in this case **a**, and its magnitude is the difference between those of **a** and **b**.

Remembering that the sum of a number of vectors is defined in this way, we find that Newton's Laws may be expressed as follows: the total momentum of a system of particles is conserved; this is called the *Law of Conservation of Momentum*. For an individual particle, it clearly says that the velocity is constant – just Newton's First Law. For collisions between two particles, and since acceleration is defined as rate of change of velocity, Newton's Second Law says that the force on each particle is equal to the rate of change of momentum of that particle. But his Third Law tells us that the forces on the two particles are equal and opposite, so the rates of change of the momenta of the particles are equal and opposite, or, remembering the way momenta are added, the rate of change of total momentum is zero – momentum is conserved.

Those readers skilled in snooker or pool will recognise the part that conservation of momentum plays in collisions between the balls. I should point out, though, that while this law puts conditions on the velocities of objects following a collision, it cannot totally determine them; in addition one needs to consider the behaviour of energy during the collision, as will be seen shortly.

At this point I shall explain one other very important aspect of working with vectors. The addition formula displayed in Fig. 2.2(a) is simple geometrically, but usually awkward to express algebraically. A nice way to handle it involves *components* of vectors. In Fig. 2.3(a) a two-dimensional example is shown. Vector

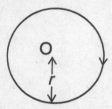

Fig. 2.4. For a particle performing a circular orbit, the magnitude of the angular momentum is the product of its mass, the radius r and its speed v. When the direction is that shown in the figure, the direction of the vector is *into* the paper; when the direction of the orbit is reversed, so is that of the vector.

a is of length 2 and bisects the x- and y-axes; the x- and y-components of this vector, a_x and a_y, are shown in this figure, and are each equal to $\sqrt{2}$. In Fig. 2.3(*b*), another example is shown; the length of the vector is 5, and its components are 3 and 4, respectively. Components of vectors are not themselves vectors. The nice thing is that, in the vector addition of Fig. 2.2(*a*), it is quite easy to work out that c_x is the sum of a_x and b_x; c_y is the sum of a_y and b_y. This makes use of vector components very convenient, and they will be used a good deal in later chapters of this book.

Another very important quantity is *angular momentum*. For the case of a particle travelling in a circular path at constant speed, the magnitude of the angular momentum is defined to be the product of the particle's mass, its speed and the radius of the circle.

I have said this is its *magnitude*, because angular momentum is defined as a vector, so a *direction* must also be specified. It is, of course, far from obvious what the direction should be defined as; after all, the direction of motion of the particle itself is constantly changing. The rather surprising answer is that it is convenient to define the direction of the angular momentum vector to be at right angles to the plane of the orbit. For the case of Fig. 2.4, the direction is *into* the page; if the direction of rotation is reversed, so is the direction of the vector. Though it may seem unlikely, with this definition virtually all the usual properties of vectors are obtained.

What I have described so far is *orbital angular momentum*, due to motion in a particular path or orbit. Classically there is also another type, *spin angular momentum*; the Earth spins about its axis, and sports players are well aware of the uses and perils of spinning balls. While angular momentum is very important classically, in some ways it is even more important in quantum theory; both orbital and spin angular momentum play central parts in the quantum theory of atoms.

I shall now discuss another extremely important quantity in both classical and quantum physics – that of *energy*. While general ideas of the part played by this quantity go back to Newton, a full understanding was not obtained till the nineteenth century, because, while energy played an important part in mechanics, it also straddles every other area of science – electricity, magnetism, light, heat, chemistry and so on.

I shall start from the basic point that there is *conservation of energy*. Though energy is readily transformed from one form to another, the total amount of the different forms is constant.

The most natural type of energy to mention first is *kinetic energy* – energy due to motion. The amount of kinetic energy associated with a moving particle is equal to half the mass times the square of the speed. It should be noted that, though the motion itself has, of course, a direction associated with it, kinetic energy, like all other forms of energy, is a scalar; *it is entirely specified by its magnitude.*

Now consider transformations. A rolling ball may roll up a hill, transforming kinetic energy into what is called *gravitational potential energy*; when it rolls down again, the reverse transformation takes place. A useful formula is that when a body of mass m is raised in height by an amount h, the increase in its gravitational potential energy is the product of m, the acceleration due to gravity, g, and h. Regarding potential energy as stored or hidden energy, other examples are a stretched elastic string and a squeezed elastic ball; the string or ball may revert to its usual length or shape, and, as it does so, potential energy is transformed into kinetic energy.

Energy is associated with electricity, magnetism and light; some details will emerge in this and later chapters. Chemical reactions may produce or require energy. This is discussed classically in terms of *chemical energy*; an explanation of this quantity in rather deeper terms is given in the following chapter.

Here I want to stress the importance of heat, or thermal energy. When we boil a kettle, electrical energy is transformed into heat. More fundamentally, in nearly every mechanical process, the end-product is heat energy. A ball rolling along a flat though not *perfectly* smooth surface gradually slows down as kinetic energy is 'degraded' to heat. When you get up in the morning, you take in chemical energy with your breakfast; during the day this energy is transformed into kinetic and possibly other forms of energy at various times, but by the end of the day it is all 'wasted' as heat. I shall return to this topic, and a better definition of the word 'heat', later in this chapter.

For the moment, I consider further the collision process, taking the simplest possible example – two particles of equal mass, m, each travelling at speed v, collide head-on. Remembering the directional nature of momentum, the total initial momentum must be zero. By conservation of momentum, the total final momentum must also be zero, which tells us that the particles bounce off each other with equal speeds (which may be zero). Momentum alone, though, cannot tell us what the speeds are.

Each particle initially has kinetic energy equal to half the product of m and the square of v; since there is *no* directional element in kinetic energy, the total is just m times the square of v. To complete the analysis, we need to know the nature of the colliding particles. If they are rubber balls, very little energy is lost as heat in the collision, so to a good approximation the final kinetic energy will be the same as the initial value. The balls will bounce off with speeds the same as before the collision, but of course directions reversed. This is called a perfectly *elastic* collision. The other extreme is if the balls are made of putty. In this case they coalesce, and *all* the initial kinetic energy is lost as heat; this is a totally *inelastic* case. All intermediate cases are possible, in which a fraction of the initial kinetic energy is lost as heat, and the balls bounce off with any speed between zero and v.

I now wish to make one final point about energy. Some types of energy – kinetic energy and energy of stretching, for example, can take positive values only. Other types, such as that due to the interaction of electric charges, may be positive or negative (as will be explained shortly). So some *systems* are restricted to positive values of energy, whereas others may have positive or negative energy. However, once this restriction is taken into account, one would expect that, classically, any energy would be allowed. One would never even contemplate the possibility that only certain specific values would be permitted.

Let us look at an example, one that will be particularly important later on, the *simple harmonic oscillator* (SHO), examples of which are a pendulum in a clock, and a mass vibrating on the end of a stretched spring. Imagine such a system left untouched for a time, so it is stationary (or in an *equilibrium* position). If the pendulum is then displaced to one side, the mass pulled up or down, either will continue to oscillate about the initial equilibrium position, as shown in Fig. 2.5. On this sketch, T is called the *period* of the oscillation, and the *frequency*, f, is just the reciprocal of T. Of course, to be more precise, the *amplitude* of the oscillations should decrease slowly in time, and the oscillations should eventually cease. I have assumed, though, that this *damping*, as it is called, takes place so slowly that I can neglect it for the length of my diagram, and have therefore kept the amplitude constant at a. On the vertical axis is what is called the *displacement*, y, and it is this quantity which oscillates in time between a and $-a$.

What technically distinguishes an SHO from any other oscillator is that, if the displacement at any particular moment is equal to y, the resulting force

Fig. 2.5. Shown is the displacement of a simple harmonic oscillator as a function of time. The amplitude is given by a, and the period by T.

back towards the equilibrium position must be proportional to y. However, what is really important is that, provided that oscillations are small, the behaviour of the great majority of oscillating systems approximates to an SHO. This is good because the SHO is very easy to analyse mathematically. (Technically, the curve in Fig. 2.5 is sinusoidal.) Classically it would be taken for granted that the SHO could have any (positive) value of energy. One of the great shocks of quantum theory is that, instead, its energy is restricted to certain discrete values; it is *quantised*.

The opposite to the quantised case is the *continuous* one, for which all values are allowed. In quantum theory, some types of system display the first type of behaviour, some the second, and others display both in different energy ranges. Neither is energy the only quantity that may be quantised; quantisation of the angular momentum (orbit and spin) of atoms is very important, for example. All will be revealed in Chapter 4.

The behaviour of waves

We are all familiar with waves at sea, waves on strings (as in violins), sound waves, and so on. A wave may be thought of as an oscillation in both time and space. Consider standing in the sea with fairly robust waves coming in. At any particular point, the water goes up and down in time rather like Fig. 2.5. However, if we change our point of view, and restrict ourselves to a particular instant of time, but look out to sea, we will see the water height behaving like Fig. 2.6. On this sketch I have marked the wavelength, λ (lambda), the distance from one peak to the next; the amplitude, a, and displacement, y, are as on Fig. 2.5.

I shall immediately state two important facts about waves. (To avoid confusion, I should mention that I am thinking of cases, like the sea-wave one above, where the medium in which the wave travels is of large extent, so one may

Fig. 2.6. The displacement of a waveform as a function of position. The amplitude is represented by a, and the wavelength by λ.

think of a long wave-train moving about without meeting any boundaries. This is often called a *travelling wave*. I shall look at the opposite *standing-wave* case, for which boundaries are all-important, at the end of this section.) First, it is clear that waves move at a certain speed. One of the most famous equations of physics says that the speed of the wave is equal to the product of its wavelength and its frequency.

Secondly, it is also clear that the wave carries energy. The *intensity* of the wave is defined as the rate at which energy is carried past a particular point. This could not be related directly to the value of y, because the average value of that in time or space is zero. It makes good sense that it is proportional to the *square* of y, and so, averaged over space or time, to the square of the amplitude a.

We must now ask what are the characteristics of a wave? How do we know that a particular phenomenon is wavelike? This may sound a strange question for sea-waves, or waves on a string, where the wavelike nature is obvious (though less so for sound waves). Historically, though, there have been important cases where the nature of the phenomenon was much less clear – light, the nature of which appeared to have been cleared up in the nineteenth century, but had to be reconsidered in the twentieth, and then a series of discoveries towards the end of the nineteenth century, cathode rays, α-rays, β-rays, γ-rays (alpha-, beta-, gamma-rays), X-rays. The suffix 'ray' meant, in effect, that scientists did not know what they were.

The two chief contenders were waves and particles. (Indeed, nineteenth-century physicists would have said that they were the only possibilities.) To see these competing, imagine two people, one at each end of a swimming-pool and facing away from each other. Initially the surface is calm. Person A can then interact with person B in two ways, either by throwing an object (a particle or number of particles) in the appropriate direction, or by splashing so that a wave travels towards B. We can say A transmits *energy* to B by either means, or, even more significantly, A transmits *information* to B; for instance, the arrival of the

Fig. 2.7. A plane wave travelling from the left is incident on a narrow slit, and displays diffraction (bending) on the far side.

object or wave could be the (pre-arranged) signal giving *information* that it is time to leave the pool.

How, then, *can* one distinguish wave from particle? An obvious way is that a wave can travel round corners, or *diffract*. Figure 2.7 shows a water wave constrained to move through a narrow slit, and exhibiting diffraction on the far side. Particles, on the other hand, move in straight lines, gradually falling under gravity. Since we can hear round corners, it appears certain that sound, for example, must be a wave.

Another characteristic of waves is *interference*. I have not yet introduced the last basic property of waves – *phase*. In Fig. 2.8(*a*), the two waves have crests at the same values of *x* and troughs at the same values; it is said that they are *in phase*. In Fig. 2.8(*b*), crests for one wave coincide with troughs for the other; the waves are *totally out of phase*. (For this terminology to make any sense, the two waves must have the same wavelength, and so the same frequency.)

What happens when the two waves meet at the same point? It turns out that we must add the displacements together. In Fig. 2.8(*a*), we obtain a wave with displacement twice that of each of the starting waves; this is called *constructive interference*. In Fig. 2.8(*b*), the displacements sum to zero; this is called *destructive interference*. These two examples are extreme cases; one may have anything in between totally in and totally out of phase. Again, if, unlike the case in Fig. 2.8, the amplitudes are not equal, the totally out-of-phase case will give a reduced displacement, but not precisely zero.

The interference phenomenon would seem to be an ideal way of identifying a wave; it is difficult to see how a particle could exhibit such effects. In a general way, such phenomena may easily be seen, say on a calm swimming-pool surface. Disturbances may be created in two well-separated regions; as the waves spread out and meet, a fairly complicated pattern results from interference. For the case of light, an accurate experiment is described on page 46.

Waves may be of two different classes. In a water wave, the *direction of travel* of the wave is, of course, along the surface of the water. The *displacement* is that

Fig. 2.8. (a) y_1 and y_2 are two waves of equal amplitude and wavelength; they are *in phase*. Their sum y_s exhibits *constructive interference*. (b) y_3 and y_4 are two waves of equal amplitude and wavelength; they are *totally out of phase*. Their sum y_s exhibits *destructive interference*.

of the water surface above or below the undisturbed level; it is *at right angles to* the direction of travel. Similarly, on a string, the direction of the wave is along the string, and the displacement *at right angles to* that direction. These waves are called *transverse*.

Sound waves are different. A sound wave is a pressure wave. In the vicinity of a crest, the density and hence the pressure are higher than for the undisturbed case and, in the vicinity of a trough, they are lower. To achieve this, particles are oscillating about their natural positions, and their motion is *along* the direction of travel of the sound wave. Because of this, a sound wave is called a *longitudinal* wave.

Fig. 2.9. The two types of plane polarisation for *transverse* waves propagating in the z-direction: (a) shows polarisation along the y-axis, or vertical polarisation, which we call W_1; (b) shows polarisation along the x-axis, or horizontal polarisation, or W_2.

Transverse waves can exhibit one property not shared with longitudinal waves. This is called *polarisation*, and I shall explain it in terms of waves on a string. Motion of the string itself must be at right angles to the direction of travel of the wave along the string, that is, at right angles to the string. That still leaves two possible directions, themselves mutually at right angles, as shown in Fig. 2.9. These two geometries represent two different *polarisations* of the transverse wave, which may be described as polarisation along the y- and x-axes, or vertical and horizontal polarisation (or W_1 and W_2 in what follows). This type of polarisation is called *plane polarisation*.

We can obtain a different type of polarisation by adding together a wave of W_1-type, with maximum displacement for a particular value of z, z_0, occurring at $t = 0$, and one of W_2-type with maximum displacement at the same point, z_0, occurring at $t = T/4$, where T is the period of the wave. A little thought may convince us that, at z_0, the displacement is constant in magnitude, and

Fig. 2.10. The first three modes of vibration of a string clamped at each end.

rotates in the xy-plane in an anti-clockwise direction; this is called *circular polarisation*. There is an alternative form of circular polarisation, where the rotation is clockwise; this is obtained by taking W_2 with its maximum at z_0 at time $-T/4$ rather than $T/4$.

(I would mention that, although water waves are transverse, the water surface itself is two-dimensional, so in this case there is only *one* direction at right angles to both the direction of travel and the water surface, and hence only one polarisation, which is plane.)

It is obvious that polarisation is another characteristic of waves, but also gives the information that the waves involved are transverse. (Experimentally, the phenomenon is manifested when different polarisations behave differently at interfaces between two different media.)

Lastly in this section I want to consider *standing waves*. In contrast to travelling waves, here boundaries are important. The best example is a violin-string, which is clamped at both ends. Again in contrast to travelling waves, only certain wavelengths are permitted. Figure 2.10 shows that the length of the string must be equal to a whole number of half-wavelengths. These patterns do not move along the string, but just oscillate at the appropriate frequency.

For each *mode* (as the vibration pattern is called), the frequency is related to the wavelength; their product is equal to v, the speed of the wave, which depends only on the properties of the string, its mass and tension. Thus, if the length of the string is l, the mode of Fig. 2.10(a) has wavelength $2l$, and frequency $v/(2l)$; that of Fig. 2.10(b) has wavelength l, and frequency v/l. Note the existence of *nodal points* (or *nodes*), where there is no motion of the string. The end-points are not counted as nodes, so there are no nodes in Fig. 2.10(a), one in Fig. 2.10(b), two in Fig. 2.10(c), and so on. (Please don't confuse 'modes' with 'nodes'!)

All this is, of course, very important for music. Figure 2.10(a) represents the *fundamental*, and Figs. 2.10(b) and so on represent *overtones*. Merely plucking a

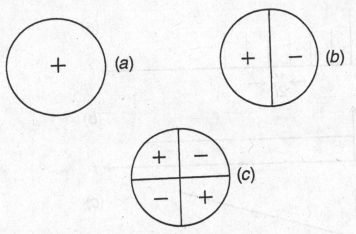

Fig. 2.11. Three of the simpler modes of vibration for a drumskin.

violin string does not give just one of the modes of Fig. 2.10, but a sum of them, a particular combination giving a note of its own individual character.

Other instruments may be analysed in similar, though usually more complicated, fashion. I shall give a very sketchy account of a drum, shown in Fig. 2.11. I consider the skin of the drum to be circular. Again different modes of vibration may be excited. Because the drum surface is two-dimensional, I cannot draw the displacement for the various modes in detail as in Fig. 2.10. Instead I mark with a plus sign regions where the displacement is 'up' at a particular time, and with a minus sign those where it is 'down'. The simplest mode is shown in Fig. 2.11(a); the whole surface vibrates together. In Fig. 2.11(b), the right-hand side is 'up' when the left-hand side is 'down', and vice versa. Figure 2.11(c) is more complicated still; there are four distinct regions. Note the presence of *nodal lines* (as distinct from nodal points in the case of the string) where there is no oscillation; there are none in Fig. 2.11(a), one in Fig. 2.11(b), and two in Fig. 2.11(c).

Such considerations are very important in many branches of science. Here I stress one aspect that was crucial at the very beginning of quantum theory. I want to discuss the frequency values of the various modes. For the wave on the string case, if we continue from Figs. 2.10(a), (b), (c) adding in one nodal point each time, we obtain the values of frequency, f, shown on Fig. 2.12(a). These values are equally spaced, the spacing being given by $v/(2l)$.

This is fine, but for some purposes the actual values of f are not as useful as the density of modes in a particular range of f. Because of the equal spacing in Fig. 2.12(a), the density, ρ (rho), shown in Fig. 2.12(b) is constant for all values of f. When we go to the two-dimensional drum case, though, the density of modes increases with f; it is actually proportional to f, as shown in Fig. 2.12(c).

Fig. 2.12. Part (a) shows the frequencies of the modes of vibration of a string of length l; (b) shows the discrete frequencies of (a) smoothed to give a density, ρ, of modes in a given range of frequency; (c) gives the same information as (b) for the two-dimensional drumskin case; the density, ρ, is proportional to the frequency f.

In Chapter 4, I shall consider an analogous three-dimensional case, and, through much of that chapter, study of modes of vibration of various systems will be very important.

A sketch of electricity and magnetism

In this section I shall present an outline discussion of those aspects of electricity and magnetism which will be most important in later chapters.

Magnetism has, of course, been known for many centuries, through the compass. We often discuss magnetic effects in terms of *poles*, a bar magnet having a 'north' pole and 'south' pole at either end, and the Earth having magnetic poles in the regions of its geographical poles.

While this is convenient, modern science considers that single poles do not exist, the basic unit of magnetism being the *magnetic dipole*, which behaves as a north and a south pole some distance apart. (A prediction by Paul Dirac, one of the founders of quantum theory, that a single pole, the so-called *magnetic monopole*, could exist independently, has been taken seriously, but none have been found.)

An understanding of even the most basic facts about electricity was more difficult to obtain. We are all familiar with such phenomena as a rubbed balloon clinging to the wall, sparks occurring between inner and outer garments during dressing, hair crackling while it is being combed, and so on. We start to interpret these observations in today's terms by noting that the basic constituents of (normal) matter are protons, electrons and neutrons. (For more details, see the last section of this chapter.) We say that a proton has a positive electric charge, an electron an equal negative charge, and a neutron no charge at all. Usually a body has equal numbers of protons and electrons, and is electrically neutral, but when two bodies are rubbed together, a number of electrons may become attached to the 'wrong' body, giving it a net negative charge, and the other body an equal net positive charge. (Note that this proton–electron model makes obvious *conservation of electric charge*; the total charge in a system remains constant.)

There is a force between charged bodies, called the *electrostatic force*. In one way this is analogous to the gravitational force; it is an inverse-square relation. The magnitude of the force between two charges is proportional to the product of the charges and inversely proportional to the square of the separation; this is known as Coulomb's Law, after Charles Coulomb, who was the first to demonstrate it explicitly in the 1780s.

In another way, the electrostatic force is more complicated than the gravitational, because there are two signs of charge – positive and negative. While masses *always* attract one another, it is found that, whereas unlike charges (one positive, one negative) attract, like charges (two positives or two negatives) repel.

(The balloon-on-wall case is rather more complicated. The charged balloon attracts towards itself particles in the wall of the opposite charge, and repels those with like charge. It is then itself attracted to the net opposite charge in the adjacent region of the wall. This is called *charging by induction*.)

Just as for the gravitational case, it is convenient to define an *electrostatic field* around a charged particle. We may use the language that it is this field that interacts with a second charged particle. Similarly we may define a *magnetic field* produced by a magnetic dipole.

I shall now say a little more about the energies associated with electrostatic and gravitational interactions. We need to define a zero of energy for these interactions, and we say that the energy is zero when the charges or masses are infinitely far apart. Now imagine two masses starting off a *finite* distance apart. Since they attract each other, we must *push* them further apart, and so give them energy to move them an infinite distance apart where they have zero energy. Logically, then, they must start off with *negative* energy when they are a finite distance apart. The same goes for two *unlike* charges, which also attract each other. Since two *like* charges repel each other, however, they must be provided with energy to push them *from* an infinite distance *to* a finite distance apart, so in the latter position they have *positive* electrostatic energy.

So far this section has contained (apart from a little gravitation) an outline of electrostatics. I have not yet mentioned the crackle in the hair, the spark between pullovers, which is an *electric current*, just a *flow* of electric charge. In these cases there is a flow of electrons from the negatively charged to the positively charged body, and this serves to restore electric neutrality. Lightning is a similar phenomenon; in the mid eighteenth century, Benjamin Franklin brought down lightning to Earth using a kite, and this led to the use of lightning conductors – metal rods on the sides of buildings to conduct or lead the lightning to the ground.

Most of the account of electricity so far has been of processes that are uncontrollable, or at least very difficult to control – they are interesting, but difficult to analyse, and even more to make use of. Gradually this changed. Following the distinction between bodies through which electricity runs freely (conductors, in modern language) and those which may be used to store it (insulators), the Leyden jar, or condenser, or (the modern term) capacitor, was invented in the mid eighteenth century. This consisted of two conducting plates with an insulating medium between them, and it could store large quantities of electric charge.

Then, through the last quarter of the eighteenth century and the first of the nineteenth, the subject of current electricity was developed. The most important discovery was that of Alessandro Volta in 1800. His *Voltaic pile* consisting of repeated sandwiches of zinc, copper and paper was the first example of a *battery*; between the bottom zinc and the top copper, there is developed what is called a voltage difference.

If point A is at a higher voltage than point B, the voltage difference being V, this means that it requires energy to move a positive charge q from B to A, the amount of energy being the product of q and V. If it is possible, a positive charge will move of its own accord from A to B, a negative charge from B to A. If a conducting wire, through which the electrons are free to move, is provided between the bottom zinc and the top copper, electrons will flow from low potential to high; this constitutes the electric current. (For historical reasons, the current is still spoken of as being a flow of *positive* charge *from* high *to* low potential.)

Current electricity is, of course, of great importance in heating, lighting, and so on. Here I concentrate on its connection with *magnetism*. In 1820, Hans Christian Oersted showed that a magnetic needle could be deflected by an electric current. From this, André Ampère at once deduced that, if a current-carrying wire acted as a magnet, two such wires should deflect one another, and showed experimentally that two parallel currents attract each other, whereas two anti-parallel currents repel each other.

For any particular geometry of electric current, it is possible to calculate the resulting magnetic field. For example, if we consider a long straight conductor, the direction of the magnetic field is tangential to a circle centred on the wire, as in Fig. 2.13. A very interesting case is a coil of conducting wire, for which it

Fig. 2.13. The magnetic field B due to a long straight conductor carrying current i.

Fig. 2.14. The magnetic field B due to a coil of conducting wire carrying current i. The field is exactly that of a magnetic dipole.

turns out that the magnetic field, shown in Fig. 2.14, is exactly the same as that from a magnetic dipole.

Ampère suggested that, in a sense, the logic of the previous sentence may be inverted. Perhaps the origin of all magnetism was just circulating currents at the atomic level. Each would give a magnetic field corresponding to a dipole, and this would explain why the dipole, rather than the single pole, is the basic unit of magnetism. This corresponds very closely to the explanation given by modern atomic physics, the Bohr model of the atom corresponding almost exactly to Ampère's ideas, and, though the more rigorous quantum theory gives a less concrete picture, the basic connection with magnetism has not really altered.

The work of Oersted, Ampère and others gave rise to the electric motor, in which the force on a current-carrying conductor in a magnetic field is exploited.

The next important step was taken by Michael Faraday [19, 20], one of the greatest experimental physicists of all time. If, as Oersted had shown, 'electricity produced magnetism', why should the reverse not be true? In 1831, after a decade of attempts, Faraday succeeded in showing that it was a *change* of magnetic field in the vicinity of a closed electric circuit that could induce a current, and the effect he demonstrated is called *electromagnetic induction*. The change in magnetic field itself could be caused by moving a permanent magnet near the circuit, or by switching on or off a current in a second circuit near the first.

Electromagnetic induction is behind the dynamo, where motion of a conductor in a magnetic field creates an electric current. Thus heat energy of coal or oil may be transformed into electrical energy.

Faraday also developed a pictorial but highly illuminating approach to electromagnetism. While I have explained the ideas of electric and magnetic field in rather a formal way as a means of calculating forces, Faraday attributed much more physical reality to them. If, for example, a card is placed on a magnet, and iron filings are scattered on it, they form loose chains running from one pole to the other. These provide a physical picture of the magnetic force as *lines of force*, and Faraday extended the picture to the electric field, which he also imagined as continuous lines of force starting at positive charge and ending at negative.

Faraday was able to interpret many electromagnetic phenomena as resulting from the interaction of these lines of force. He thus located these phenomena, and the various contributions to the energy of the system, in the medium, rather than at the coils and magnets themselves. While Faraday's work was not mathematical, Maxwell was able to build a rigorous mathematical system using the same ideas. This will be discussed in the following section after a detour to consider the nature of light.

Light, from Newton to Maxwell

Was light a wave or a particle? (Such, as I said earlier, seemed the only options for a long time.) If it were a wave, it would seem that diffraction should enable us to see round corners. For this reason, Newton favoured a particle-like model, though the experiment usually known as *Newton's rings* and shown in Fig. 2.15 led these to end up as unusual particles, subject to 'fits'. (Today we interpret these experiments as indicating interference between light that has travelled through the air-film and that which has been reflected at the surface of the lens, interference that may be constructive or destructive for a given wavelength, depending on the thickness of the air-film at that point. We thus deduce that light is behaving in a wavelike manner; I won't say 'is a wave' because that would have to be withdrawn in Chapter 4.)

It is easy to explain the lack of obvious diffraction, which is a result of the very short wavelength of light. White light consists of wavelengths between 3.8×10^{-7} m and 7.5×10^{-7} m, the violet and red ends of the *visible spectrum*.

Fig. 2.15. The so-called 'Newton's rings' experiment, in which the curved surface of a lens lies on a flat glass surface. Interference may be detected between light that is reflected at the glass–air surface at P and that which travels through the air along the path PRS. The nature of the interference depends on the thickness of the wedge of air at that point. If light of a specific wavelength is used, a series of light and dark bands will be seen as one moves out from the point of contact of lens and flat glass surface.

It was not till the end of the eighteenth century that Thomas Young performed an experiment that, at least in retrospect, showed clearly that light exhibited the phenomenon of interference. (At the time, interpretation of his experiment was highly controversial.) This was the famous *Young's double-slit experiment*. In Fig. 2.16, *monochromatic* light, that is to say light of a given colour, or a given wavelength, λ, from source L, falls on a screen in which two slits A and B have been cut. The light which travels through the slits reaches a screen beyond them. At point P, such that AP and BP are equal lengths, the light that has passed through the two slits must be in phase, and constructive interference results; a bright image is seen on the screen.

However, if we move to Q, where BQ is greater than AQ by exactly $\lambda/2$ (and similarly a point Q′, where Q′P is equal to QP; AQ′ is greater than BQ′ by $\lambda/2$), the light that has passed through slit A is totally out of phase with that which has passed through slit B. There is therefore destructive interference at Q and Q′, and a dark region. Moving up from Q, down from Q′, we meet points R and R′, where AR and BR, AR′ and BR′, differ by λ. We are back to constructive interference, and bright regions on the screen.

Overall, as we move across the screen, we see a whole series of alternate bright and dark regions. This seems clear evidence that light is behaving as a wave *in this experiment*. (During the nineteenth century, of course, people would have spoken much less cautiously; they would have said that the experiment 'shows that light is a wave'.) Mathematical analysis of the experiment enabled a value of the wavelength of the light to be obtained from the distance between successive bright regions.

Fig. 2.16. Young's double-slit experiment to demonstrate the interference of light. Light from source L passes through slits A and B. Since distances AP and BP are the same, constructive interference takes place at P. Q and Q′ are points such that pathlengths from A and B differ by $\lambda/2$, and there is destructive interference at these points. At R and R′, the difference in pathlengths is λ, and we are back to constructive interference.

More accurate values can be obtained if the double slit is replaced by a diffraction grating, a sheet of glass on which very many lines have been drawn mechanically. The diffraction pattern formed is of the same general nature as that produced by the double slit, but the maxima are very much sharper. The diffraction grating is used in atomic spectroscopy, a discipline of central importance in the development of quantum theory. (See Chapter 4.)

What kind of wave was light? A clue to one aspect was given by experiments involving calcite; light passing through this material is broken into two distinct rays travelling in different directions – *double refraction*. This phenomenon was explained by Augustin Fresnel in a thorough account of the wave nature of light published in the decade after Young's work. He assumed that light was a *transverse* wave, and the two rays in the double refraction corresponded to the two modes of plane polarisation. Everything seemed thoroughly convincing; today everybody has heard of 'polaroid' sunglasses, which preferentially transmit one polarisation, and thus reduce glare.

It was possible to measure the speed of light, which is one of the most important constants of physics, always known as *c*. During the second half of the nineteenth century and into the twentieth, progressively more accurate values were obtained by Armand Fizeau, Léon Foucault and Albert Michelson. The value accepted today is close to 3×10^8 m s^{-1}. (This is the speed of light in a vacuum; in air the speed is very slightly lower, and in other media it is considerably lower.)

Any further understanding of the nature of light had to wait till Maxwell's Theory of Electromagnetism of 1864. Following Faraday, though proceeding more mathematically, Maxwell [21, 22] wrote down a set of four equations for the electric and magnetic fields at any point. Three of these equations were based on the contributions of Coulomb, Ampère and Faraday himself which were described in previous paragraphs. (The fourth just expresses the fact that magnetic lines of force must be continuous, because there are no free poles for them to start or end on.)

Maxwell noticed an important inconsistency in the set of equations. Ampère's Law as discussed so far was correct only for the time-independent case, and Maxwell had to add an important time-dependent term. His final set of equations related the spatial variation of electric field **E** and magnetic field **B** at a particular point to their time-dependence and also to the density of electric charge at that point. These equations, the very famous *Maxwell's Equations*, should be regarded as the final and complete statement of classical electromagnetism; as such they are of comparable stature to Newton's Laws for mechanics.

A most exciting feature of these equations was that they predicted a new type of wave – the *electromagnetic wave* (or electromagnetic radiation). Such a wave would be produced whenever a charged particle is accelerated. The equations gave a value for the speed of such a wave (in a vacuum) and this value turned out to be precisely the speed of light as measured experimentally. It thus seemed likely that light was just such an electromagnetic wave, or, to put this a little more precisely, what we call light is that part of the electromagnetic spectrum (with wavelength between about 3.8×10^{-7} m and 7.5×10^{-7} m) to which our eyes are sensitive. We may express this equally in terms of frequencies, remembering that, for any wave, the product of wavelength λ and frequency f is the speed. For any electromagnetic wave, the speed is the same, c, so the larger λ, the smaller f, and so on.

It is well known that, outside the visible spectrum, there are waves detectable by various means, though our eyes are not sensitive to them – ultra-violet (UV) and infra-red (IR). These are part of the electromagnetic spectrum. It was considered good further evidence for Maxwell's theory when, in 1879, Heinrich Hertz succeeded, by vibrating charged particles at an appropriate frequency, in stimulating waves of the nature predicted by Maxwell, but of much greater wavelength

Fig. 2.17. The spectrum of electromagnetic radiation, showing the approximate bounds of each class of radiation in terms of both wavelength and frequency.

than light. These were the first radio waves. For each of these classes of electromagnetic wave, I have indicated ranges, in terms of both wavelength and frequency, on Fig. 2.17.

The microwave (M/wave on Fig. 2.17) region (with wavelength a few centimetres) was opened up in the Second World War for radar purposes; nowadays it may be better known for ovens. I have indicated X-rays and γ-rays on this diagram to represent present understanding. It was some time after their discovery, though, before their place in the scheme was confirmed, and in the meantime considerable confusion was caused, as explained later in this chapter.

Maxwell's work also made clear the nature of the electromagnetic wave. As shown in Fig. 2.18, it consists of an electric field and a magnetic field, mutually perpendicular, each perpendicular to the direction of propagation of the wave, and each oscillating at the frequency of the wave. This corresponds to one particular (plane) polarisation; for the other, the directions of the fields are interchanged.

Once it was recognised that light was behaving as a wave, one thing seemed clear; it must be a wave *in a particular medium*. It just seemed obvious that a wave requires a medium; one could not even imagine a wave on a string without the string, or a water wave without the water! Sound is also a wave *in a medium*, as proved by a well-known demonstration in which a bell is set ringing in a glass jar from which the air is gradually evacuated. The bell can still be seen vibrating, but no sound emerges, because sound cannot travel through a vacuum.

Fig. 2.18. The nature of the electromagnetic wave. The direction of propagation is shown by v. The solid lines show E and B at one time, and the dashed lines show them half a period later.

For electromagnetic waves, the name dreamed up for this postulated medium was the *ether*, and nineteenth-century scientists spent (wasted, one has to say, in retrospect) a lot of time studying its properties. They never made much sense of them and it really took both relativity and quantum theory to clarify the situation. Einstein's theory of relativity showed that there could be no ether; it took his (quantum-theoretical) idea of the photon to suggest how you could do without it.

Heat and thermodynamics

I have dealt with the two really fundamental areas of classical physics – mechanics and electromagnetism, based on Newton's and Maxwell's Laws, respectively. The science of heat is not based on new fundamental principles; it is really just an application of Newton's Laws. However, it deals with systems containing many particles; complexity leads to the subject's own generalisations and laws, although these are statistically rather than fundamentally based. The area still has major difficulties of principle, so conceptually it has required as much study by leading physicists as the more fundamental areas of research [23].

I shall start from the realisation that heat *was* a form of energy (rather than 'caloric', a weightless fluid, a belief that lasted well into the nineteenth century). At the very end of the previous century, Benjamin Thompson (later Count Rumford), who argued from the high temperature of the brass chips produced in boring cannon, and Sir Humphry Davy, who tried to melt lumps of ice by rubbing them together, had suggested that heat was energy, but it was not until the quantitative experiments of James Joule in the 1840s that the idea became generally accepted. These experiments included measuring the temperature rise caused by expending a known amount of energy stirring water in a vessel. Thus Joule, and also Hermann von Helmholtz, introduced the idea of conservation of energy, which was later much developed, and has been discussed earlier in this chapter.

We should really be rather careful how we use the word 'heat'. If we wish to be consistent, we should talk only about heat as *energy in transit*. When I heat water to boiling, or heat air in a balloon, I transfer heat to it. But we should never talk about 'the heat energy of a system' for the very good reason that such a quantity cannot be defined! Let us imagine a fixed mass of gas, in a balloon say, starting at one state, A (which I specify by pressure P_1 and temperature T_1), and ending at another state, B (specified by a different pressure and temperature, P_2 and T_2). We may arrange this process in many different ways. We may heat the gas to temperature T_2, then adjust the pressure to P_2 at fixed temperature T_2; or we may increase the pressure to P_2, then heat at that pressure till the temperature is T_2; or follow an infinity of other more complicated routes.

It turns out that each route requires a different amount of heat, so we must not claim to have 'put a particular amount of heat energy into the system' to change it from state A to state B.

There *is* a quantity that tells us how much energy has gone into the system itself; this is called the *internal energy of the system*, and written as U. Given the state of the system (pressure and temperature, say, for a fixed mass of gas), we *can* talk about 'the internal energy of the system' (but it is *never* correct to talk about 'the heat of the system'). What happens, for example, when a gas is heated at constant pressure is that the gas expands, and transfers energy to its surroundings, or, in more traditional and probably more meaningful language, *does work on* its surroundings. The heat energy transferred to the gas is divided between the increase in internal energy of the gas and the work done by the gas in expanding.

I shall now discuss the first of the theoretical methods built up to discuss heat and work. *Thermodynamics* was largely developed in the nineteenth century; it was based initially on information obtained from heavy industry – the efficiency of the various means of obtaining energy from coal, for example. This empirical information was codified as the Laws of Thermodynamics. The First Law related to the conservation of energy, and the account of heat, work and internal energy I have just sketched.

The famous (or infamous) *Second Law of Thermodynamics* established further limitations on what machines could do. One statement of this is as follows – it is impossible to take energy continuously from a (high-temperature) source, and use it to do work, without depositing some energy in a sink, which must be at a lower temperature than the source. It would be nice if we could break this law;

we could, for example, steam across the ocean by extracting energy from the sea. But common experience (of which the Second Law is a distillation) tells us that it can't be done. There is no available sink at lower temperature than the sea.

An alternative statement of the Second Law is that you cannot transfer heat from a cooler to a hotter body without providing energy. (In contrast, of course, heat will automatically flow from hotter to cooler.) A fridge, unfortunately, *does* require to be plugged in, and contributes to the electricity bill. The two statements of the law are equivalent, though it takes a little algebra to demonstrate this point.

After very penetrating insights by Sadi Carnot as early as 1824, these statements were articulated by Lord Kelvin and Rudolph Clausius in the 1850s. The subject of thermodynamics developed through the remainder of the nineteenth century (when Max Planck, later initiator of the quantum theory, spent the earlier decades of his career on such work) and into the twentieth (when a Third Law of Thermodynamics was announced, which is discussed in the following section).

The theory was a major achievement of science; it refrained from any discussion of the microscopic nature of the substances involved in any process – whether they are atomic or not, and used only macroscopic properties – pressure, temperature and so on. In spite of this, it produced, with great mathematical elegance, results of enormous generality and detail concerning the behaviour of substances and the potential efficiencies of industrial processes. In particular it made a great contribution to understanding the direction and nature of chemical processes, both in the laboratory and in industry. Its strength was that it was independent of any (possibly temporary) atomic model of the substances; of course, in turn, it was limited in what it could say about the nature of substances, as distinct from how they behaved.

The only aspect of thermodynamics I shall discuss here is the quantity called *entropy*, and the *arrow of time* [24]. In my previous discussion of energy, I implied a *direction to physical processes*, in the sense that useful energy, by which I mean chemical energy (fuels or food) and potential energy, is lost as heat, and ends up as what I now call internal energy. The Second Law reinforces this – heat flow to a sink at a low temperature is an inevitable part of physical processes.

The idea of entropy sums up this idea mathematically. Technically, its definition is such that, if an object at temperature T receives an amount of heat energy Q, its entropy goes up by an amount Q divided by T. It is difficult to get a feel for this quantity, though it may become a little easier to appreciate from a statistical point of view in the succeeding sections. Its importance, though, is that the total entropy of the Universe constantly increases. Some parts may

Fig. 2.19. The relationship between pressure and temperature for a fixed mass of a gas, as given by what is known as Charles' Law or Gay-Lussac's Law. The straight line cuts the temperature axis at −273 °C, the absolute zero of temperature.

exhibit a decrease in entropy, but that is always at the expense of other parts that exhibit a larger increase. Again there appears to be a clear 'arrow of the Universe' from past to future. This increase of entropy is an alternative statement of the Second Law of Thermodynamics.

Kinetic theory

I now move on to the second approach to these matters, an approach based wholly on the properties of the atomic constituents of the systems. While the method obviously relies on getting the atomic model correct, it would then seem to be a straightforward application of Newton's Laws. In principle this is indeed the case, but in practice one is talking about very large numbers of atoms, and is forced into methods that are, implicitly or explicitly, statistical, and the branch of science is called *statistical mechanics* or *statistical thermodynamics*.

As an introduction to this topic, in this section I shall discuss *kinetic theory*, which may be regarded as a statistical approach to the behaviour of gases. Knowledge of the properties of gases developed over a long period of time. What is known as 'Boyle's Law' or 'Mariotte's Law' was enunciated in the mid seventeenth century – for a fixed mass of gas at a constant temperature T, the pressure P is inversely proportional to the volume V. What is known as 'Charles' Law' or 'Gay-Lussac's Law' was discovered towards the end of the eighteenth century, and it is shown in Fig. 2.19. If the volume of a fixed mass of gas is kept constant, a graph of P against T is a straight line. While experiments on temperatures much below zero on the centigrade scale were not possible, if the graph is extended it cuts the T-axis at −273 °C. If P is kept constant, a graph of V against T has much the same form as Fig. 2.19, again cutting the T-axis at −273 °C. This led Kelvin to propose the use of an *absolute* or *Kelvin scale of temperature* obtained by adding 273 to the temperature in °C. So −273 °C is 0 K; 0 °C is 273 K; 100 °C is 373 K.

(Here K stands for Kelvin; the omission of the degree sign is standard, though, I would suggest, regrettable.) And $-273\,°C$ or $0\,K$ may be regarded as an *absolute zero* of temperature.

From now on, I shall always work with temperatures *in Kelvin*. When this is done, the gas laws above may be combined to say that the product of P and V is proportional to T. This is called the *general gas equation*, which is actually only an approximation, but a good one, particularly when the gas is not very dense.

Can this equation be explained by a simple molecular model of a gas? (I use the term 'molecular' rather than 'atomic' because molecules, not (necessarily) atoms, are the basic units of a gas; the relation between atoms and molecules will be explained towards the end of the chapter.) This was achieved by *kinetic theory*, which was developed by Clausius and Maxwell in the 1850s, though Daniel Bernoulli anticipated many of its features a century before.

Just about the simplest model of a gas we may think of is one in which (i) molecules are taken to be of zero size, so they do not collide with each other; (ii) they exert no forces on each other; and (iii) they move around the container travelling in straight lines but in random directions, and making perfectly elastic collisions with the walls, these collisions constituting the pressure of the gas on the walls. Straightforward analysis then leads to Boyle's Law. (The general idea is that, if V is increased, but the distribution of speeds of the particles is unchanged, they will have fewer collisions with the walls, so the pressure will be reduced.)

To recover the full gas equation above, we require the additional important idea that the average energy of the molecules is proportional to T (in, of course, Kelvin). Thus absolute zero would seem, classically, to be the temperature at which all motion ceases. This is not the case quantum-mechanically, where many systems must have non-zero energy even at the lowest temperatures. In any case, there is a *Third Law of Thermodynamics*, announced by Walther Nernst and developed by Francis Simon in the first half of the twentieth century, which says that it is impossible to reach absolute zero!

Many other properties of gases can be deduced from this simple form of kinetic theory, and improved assumptions (for example, a non-zero volume for molecules and a non-zero force between them) yield adjustments to the gas equation that are generally in line with experiment. The aspect I want to stress here is the distribution of speeds of the molecules. Maxwell showed that there was a natural 'Maxwellian' distribution (for all cases except the artificial one just described where no collisions between atoms are permitted). If the molecules happen to start with a different distribution, collisions rapidly change it to the Maxwellian form. In subsequent collisions, individual molecules will gain or lose energy, but statistically the distribution will be maintained.

Fig. 2.20. The Maxwellian distribution of speeds of molecules in a gas at a particular temperature. The shaded area gives the probability of the molecule having speed between v_1 and v_2. Since the probability of the molecule having *any* speed must be unity, the total area under the curve must also be unity. The most probable speed is given by v_m, and the average speed, which is slightly higher, by v_{av}.

The distribution is shown in Fig. 2.20, where the proportion of molecules with speeds between v_1 and v_2 is given by the shaded area of the curve, the total area under the curve being unity. We notice that there is a most probable speed v_m, but we find molecules with all speeds from zero to very high values. The average speed, v_{av}, is close to v_m, though they are not equal.

Also of interest is the distribution of values of *one component* of the velocity – of v_x, v_y v_z. This is shown in Fig. 2.21. (It should be noted that, unlike v, v_x may be positive or negative.) While Fig. 2.20 has a minimum at v zero, Fig. 2.21 has its maximum (most probable value of v_x) at v_x equal to zero. (The connection between Figs. 2.20 and 2.21 seems obscure at first sight. The point is that there is only one way of making v zero; v_x, v_y and v_z must all be exactly zero. On the other hand, there are very many combinations of v_x, v_y and v_z that result in values of v around v_{av} and v_m.)

Recognising that the kinetic energy associated with v_x is given by half the mass of the molecule times the square of v_x, it is easy to find a much more general relation, and an extremely important one, which is displayed in Fig. 2.22.

Here I am plotting the probability of this energy, E, taking particular values. Note that the highest probability is for E zero, but there is a non-zero probability of E having any positive value, though the probability decreases as E increases.

Fig. 2.21. The distribution of values of *one component* of the velocity of a gas molecule at a particular temperature. The shaded area gives the probability of the molecule having an x-component of velocity between v_{x1} and v_{x2}. Unlike the case of Fig. 2.20, the highest probability is for v_x equal to zero, and v_x may take positive or negative values.

Fig. 2.22. The distribution of values of energy associated with one component of the velocity of a gas molecule. E must be positive, and $P(E)$ takes its highest value for E equal to zero. This curve is the important Boltzmann distribution, and applies for the energy corresponding to any 'degree of freedom' of any physical system. The term 'degree of freedom' is explained in the text.

Statistical mechanics

I have now bridged the gap from the rather specific kinetic theory models to more general *statistical mechanics*. Figure 2.22 is known as *Boltzmann's distribution*. In Fig. 2.23, I show its form at three different temperatures – T_l, T_m and T_h (with subscripts denoting low, medium and high, respectively). For each case the distribution has the same general shape (and, since it represents a probability,

Fig. 2.23. The Boltzmann distribution, as explained in the caption of Fig. 2.22, for three temperatures, T_l, T_m and T_h. kT_l, kT_m and kT_h are measures of the widths of the probability distributions, and these increase as the temperature increases. However, and rather surprisingly, the highest value of probability remains at E equal to zero.

in each case the area under the curve must be unity), but as T rises the distribution becomes flatter, extending out to higher values of E (though, almost paradoxically, the maximum is always at E equal to zero). In fact, as I said earlier, the average value of E is proportional to T, the constant of proportionality being half the important *Boltzmann's constant*, which is always written as (small) k. The value of k is 1.38×10^{-23} J K^{-1}, where J is joules, the unit of energy, and (large) K refers to Kelvin.

The name refers to Ludwig Boltzmann [25], for 40 years one of the chief advocates of atomic ideas, and the main builder of statistical mechanics. His leading opponents were Ernst Mach, Wilhelm Ostwald and Pierre Duhem, the first because of the *positivist* desire to keep hypothetical quantities out of science, the others because of their support for the rival *energetics*, a form of generalised thermodynamics. The debates were often bitter, and Boltzmann committed suicide in 1906, ironically just as the existence of atoms was becoming firmly established. (Strangely, the so-called Boltzmann constant was never used by Boltzmann himself; it was first used by Planck in the course of the work that introduced the quantum.)

Boltzmann's distribution is so important because it is a probability distribution for each *degree of freedom* of *any* physical system. The idea of degrees of freedom is best explained by giving examples. A gas molecule moves in three dimensions, and has only kinetic energy. It has therefore three degrees of freedom, and its average energy at temperature T is $\frac{3}{2}kT$, that is $\frac{1}{2}kT$ per degree of freedom. (This does not include any rotational or vibrational energy; for this question see Chapter 4.) A simple harmonic oscillator in one dimension has one degree of freedom for its kinetic energy and one for its potential energy, so its average energy at temperature T should be kT. This *Law of Equipartition of Energy* is a very general result of *classical physics*, applying, for example, to modes of electromagnetic radiation, as will also be shown in Chapter 4. In retrospect at least, the failure of this law in some circumstances is just about the clearest evidence of the need for a quantum theory.

I shall now briefly review the statistical nature of our kinetic-theory model of a gas. Pressure is seen to be a statistical concept; it is caused by the random collisions of gas molecules with the walls of the container. Similarly, the concept of temperature is statistical; we may use the term only because there is a large number of molecules in the gas, with energies distributed according to Boltzmann's distribution. The temperature itself is related to the distribution, not directly to individual molecules. I would stress, though, that, although *our approach* is statistical in nature, the model in terms of positions and velocities of individual molecules is entirely deterministic. This is another point where at least standard interpretations of quantum theory differ.

An important part of Boltzmann's work was the discussion of entropy, a concept derived from thermodynamics, in statistical terms. To explain the fairly subtle argument involved, let me start with an example. Suppose that I have N molecules (a large number), which I shall put into a container that has a partition in the middle. If I wish to have all N on the left of the partition, clearly there is only one way in which I can choose the molecules to achieve this. If, on the other hand, I wish to have $N/2$ on the left and $N/2$ on the right, and I don't care which molecules go on which side (technically I may say that the molecules are *indistinguishable*), there will be very many ways to do it.

If I define W as the number of *microstates* (where particular molecules are on each side) corresponding to a *macrostate* (where only the numbers of molecules on each side are relevant), W will be one in the first case, but very large in the second. We may say that, if the molecules are put into the container at random, the probability of the second case emerging is very much greater than that of the first. So W may relate to the intrinsic *probability* of the macrostate. Alternatively, since we may say that the second case is much less ordered than the first, W may be called a measure of the *disorder* of the system.

Now suppose that we start off with all the molecules in the left-hand side, but otherwise with positions and velocities random, and raise the partition. We would expect that after a very short time we would be fairly close to the second case, roughly equal numbers of molecules on each side of the container. It thus seems that W has a tendency to rise, and in fact this is the case for more general situations. The statistical quantity W thus seems to behave in a similar way to the entropy of thermodynamics, and the precise link was made by Boltzmann. I shall write it in the form that the entropy, S, is equal to the product of k, Boltzmann's constant, and the logarithm of W. (As W increases, so does its logarithm, so S does as well.) This equation is the crucial bridge between thermodynamics and statistical mechanics, and it was inscribed on Boltzmann's grave.

Around 1870, Boltzmann himself constructed a famous theorem that claimed to show explicitly that W, and hence the entropy, S, always increased (or, to be exact, never decreased). Such would have been an amazing triumph; it would have established, directly from the atomic basis of matter, that there was an arrow of time pointing in the direction of increasing entropy, that physics was thus inherently irreversible. (The state of the Universe to which such processes would lead is the *heat death* where everything is at the same temperature and disorder is complete. This seemingly inevitable fate caused scientists considerable disquiet in the period before they perfected their own ways of providing a much speedier end for the Universe.)

But Boltzmann's theorem, which involved delicate approximations, could not be entirely correct, because, as Johann Loschmidt and Ernst Zermelo independently pointed out, the result itself is certainly wrong. Newton's Laws themselves are time-reversible. What this means is that, if a film of a number of particles colliding or interacting via Newton's Laws is shot, there is no way of distinguishing the film being shown forwards from its being shown backwards; in both cases Newton's Laws are obeyed. Again, if the motion of every particle in the Universe were suddenly reversed, Newton's Laws would still be obeyed, but the entropy must start to *decrease*.

Boltzmann's theorem more properly says that, in the fairly low-entropy state which the Universe is in at the moment (and which is necessary if life is to exist), it is extremely *probable* that W, and hence S, will increase. Such issues remain controversial; a detailed account is given by Coveney and Highfield [24].

Indeed, this is only one aspect of the general point that Boltzmann's work was by no means an established part of classical physics by the turn of the century.

During the first decade or so of the quantum theory, right up to Bohr's major work, the principal steps were taken by Planck and Einstein, and were statistical in nature and based on Boltzmann's theories. Yet Planck had, till recently, been a determined opponent of Boltzmann, and Einstein, though a strong believer in statistical methods, disagreed with Boltzmann on central issues such as the methods of counting W. It all made the development of quantum theory particularly troublesome.

Lastly in this section I want to mention the *ensemble* concept. This was largely the work of Willard Gibbs, who should be regarded as co-founder of statistical mechanics with Boltzmann, though Gibbs' isolation in America meant that he had little immediate influence on the European scientists.

To explain the use of an ensemble, let us consider the microstate of the actual system we wish to consider – a detailed listing of the position and velocity of *each* molecule. From this we may calculate the energy of the system. Now we take a jump and consider *all* possible microstates with the same number of molecules and the same energy. We are thus considering a very large number of microstates, with these two points in common, but often with very different behaviours of the individual molecules. This hypothetical collection of systems is called an *ensemble*. What Gibbs first showed is that one can do a vast amount of mathematical analysis on the ensemble itself; one can obtain average values of macroscopic quantities such as pressure and temperature for an ensemble of gas systems.

Such would not immediately seem to be of much use; after all, we want the pressure of the system we started off with, not the average of an ensemble of hypothetical systems. The other crucial point in working with ensembles, though, is that, provided that there is a large number of particles in each system, the overwhelming majority of members of the ensemble will have (macroscopic) properties very close to those of the ensemble average. This is rather like the problem of putting 1000 molecules, randomly and independently, on one or other side of a partition. Across all arrangements of the molecules, or all members of the ensemble, not only is the average number of molecules on each side equal to 500, but in addition we are overwhelmingly likely to be close to the 500 : 500 distribution.

To discuss this more precisely, we may mention *fluctuations*. (Einstein was the master of this technique, as we shall see in Chapter 4.) If the average pressure across the ensemble is P, then the average *fluctuation*, the average departure from P across the ensemble, measured as a fraction of P, is inversely proportional to the square root of the number of particles; it becomes smaller and smaller as the number increases.

The above is no more than an entry to the study of ensembles in classical statistical mechanics, a discipline with very many important applications across a wide area. I particularly discuss it here because, as will be explained in Chapters 6 and 8, Einstein and others have argued that quantum theory, even for systems naturally considered individual, is inherently a theory of ensembles.

The chemical atom

In this chapter, I have already said much about atoms, mostly, following from Boyle and Newton on to Clausius, Maxwell, Boltzmann and Gibbs, treating them as small hard spheres in motion, and examining their average behaviour in order to obtain values for the macroscopic properties of the systems they compose.

Another approach concentrates on the specific properties of individual atoms in chemical combination [26]. John Dalton is usually spoken of as the founder of (chemical) atomic theory, at the beginning of the nineteenth century. He postulated different atoms for each chemical element – hydrogen, oxygen, carbon and so on. The atoms themselves were taken to be indivisible and so small that it seemed impossible that one could ever experience a single atom. Atoms of a particular element had a characteristic mass. (Note that this brief statement justifies the two basic premises of modern chemistry – Boyle's characterisation of the chemical element as a substance that cannot be further decomposed and Antoine Lavoisier's statement of *conservation of mass*. With Dalton's use of atoms of a fixed mass, the latter requires atoms to be indestructible.)

Dalton made the important assumption that, to form a given chemical compound, *small numbers* of atoms of the constituent elements combine to form a *molecule* of the compound, this being the definition of the molecule in Dalton's theory. He assumed, for example, that one atom of hydrogen would combine with one of oxygen to give one molecule of water – HO. (Readers of exceptional chemical knowledge may recognise that this is not quite right, but please read on.) This assumption showed that water would always contain the same ratio of masses of hydrogen and oxygen, an example of what is known as the *Law of Constant Composition or Definite Proportion*, usually attributed to Joseph Proust just before Dalton's own work.

This argument suggests that the ratio of the masses of oxygen and hydrogen in any quantity of water should be the same as the ratio of the masses of individual oxygen and hydrogen atoms. Extension of this principle would give the masses of atoms of every element in terms of that of hydrogen, hydrogen being a convenient marking-point because its atoms have the lowest mass. It seemed to Dalton, and to everybody else for a century or so, that one could never hope to move beyond these *relative* atom masses to actual atomic masses. (So, in this book, atomic mass usually just means relative atomic mass.)

Now let us consider cases where two elements form a number of different compounds. An example of Dalton's ideas here was that a molecule of carbon monoxide should contain one atom of carbon and one of oxygen, while a molecule of carbon dioxide should contain one of carbon but two of oxygen – CO and CO_2. (Here I am using modern atomic notation, essentially due to Jöns Berzelius, which is more practical than the patterns proposed by Dalton, though not as attractive.) This idea (correct, this time in detail) suggested that the mass of oxygen in the two compounds combined with a particular mass of carbon should be in the ratio of 1 : 2, an example of the *Law of Multiple Proportions*. Dalton deduced this law from his theory, and then verified it experimentally.

Though clearly of tremendous potential power, Dalton's methodology is threatened by the problem mentioned for water above. How can one find out how many atoms of hydrogen and oxygen combine? Without this knowledge, no reliable atomic masses may be calculated. Berzelius used a number of ad-hoc rules, and produced a good table of atomic masses by 1814, and an excellent one, differing only in a few places from today's, by 1830. The most important of these ad-hoc rules was that of Pierre Dulong and Alexis Petit, which said that the product of atomic mass and thermal capacity is constant. (The thermal capacity is the amount of heating required to increase the temperature of a fixed mass of the element by 1°.)

A more rigorous development came from Joseph Gay-Lussac's Law of Combining Volumes of 1808. He showed that gases combine in definite and simple ratios by volume; for example, hydrogen and oxygen combine in the proportions 2 : 1 by volume. One may add the 1811 Law of Amadeo Avogadro, that, at a definite pressure and temperature, a given volume of *any* gas contains the same number of molecules, and deduce, as Berzelius did, that the formula for water is H_2O.

There remained a fundamental ambiguity, however, namely the distinction between atom and molecule for a gaseous element. This problem had actually been solved by Avogadro, but general acceptance did not come till the work of Stanislao Canizzaro in 1858. If one's starting assumption is that the molecule of many gases, including hydrogen, oxygen and nitrogen, contains *two* atoms, many hitherto rather confusing facts fit into place.

Chemical formulation could now proceed on the dual assurance that atomic theory was sound and that correct atomic masses were known; in the second half of the nineteenth century, structural chemistry flourished. Though based securely on atomic notions, it did not suffer at the hands of the positivists quite as much as Boltzmann's statistical mechanics; it was always possible to use the slogan that atomic concepts were undeniably useful, but might not represent ultimate reality.

Once atomic masses were known, it was interesting to see whether any pattern emerged. As early as 1815, William Prout observed that many elements had atomic masses almost exact multiples of that of hydrogen. For example carbon, oxygen and nitrogen had (relative) atomic masses of almost exactly 12, 16 and 14, respectively. Perhaps all atoms consisted of combinations of hydrogen atoms. However, many other atoms did not obey the rule, the atomic mass of chlorine, for example, being close to 35.5.

In 1866, John Newlands put forward the idea that, when the elements are written in order of atomic mass, there emerges a periodicity, which he described as *octaves*; every seventh element, he claimed, had similar properties. He was ridiculed for his ideas, and in retrospect it may be seen that both he and Prout had systems that were too rigid.

The first successful *Periodic Tables* were produced by Lothar Meyer and Dmitri Mendeleev around 1870. As with Newlands, elements were ordered by atomic mass; now, though, successive 'periods' could be of different length. Hydrogen was on its own, then followed two of Newlands' octaves, but then further periods of greater length (which could, of course, be interpreted as octaves with other groups of elements – *transition elements* – interposed at various places). As a result, elements at corresponding places in successive periods are, in all cases, similar – the halogens (fluorine, chlorine, bromine, iodine); the alkali metals (lithium, sodium, potassium); nitrogen, phosphorus and arsenic; and so on.

There were gaps, where Mendeleev in particular was prepared to predict missing elements, together with many of their properties; several such elements were eventually found. He also suggested, on the basis of apparent irregularities of ordering, that some atomic masses had been measured incorrectly. In some cases he was right – but not in all. The order of cobalt and nickel, for example, really *should* be the reverse of that given by atomic mass; it is a related but different parameter, *atomic number*, that actually provides the correct order. This was shown by Henry Moseley, but not for another 45 years.

In the last decade of the nineteenth century a great surprise was a new family of elements, with a member in each period of the Periodic Table. In a series of experiments, Lord Rayleigh and William Ramsey showed that, if all known gases were removed from the atmosphere, other gases remained, which were successively discovered as argon, helium (previously identified in the Sun by spectroscopic means, for which see Chapter 4), krypton, neon and xenon. (A final radioactive member, radon, was discovered by Ernest Rutherford.) These gases had remained hitherto unnoticed because they were extremely unreactive;

they have often been called 'inert gases', though now the term 'noble gases' is preferred.

So in today's Periodic Table, the first period consists of just hydrogen and helium. The second and third have eight members – Newlands' octaves plus a noble gas. As transition series begin, the next periods have 18, 18 and 32 members, and the seventh period remains incomplete.

Dividing the 'indivisible'

So far everything I have said has referred to whole atoms: the word 'atom' *does* mean 'indivisible'. I now want to turn to those discoveries around the turn of the century which emphasised that the atom had internal structure [27].

Before this, though, I shall briefly discuss *electrolysis* or *electrochemistry*. This phenomenon was discovered by Anthony Carlisle and William Nicholson immediately after Volta's invention of the battery in 1800. When an electric current was passed through water that had been made conducting by the addition of a little salt, hydrogen was produced at the *cathode*, oxygen at the *anode*. (The anode and cathode are the two *electrodes*, or plates at which the current enters the liquid, the current flowing from anode to cathode.)

Other substances were soon electrolysed, or broken down into constituent elements; Humphry Davy discovered new metals in this way – sodium and potassium. From such experiments came the electroplating industry in which a coat of tin or silver, for example, is deposited onto a cheaper metal used as an electrode.

It was Faraday who built up a quantitative approach to electrochemistry. As seems reasonable, the amount of hydrogen, or other element, liberated was proportional to the amount of electric charge that had flowed through the circuit. The amounts of different elements liberated by the same amount of charge were closely related to the atomic masses. This led Helmholtz to suspect the existence of a fundamental unit of electric charge, which would liberate one atom of any element. Thus when Faraday divided the amount of charge that had flowed by the mass of hydrogen produced, he was obtaining a value of e/M, where e is the unit of charge and M is the mass of a hydrogen atom.

(To put things a little more exactly, and in today's terms, it is known that hydrogen exists in solution in part as hydrogen *ions*, atoms that have lost one electron, and thus are positively charged as H^+. The electric current effectively supplies an electron at the cathode to produce an H atom, which is then liberated (two atoms becoming a molecule). Thus *e*, in Faraday's ratio, is explicitly the charge on the hydrogen ion; −*e* is the charge on the electron.)

I now move on to electric currents in gases, specifically gases at low pressures, where in the second half of the nineteenth century it was found that the current flows in straight lines, and may be blocked by solid obstacles, producing 'shadows' on the phosphorescent walls of the vessel. The flow was chiefly of negatively charged particles *from* cathode *to* anode, and was called a *cathode ray*.

An important question was posed: are cathode rays particles or waves? Electromagnetic theory says that, just like a wire carrying a current, a beam of charged particles should be deflected by a magnetic field, and it was found that cathode rays were so deflected. Also Jean Perrin found that a conductor became negatively charged when cathode rays fell onto it. These experiments seemed very difficult to understand if cathode rays were waves, very easy if they were particles, so the latter was assumed. (Like so much in this chapter, this seemed little more than common sense at the time, but will require revision in Chapter 4.)

In 1897, J. J. Thomson and Emil Wiechert independently performed experiments in which cathode rays were deflected by magnetic and electric fields simultaneously, but in opposite directions; the strengths of the fields were varied till the particles travelled in a straight line. In a second experiment only an electric field was present. Together, these experiments gave a value for the speed of the particles and, much more significantly, their charge/mass ratio, which turned out to be many times (1836 is today's value) the *e*/*M* obtained by Faraday. (The ratio was negative, since the charge on the particles was negative.)

Two obvious possibilities were that the charge was 1836 times higher than the one Faraday was considering, *or* the mass 1836 times lower. The latter perhaps seemed unlikely since atoms were supposed to be indivisible. Meanwhile, experiments to determine the value of the basic unit of electric charge were carried out by C. T. R. Wilson, Thomson and John Townsend. These experiments used droplets of water collected around singly charged particles and were not very accurate, but they did indicate that there was one basic unit (not two quantities differing by a factor of 1836). Also Hertz showed that cathode rays could penetrate thin metal foils to an extent that seemed unlikely if they were of atomic size.

Thus physicists became convinced that cathode rays consisted of particles with a mass 1/1836 times that of a hydrogen atom, and with a charge −*e*, where *e* is the basic unit of charge. It seemed reasonable that the particles were constituents of the atom, and they became known as *electrons*. An accurate value for *e* was not produced till the work of Robert Millikan in 1911; modern values make *e* equal to 1.6×10^{-19} coulombs, and *m*, the mass of the electron, equal to 9.11×10^{-31} kg, with the mass of the hydrogen atom 1836 times as high.

Another important discovery in this period was that of X-rays by Wilhelm Röntgen in 1895. These were produced in the vicinity of a high-voltage cathode-ray tube, when the high-energy electrons struck the glass wall. They could pass through low-density material, though not that of high density (and hence their medical use); they could darken photographic plates and make phosphorescent material luminous. Röntgen showed that they travelled in straight lines, and were undeflected by electric or magnetic fields, so they must be uncharged. If they were electromagnetic waves they would have to be of very short wavelength to explain their penetration.

I shall now explain how what may be called X-rays' place in classical physics was ascertained. In 1899, Hermann Haga and Cornelis Wind produced a diffraction photograph for X-rays indicating that they were wavelike with a wavelength of about 10^{-10} m. In 1904, Charles Barkla demonstrated that X-rays could be polarised, and were thus a transverse wave, probably electromagnetic.

These experiments were actually not totally convincing, but conclusively, in 1912 and on the basis of a suggestion of Max von Laue, interference of X-rays was demonstrated by Walther Friedrich and Paul Knipping. Since no man-made system could be of small enough dimension, von Laue's idea was to use a naturally occurring crystal of, in this case, zinc sulphide. A perfect crystal consists of a regular array of atoms and typical inter-atomic distances are between 10^{-10} m and 10^{-9} m. Thus it seemed that such a crystal could do for X-rays very much what a diffraction grating does for light, produce bright spots in particular places on an X-ray photograph. This is indeed what was found.

It was William Henry Bragg, and his son William Lawrence Bragg, who, in a sense, reversed the process to study the crystal rather than the X-ray itself. Once the basic mechanism of *X-ray diffraction* had become known, and the Braggs had simplified von Laue's efforts in this direction, it became fairly straightforward in principle, though increasingly complex in practice, to determine the structures of more and more complicated crystal structures. X-ray diffraction became practically a scientific discipline in its own right, *X-ray crystallography*, providing valuable information to physics, chemistry, metallurgy and molecular biology. For the last of these areas, I might particularly mention its contribution towards the determination by James Watson and Francis Crick of the double-helix structure of DNA [28].

From the point of view of this book, though, the importance of von Laue's work was to remove all doubt that X-rays were another aspect of electromagnetic radiation, with their own distinct range of wavelengths and frequencies (already inserted on Fig 2.17).

Yet so far, while I have told the truth, I have been extremely economical with it. A nice account of the wider issues is given by Wheaton [29]. Increasingly from 1904 on, W. H. Bragg had been able to marshal evidence that X-rays must have a particle-like nature. Bragg and Barkla had a vigorous controversy over the wave/particle question, particularly in 1908. I would stress here that, when X-ray diffraction was invented in 1912, Bragg's evidence did not go away. It is actually evidence of the need for a quantum theory, and is thus dealt with in Chapter 4. Here I just note that it is a good example of classical and quantum theories being developed simultaneously.

Another important discovery around the turn of the century was *radioactivity*. Henri Becquerel discovered in 1896 that compounds of uranium emitted radiation continuously and spontaneously. Like X-rays, this radiation was highly penetrating, darkened photographic plates, caused phosphorescence, and made gases conductors of electricity. The origin of the radiation was soon found to lie in the atom itself, and Marie and Pierre Curie sought other radioactive elements; their most important discovery, after immense labours of purification, was the element *radium*, thousands of times more radioactive than uranium. In fact all atoms more massive than bismuth (atomic mass 209) are radioactive.

Much of the intricate work of analysing radioactivity was carried out by Ernest Rutherford. There were, in fact, three distinct kinds of rays, α-, β- and γ-rays; γ-rays were the most penetrating and α-rays the least. The method of Thomson and Wiechart described above showed that β-rays were electrons travelling with speeds often approaching that of light. Similar experiments on α-rays showed that their ratio of charge to mass was much lower than that of the electron. When α-rays were trapped in a chamber, helium was formed, and it turned out that α-particles (as they are always called) are just helium atoms that have lost two electrons, so they are doubly ionised, and hence doubly charged, helium atoms, He^{++}. Finally, γ-rays were discovered to be electromagnetic radiation of very short wavelength; γ-rays and X-rays differ in their *origin* rather than their *nature*, so γ-rays were subject to the wave/particle problem mentioned above for X-rays.

I shall now carry the story of the atom beyond the official limit of this chapter into the twentieth century. It could be assumed that electrons were components of the atom, but where was the rest of it – nearly all the mass, and the positive charge to neutralise that of the electrons? It seemed likely that this was spread over most of the atomic volume. In 1911, Rutherford asked two of his research workers, Hans Geiger and Ernest Marsden, to check this by bombarding atoms with high-energy α-particles. While most of the α-particles were deflected very little, a small number suffered very large deflections. For Rutherford it was as amazing as if you had fired a shell at a sheet of tissue paper and it had come back and hit you.

Clearly the positive charge of the atom must be collected in a very small volume or *nucleus*; the α-particles will rarely come near this volume, but if they do so, they will experience strong electrostatic repulsion. In fact, the radius of the nucleus is of order 10^{-13} m, compared with about 10^{-10} m to 10^{-9} m for different atoms. Thus the *volume* of the nucleus occupies only around 10^{-15} that of the atom. Detailed analysis of α-scattering data confirmed this picture, and no subsequent doubt has been expressed on the correctness of Rutherford's *nuclear atom*.

The nucleus of the hydrogen atom is just the *proton*, another important constituent of all atoms. The proton has a charge equal to that of the electron, but positive, and, since the hydrogen atom consists of one proton and one electron, the mass of the proton must be about 1835 times that of the electron. The electron could be pictured as orbiting the nucleus, rather as a planet orbits the Earth, but problems were raised by this idea, which cannot be resolved till Chapter 4.

More crucial information about the atom emerged in the 1913 work of Moseley. Five years before, Barkla had put beyond doubt that X-rays were produced in two very different ways. I have already mentioned the radiation produced when electrons are suddenly decelerated at the walls of the tube – *braking radiation*. This consists of a continuous range of frequencies, analogous to white light. The other type, present whenever these X-rays are scattered, is *characteristic secondary X-rays* (named K, L, M, . . .), and they occur at a few sharp wavelengths, which are dependent on the scattering material. Characteristic X-rays are analogous to the light at specific frequencies produced in atomic spectroscopy – for which, yet again, see Chapter 4.

Moseley did not need to understand the mechanism entirely. He merely measured the K, then the L wavelengths, for as many elements as possible, and discovered that a graph of wavelength against what he called *atomic number* gave a straight line. Atomic number was (close to) the number of the element when they are all listed in order of atomic mass. However, to obtain the straight line, some gaps had to be left, and some changes in order made – just the old changes familiar from the Periodic Table, cobalt before nickel and so on.

It was clear that atomic number was more fundamental than atomic mass. It is, in fact, the number of electrons, and also of protons, in the particular atom, and it is always written as Z. So hydrogen has, as already said, one proton and one electron. Helium has two of each, so a doubly ionised helium atom, the α-particle, is just the helium nucleus.

Protons and electrons cannot, though, be the only constituents of atoms other than hydrogen. If they were, the atomic mass, the ratio of the mass of the atom to that of hydrogen, would just be equal to Z; in fact it is around $2Z$ for the smaller atoms (after hydrogen) and rather higher for larger atoms. The complete elucidation did not come till 1932, when James Chadwick identified the *neutron*, though the solution had been anticipated before then. The neutron is an uncharged particle, with mass very nearly equal to that of the proton (about 0.1% higher, in fact).

Apart from hydrogen, nuclei of all elements contain protons *and* neutrons. The helium nucleus, for example, contains two of each, so its atomic mass is very nearly equal to 4. (Such figures are not exact; there will be relativistic corrections because of binding energy, for which see Chapter 3, as well as different isotopes playing a role, for which see below.) Similarly, carbon, nitrogen and oxygen atoms have six, seven and eight neutrons, protons (and electrons), respectively, and atomic masses very close to 12, 14 and 16. An arsenic atom contains 33 protons (and electrons) and 42 neutrons, and its atomic mass is close to 75.

This is taking us near Prout's Rule, according to which all atomic masses should be whole numbers. What, then, about the exceptions to Prout's Rule, such as chlorine, with atomic mass about 35.5? The answer is that, while, for a given element, the number of protons in a nucleus is fixed (it is having 17 protons and electrons that *makes* an atom chlorine), the number of neutrons is not, in most cases. Some chlorine atoms have 18 neutrons, some 20; these are called different *isotopes* of chlorine. Writing an isotope as (Z, A), where A is the sum of the numbers of protons and neutrons, the two isotopes of chlorine are $(17, 35)$ and $(17, 37)$. Since the first is roughly three times as naturally abundant as the second, the atomic mass of chlorine is about $(3/4 \times 35) + (1/4 \times 37)$ or 35.5.

Even hydrogen, contrary to what I have implied so far, does have different isotopes. *Deuterium*, the nucleus of which contains one neutron, is naturally 0.015% abundant. The nucleus of *tritium* contains two neutrons, but it is highly unstable to radioactive decay, so tritium has no natural abundance.

Isotopes had actually been known long before 1932. In 1903, Rutherford and Fredrick Soddy constructed *radioactive series*, consisting of sequential decays between particular isotopes of different elements, and from 1910 on, J. J. Thomson and Francis Aston developed the mass spectrometer, in which beams of positively charged ions passed through superimposed electric and magnetic fields, and different isotopes followed different paths.

Conclusion

With the exception of the recent foray up to 1932, this chapter has just about taken physics up to the end of the nineteenth century. There was a great deal of solid, and it seemed irrefutable, achievement – Newton's and Maxwell's syntheses in particular. There were acknowledged difficulties, some already mentioned, but it seemed unlikely that they could not be handled by suitable minor adjustments. Such did not turn out to be the case!

3

A glance at relativity

The idea of relativity

Probably all of us have had the experience of sitting in a stationary train, and suddenly seeing a train on an adjacent track moving (or appearing to move) past our own. The doubt is there because it might not immediately be quite clear whether we are indeed still stationary and the other train is moving, to the right, say, or whether *it* is stationary and *we* have begun moving to the left.

In fact, all our normal actions in the train – sitting, eating, drinking, walking, may be carried on in exactly the same way, irrespective of whether our train is moving or not – at least so long as the motion is at constant speed in a straight line. Could this be raised to the status of a principle – 'All laws of physics are the same in the two circumstances'? Answering this question is the subject of this chapter [30, 31].

It may perhaps seem at first sight rather a dry formal question. In fact in the previous chapter I hinted at the important part it played in Galileo's argument on the movement of the Earth. But it has a more general importance even than this. If such a principle is adopted, it puts considerable restrictions on the laws of physics – in fact it ends up by making most of classical physics unacceptable.

Let us return to Galileo. To his opponents it seemed obvious that the laws of physics must take a different form on the surface of a moving Earth from the form taken if the Earth were stationary. On a stationary Earth, a dropped object would fall straight down; on a moving Earth it 'must' be left behind. By suggesting what would today be called 'conservation of momentum', Galileo showed that this need not be the case. A dropped object would retain its momentum parallel to the Earth's surface, and so fall at the feet of the person dropping it.

Galileo often made his point by comparison of events on a moving ship with those on a stationary one. In a more technological age, Einstein talked of trains instead of ships, and nowadays people often use spaceships in their discussion, but they all do the same job. On a smoothly sailing ship, for example, a game of deck-tennis may be played just as fairly as on land. (The player hitting in the direction of motion of the ship does not, as one might suspect, have an advantage.)

I now want to make the argument rather more specific. I shall imagine two trains A and B (and it will be helpful to think of them as very long), on rails running side by side in the x-direction. It will be supposed that every compartment of each train has a clock in it, and all the clocks in train A are arranged to read the same time; also all the clocks in train B read the same time. (Indeed, at the moment there seems nothing to prevent us just saying that *all* the clocks read the same time; later on we must be more careful.)

I wish to consider the occurrence of *an event*. An event may be a dropped object hitting the ground, or a match being struck, or anything that occurs *at a particular place and a particular time*. I want to describe it as occurring at position x and time t in train A, and at x' and t' in train B. To do so, I must imagine nailing onto the wall of one compartment in train A a *set of axes* with the x-axis in the direction of the rails, the y-axis upwards, and the z-axis perpendicular to both. The axes meet at an *origin* O, so all distances x along the train are measured from O. The same thing is done for train B, but now the axes are marked x', y' and z', and the origin is O'. (It is x and x' that will be important for us.)

Now I'll explain how the trains move. Train A in fact remains stationary, while train B moves at a steady speed v in the x-direction. I shall arrange that, at the instant when O and O' differ only in their z-coordinates, the clocks in the compartments on the train containing O and O' read zero. (Of course, at the moment, while we are fairly relaxed about time, we would be inclined just to say that *all* the clocks read zero.) The situation is displayed in Fig. 3.1 (*a*). After a time t, train B has moved along the direction of the x-axis of train A, so the value of x at O' is not zero, but the product of speed v and time t. (Equally, one might say that the x'-coordinate of O is not zero but minus v times t.) The situation at time t is shown in Fig. 3.1(*b*).

Such technical machinery may seem heavy and abstract, but it does enable everything to be explained precisely. I should mention that, more technically still, the attention is usually focussed on the actual x-, y- and z-, and x'-, y'- and z' axes; these are spoken of as *frames of reference*, so we have 'the moving *frame*' and so on. I prefer to keep things more concrete, and to think of 'the moving *train*' whenever possible.

Fig. 3.1. Axes *xyz* and *x′y′z′* are fixed to trains A and B, respectively. In (*a*), the clocks at O and O′ read $t = 0$ and $t′ = 0$. In (*b*), relative motion has taken place; we may say that the dashed frame has moved along the *x*-axis of the undashed frame *or* that the undashed frame has moved along the −*x′*-axis of the dashed frame. An event now takes place at E.

Now let us imagine an *event* taking place at time *t*. I have indicated its location on Fig. 3.1(*b*). Clearly its *x*- and *x′*-coordinates will not be the same. In fact it seems obvious that, for this event, *x* will be greater than *x′* by the product of *v* and *t*. (Or, alternatively, *x′* will be equal to *x* plus the product of *minus v* and *t*. See later.) On the other hand, it would be taken for granted that the *times* of the event in the two frames, *t* and *t′*, are equal.

These are statements of the *Galilean Transformation* for the coordinates of an event. They are extremely simple, and seem to be no more than common sense.

Unfortunately, as Einstein was to show, they are not right, merely a good approximation if *v* is much less than *c*, the speed of light.

Using the Galilean Transformation for the moment, though, we may see how not coordinates but *laws* vary between the trains (if at all). We may assume that Newton's Laws are obeyed in the stationary train A, and use the Galilean Transformation to obtain the laws of mechanics in moving train B. It is found that they have exactly the same form as in train A; Newton's Laws are recovered. We should really have expected this; as already explained, Galileo based his mechanics on the belief that the laws of mechanics should be the same on the two trains, and Newton's work was based on that of Galileo.

I must emphasise (and common experience will back me up) that this is *only* true provided that speed *v* is constant, *and* the motion is in a straight line. In vector terms (see Chapter 2), the velocity, vector **v**, of train B (or frame B) must be constant. If, in contrast, train B is accelerating away from rest, passengers will speak of feeling a force in a backwards direction. Of course, there is no force; they really mean that they are accelerated towards the back *despite* there being no force. Thus Newton's First Law is *not* obeyed in the train while it is accelerating. Similarly, if the train is decelerating, passengers will be flung to the front.

Again, if the track is bumpy, the train will be subject to intermittent accelerations in random directions; passengers will find drinking their soup particularly hazardous, and will be well aware that it is their train that is moving, not A. If the rails are curved (to the right, say) at this point, so that, even though the *speed* of train B may be constant, its *velocity* (speed *and* direction) is not, as explained in Chapter 2, again Newton's First Law is *not* obeyed in train B; passengers will be accelerated to the left.

To sum up, we have found that there are two types of frame. First, there is what we have so far been calling the 'stationary' frame, the frame of train A, and also frames moving with constant velocity. In these frames, Newton's First Law is obeyed. They are called *inertial frames*. Inertia is just another word for mass; in these frames, bodies display their inertia by requiring a force in order to be accelerated. The second type consists of accelerating frames; they are called *non-inertial frames*, and in these frames Newton's First Law is not obeyed. For some pages I shall talk only about inertial frames; non-inertial frames will return briefly in discussion of Einstein's general theory of relativity.

But now I want to make the most significant observation of this chapter. So far I have talked of 'stationary frames' and 'moving frames' (or 'stationary and moving trains'). But what really distinguishes a 'stationary frame' from one 'moving with a constant velocity'? The answer, so far at least, can only be nothing, at least nothing fundamental. We may have had at the back of our minds that the 'stationary frame' is that in which Newton's Laws are obeyed, but we have now seen that *all* inertial frames have that property.

From now on, I shall not imply that any one member of the set of inertial frames is any more fundamental than the others. What I have previously called the 'stationary' frame A must be 'the rest-frame of the Earth' or just 'the Earth's frame'. Frame B becomes not 'the moving frame', but a frame moving *with respect to* or *relative to* frame A. Specifically, it is the frame moving along the x-axis of frame A with speed *v*. But equally we may say that frame A is moving along the $-x'$-axis of frame B with speed *v*. We must not imply that either of the last two sentences is more significant than the other – and we must *never* say that an inertial frame is simply 'moving'. This is the central point of *relativity*. (The Galilean Transformation is already totally in accord with this way of thinking.)

I may perhaps here answer a few possible objections. First, does not frame A have a special significance just because it *is* the rest-frame of the Earth? The answer must be that certainly from our point of view there is every difference between being anchored securely on Mother Earth and being marooned on a space-vessel heading away from the solar system at an immense, though constant, speed. But, from the wider point of view of the Universe, more particularly from the perspective of the laws of physics, these considerations are not relevant.

It may still be asked, though, from the wider point of view, must there not be a rest-frame of the Universe (which would certainly not be expected, of course, to be identical with the rest-frame of the Earth)? Such a possibility cannot be simply ruled out – indeed the Michelson–Morley experiment described below is a brilliant attempt to find such a frame, though ultimately an unsuccessful one. While the question is really one for cosmological speculation, it may be said that, if such a frame does exist, the basic laws of physics seem not to recognise its uniqueness.

Lastly it may be argued that, since the Earth is rotating about the Sun, and spinning about its own axis, its rest-frame cannot strictly be inertial. This is true, and experiments may be performed to demonstrate the fact. In 1851, Léon Foucault set up the so-called Foucault pendulum. This must be extremely long – in a version of the experiment carried out by the British Open University, the pendulum was suspended from the Dome of St Paul's Cathedral in London. The pendulum is carefully set oscillating in one plane, and Newton's Laws would predict that it should stay in that plane if the Earth's rest-frame were precisely inertial. However, the plane of swing slowly rotates (at 12° per hour in London) as the Earth spins. Thus the rest-frame of the Earth is not quite inertial, but the great difficulty in demonstrating the fact shows that to consider it inertial will nearly always be a satisfactory approximation.

Fig. 3.2. A river in which a current flows to the right at speed v. An oarsperson rowing at speed c *relative to the water* will travel at speed $c + v$ *relative to the bank* when travelling to the right, $c - v$ when travelling to the left. A rower wishing to travel along path AB directly from one bank to the other must aim the boat *into the current*; the speed relative to the bank will thus be reduced from c to u.

Michelson and Morley; Fitzgerald and Lorentz

So far a convincing story has been built up, but of course we have discussed only mechanics; it is necessary to move on to the other fundamental area of physics – electromagnetism. We may do for Maxwell's Laws what we have already done for Newton's – consider them true in one inertial frame, and use the Galilean Transformation to find the appropriate form of the laws in another inertial frame. When we do this, we reach the conclusion, perhaps rather surprising from the point of view of our recent remarks, that Maxwell's Laws do *not* hold in any other inertial frame. This appears to indicate that we *may* define a unique rest-frame of the Universe to be that in which Maxwell's Laws apply.

From another point of view, this conclusion might not appear so surprising. We saw in Chapter 2 that one of the important results of Maxwell's theory was the prediction of electromagnetic waves, of which light is an example. It seemed clear that these must be waves *in a medium*, which is given the name of the *ether*. So the rest-frame of the Universe must be just the rest-frame of the ether, and it is naturally in this frame that Maxwell's Laws must be obeyed. Or so it seemed.

At first sight it might seem easy to find the speed of the ether relative to the rest-frame of the Earth. Imagine rowing along a river in which there is a current of speed v to the right (Fig. 3.2). If you can row at speed c *relative to the water*, then rowing to the right is easy – you can shoot along at speed $c + v$ *relative to the bank*. Rowing to the left is more difficult; relative to the bank you can only travel at speed $c - v$; if v happens to be greater that c, your most strenuous efforts will result only in the boat going backwards. (I use 'c' here, incidentally, because this quantity is analogous to the speed of light in what follows.)

It would seem that light should behave in a similar fashion, with the part of v being played by the speed of the ether in the rest-frame of the Earth. By analogy it would be expected that the speed of light, as measured on Earth, should depend on the direction of the light, varying between $c + v$ and $c - v$. One might hope that measurements could detect this effect.

Unfortunately, because c is so large, it is never possible to measure 'one-way' speeds of light. The kinds of experiment that *are* possible are two-way; light travels a distance l, from A to B say, *with* the ether perhaps, then returns from B to A *against* it. At first sight it might seem that the $c + v$ and $c - v$ would cancel out, and we would be back to a journey time of the distance travelled, $2l$, divided by c. Actually this is not *quite* the case; the time *is* slightly increased by the movement of the Earth through the ether (or the movement of the ether past the Earth – two sides of the same coin), but only to second order in v/c, that is, by a term proportional to the *square* of v/c. If we assume that v/c is small (for which see later), this will be an exceedingly small increase in time. Certainly we will not hope to detect it by a direct measurement of the time itself.

An experiment that may seem a little more feasible is to send *two* rays each on two-way paths at right angles to each other as in Fig. 3.3. At first we might suspect that the light travelling along path MM_1 in Fig. 3.3 would be unaffected by the ether, since it is travelling perpendicular to it. This is not actually the case. Thinking back to Fig. 3.2, we may realise that, to row from a point on one bank to a point directly opposite on the other, you must row *into* the current, which increases the time taken. In the optical case, too, the time to travel MM_1M is greater than $2l/c$, again to second order in v/c. The good news is that the *two* correction terms (for paths MM_2M and MM_1M) do *not* cancel out, so light travelling along the two paths will become out of phase, and there is a probability of detecting this effect by observing interference at T.

What size of effect should we expect? This of course depends on the size of v, the speed of the ether relative to the Earth, and this quantity we don't know – it's what we are trying to find, and it *could* be very large. We can, though, estimate the smallest value it *could* take. This would be if the ether happened to be at rest relative to the Sun; in this case v would be the speed of the Earth in its orbit. (Alternatively, the Earth *could* be at rest in the ether for one season of the year, but would have a speed relative to the ether of *twice* its orbital speed six months later.) If v is put equal to the orbital speed, and used with a plausible value of l, it seems that the experiment is conceivable, though extremely difficult.

In one of the greatest experimental feats of all time, Albert Michelson and Edward Morley performed this experiment in the 1880s. They shone light onto a

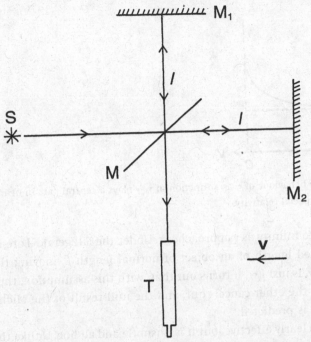

Fig. 3.3. The Michelson–Morley experiment. Light from source S reaches semi-silvered mirror M, and may subsequently reach telescope T by travelling along path MM₁MT or MM₂MT. Interference between light that has travelled along these two paths may be observed. The distances MM₁ and MM₂ are assumed to be exactly equal, and are given by l. The velocity of the ether in this frame is indicated by **v**.

half-silvered mirror M (in Fig. 3.3), thus producing the two beams, and inspected the interference pattern at T. (The instrument shown in Fig. 3.3 is called the *Michelson interferometer*.) It was necessary to rotate the apparatus through 90° so that the two wings of the apparatus interchanged. (To achieve this, the apparatus was mounted on a massive stone slab and floated on mercury.) The experiment was sufficiently sensitive that a change in the interference pattern should definitely have been expected. Yet they found – *nothing*! The Michelson–Morley experiment, one of the most famous in the history of science, gave a null result. The search for the ether had failed.

One suggested explanation of this result came fairly quickly. George Fitzgerald and Hendrik Lorentz proposed independently that, when an object moves through the ether, its length decreases – the *Fitzgerald–Lorentz contraction*. The amount of contraction is related to a quantity γ, which plays a large part in the remainder of this chapter. If β is defined as v/c, then γ is defined so that $1/\gamma^2$ is equal to $1 - \beta^2$. On Fig. 3.4 the behaviour of γ as a function of v is shown. For small v, γ rises only slightly above unity, but as v increases, γ increases very

Fig. 3.4. The behaviour of γ as a function of v. γ plays a central part in nearly every formula of special relativity.

rapidly, and tends to infinity as v approaches c. Under the Fitzgerald–Lorentz proposal, the contracted length of an object of normal length L, moving through the ether at speed v, is just L/γ. It turns out that, with this assumption, the effect of motion through the ether cancels out, and the null result of the Michelson–Morley experiment is predicted.

Thus the idea is clearly effective, but it is unsubtle and ad hoc. Unlike the earlier ideas in this chapter, it retains a rest-frame for the Universe, so the motion of the interferometer is absolute rather than relative, and the contraction proposed is a real physical one. In contrast, Einstein's ideas, which I now turn to, are subtle and elegant; they constitute an overall approach to the whole of physics, rather than an ad-hoc stratagem to solve a particular problem.

Relativity: the special theory

The Michelson–Morley experiment is a good motivation for *us* to consider Einstein's 1905 special theory of relativity, since the theory explains the null result in a simple way. This does not mean, of course, that the experiment was the genesis of his own ideas; much later in his life he denied this (very common) suggestion rather emphatically.

Einstein's ideas were rather more abstract, born out of years of speculation about basic physical phenomena, especially the nature of light. For example, suppose that we chase after a beam of light at, say, half its speed, $c/2$; common sense, backed up by the Galilean Transformation, tells us that its speed, relative to us, must be $c/2$. If we move in the opposite direction at $c/2$, its speed, again relative to us, must be $3c/2$. Finally, if we move in the direction of the light and at its speed, c, it must appear to us to be stationary. Rather than a wave, oscillatory in space *and* time, light is reduced in status to an oscillation in time. And Einstein refused to accept that this was conceivable.

In the light of this, and other deeply felt ideas, Einstein proposed a fundamental condition on physics – *all the laws of physics must be the same in all inertial
frames*. A particular very important aspect is that *the speed of light is c in all inertial
frames*. It will thus be inappropriate to define a rest-frame for the Universe, and
so all motion must be relative. (We must not talk of a particular inertial frame
as 'stationary' or 'moving', but, for example, as moving *relative to* another inertial
frame with a particular velocity.) Finally, with the abandonment of a rest-frame
for the Universe, the ether serves no function and may be dispensed with.

Clearly this postulate explains the Michelson–Morley result. Light travels at
speed c along *each* path of the interferometer, so there is no phase difference
between different light-paths, and hence no change in interference pattern when
the apparatus is rotated.

Einstein's own difficulty above is also put in a different perspective. For the
person running either after or away from a beam of light at speed $c/2$, the speed
of that light is still just c. And we must add in a constraint that no object –
or, at least, no object with mass – may travel at or beyond the speed of light
(relative to anything else), so the final conundrum is removed. This point is not,
as it might seem at first, an ad-hoc postulate to remove a difficulty; it emerges
from the very core of the relativistic argument. (Particles *without mass*, and we
shall learn in the next chapter to discuss light in these terms, do (must) travel
at the speed of light.)

But as well as removing some problems, Einstein's postulate seems at first
sight to cause many more! Even what has been said already seems frankly absurd.
If you chase after a light beam at, say, $(99/100)c$, surely the light *must* be moving
away from you at only $(1/100)c$. This certainly works for sound waves. Indeed, it
is quite possible to accelerate *through* the speed of sound; this is what is known
as 'breaking the sound barrier'.

Einstein insists though that, however hard you chase the light, its speed,
relative to you, remains c. The only way we can make sense of this is to accept
that, if there are two observers, one shining a torch, the other chasing after the
beam, we must *not* assume that the rate of flow of time is the same for both.

Before explaining this, let me first comment on the introduction of the idea
of 'observers'. This is common and helpful in discussions of relativity, but it
should not be thought to imply any *subjectivity*, or any 'observational process'.
When an event occurs, the observer merely notes what its values of x and t are
in the observer's rest-frame. (I mention this because, in many interpretations of
quantum theory, 'observers' are felt to play a more positive role.)

Indeed, I shall first point out that, contrary to what is often guessed by
those thinking about relativity for the first time, lack of simultaneity of events,
events occurring at different times for different observers, has *nothing* to do with

observers being at different places. It is quite true that, if a light is flashed at a particular time, observers at different distances from the source will register arrival of the light at different times. We know that the light we receive today from a far distant star was emitted many years ago. This is true – but it is also obvious, and automatically allowed for in any experimental or theoretical analysis. Relativity, on the other hand, is certainly *not* obvious.

A time (for a given event) corresponds, in fact, to a particular inertial frame, *not* to a particular observer in it. It is quite straightforward to ensure that all clocks in the same inertial frame, say frame A, read the same time. This is one thing that the constant nature of the speed of light makes easy. A pulse of light may be sent from one clock at $t = 0$, say; if a second clock is distance x from the first, this clock is set at time x/c when the light arrives. This may be repeated for all clocks in frame A. So when an event takes place, the time of the event in frame A is the time recorded on a clock at rest in frame A at, or as close as possible to, the position of the event.

When one talks of 'lack of simultaneity', one is implying not that an event occurs at different times at different places in the same frame, but that it occurs at different times in different inertial frames. This is, of course, highly unexpected from the point of view of classical physics; Einstein's boldest step was to refuse to rule it out on the grounds of 'common sense'.

It must imply that the Galilean Transformation, which says that, for a particular event, x' is equal to x minus vt, but t' and t are always equal, must be wrong. In special relativity it is replaced by the *Lorentz Transformation*. (As mentioned in Chapter 1, Lorentz produced many of the equations, but practically none of the understanding, of special relativity.) To obtain the Lorentz Transformation, *both* parts of the Galilean Transformation are adjusted. In the relationship between x and x', a γ (as shown in Fig. 3.4) intrudes, but more fundamentally, t' is not equal to t. We may write t' as dependent on t and x, or t as dependent on t' and x'.

I shall not discuss the Lorentz Transformation in detail, but I shall mention three important properties. First, when v is much less than c, it becomes very close to the Galilean Transformation, which we know is adequate for such values. Secondly, it transforms a ray of light travelling with speed c in frame A into a ray of light travelling with the *same speed, c*, in frame B; this is, of course, exactly what it was designed to do. Thirdly, we must remember that frame B has exactly the same status as frame A; mathematically the transformation from (x', t') to (x, t) and that from (x, t) to (x', t') have exactly the same form with v replaced by $-v$.

I shall briefly describe a few *results* of the Lorentz Transformation. First, imagine a stick of length L at rest along the x-axis of frame A. An observer in frame

B will observe this stick to be contracted, and to be of length L/γ. This is of the same form as the suggestion of Fitzgerald and Lorentz (F and L, for short), and is usually in the present context called the *Lorentz contraction*. However, it is important to note that what is happening in special relativity is very different from the idea of F and L. For them it was a *real* contraction caused by an *absolute* motion through the ether. In special relativity, it is an *apparent* effect due to relative motion of frame B with respect to frame A, *or* of frame A with respect to frame B. Thus it is a *reciprocal* effect; if a stick of length L lies along the x'-axis of frame B, to an observer in frame A it will appear to have a reduced length of L/γ.

A broadly similar effect is that of *time-dilation*. The most important example concerns particles that are unstable with respect to radioactive decay. Often these are formed when cosmic rays, reaching Earth from other parts of the Universe, collide with particles high in the atmosphere. A particular type of unstable particle may live, on average, a time τ, and so may be expected to travel only a distance of about v times τ, where v is its speed. If $v\tau$ is much less than the height of the atmosphere, it might be expected that very few particles would reach Earth. In fact, though, large numbers do.

The explanation lies in the Lorentz Transformation. It turns out that, though the lifetime of the particle in its own rest-frame is τ, for an observer on the Earth the apparent lifetime is γ times τ. Provided that v is large, γ will be much greater than 1, so many of the particles will live long enough to reach Earth.

Relativity in mechanics

I now move to consider how relativistic ideas affect mechanics. It will be remembered that Newton's Laws obeyed Einstein's postulate when transformed *by the Galilean Transformation* – they were the same in all inertial frames. Since the Galilean Transformation has now been replaced by the Lorentz Transformation, this suggests that, to keep faithful to the postulate, Newton's Laws too must be reduced in status to good approximations for low speeds.

It turns out that, in special relativity, the Newtonian expression for momentum, mv, must be multiplied by a factor of γ; the result is shown in Fig. 3.5. For low speeds, of course, γ is close to 1, and the Newtonian expression is regained, but when v is higher, the relativistic expression is much greater than Newton's, and tends to infinity as v becomes closer to c. Figure 3.6 shows what happens as a steady force is applied to the object, and helps to explain why v is restricted to being less than c. The momentum increases regularly throughout, but the speed increases more and more slowly, and never reaches c. (The form of γ shown in

Fig. 3.5. The dependence of the magnitude of the momentum p on speed v as given by special relativity. For low v, the relativistic expression is very close to the classical expression of mv, indicated by the straight line on the figure, but as v approaches the speed of light c, p increases more and more sharply, and tends to infinity as v approaches c.

Fig. 3.6. Acceleration under a constant force as given by special relativity. According to classical physics, v would increase indefinitely, as shown by v_1, the straight line on the figure. However, relativity shows that, as the speed increases, its rate of increase slows, and the speed itself, as seen in the lower line v_2, can never reach the speed of light c.

Fig. 3.4, and its occurrence in so many relativistic expressions, emphasises the same point.)

When one turns to energy, something more startling happens. According to special relativity, the appropriate expression is γ times the mass of the object times the square of c (γmc^2). Thus, even when the object is at rest in a particular frame, the formalism suggests that it has an energy of mc^2 in that frame. If it travels at speed v in the frame, the energy increases further. For small v, the *increase* is equal to the classical kinetic energy, though as v becomes closer to c, the energy, like the momentum, increases, much faster, and tends to infinity as v tends to c.

But what is really interesting is the rest-energy of mc^2. Its presence suggests that, even when a body is at rest, there is energy associated with it – and a large amount of energy because c is so large. Another way of speaking of this is that mass is a form of energy – mass-energy; thus instead of conservation of mass *and* conservation of energy, there is *one* conservation law for *one* quantity.

It is well known that this mass-energy is the source of nuclear power, as used in bombs and nuclear power-stations. In the radioactive decays these make use of, a small amount of mass is lost, and a large amount of kinetic energy and heat produced. Actually the same effect is also behind the energy-balance in conventional chemical reactions. The reaction-products have slightly greater or less mass than the substances taking part in the reaction (though the difference would be much too small to be detectable by a chemical balance, and thus to be detected as a violation of conservation of mass). Conservation of energy (including mass-energy) then requires a net output or intake of heat. The mass-energy corresponds to the otherwise unexplained 'chemical energy' mentioned in Chapter 2.

As a final comment on relativistic mechanics, please note that I have not found it necessary or helpful to introduce the idea of a 'relativistic mass' equal to γm. This is used in many treatments of relativity, but should be regarded, at least in my opinion, as being rather *against* the true relativistic spirit [31].

The background of general relativity

General relativity is a highly mathematical theory developed by Einstein over a number of years up to 1916. It is, though, possible to give a reasonable explanation of at least some aspects of the theory without elaborate mathematics. It starts where special relativity left off; it deals with bodies in *non-inertial frames*, bodies in particular that are falling, and hence accelerating, towards Earth (or moving in some other gravitational field). Thus general relativity is really a theory of gravity, replacing Newton's theory of gravitation. The two theories are totally different in nature, but their results are very close in a great range of circumstances.

An important step towards the theory is Einstein's *principle of equivalence*. It may be remembered from Chapter 2 that, in Newton's account of an object falling under gravity, two different masses, gravitational mass – as in the law of gravitation, and inertial mass – as in the Second Law, must be taken as identical.

Fig. 3.7. A spaceship in outer space is subject to an acceleration in the *x*-direction. The traveller will feel pushed in the −*x*-direction, and one of the *yz*-planes will seem like a floor, the other like a ceiling.

Einstein took a different approach. Consider first a spaceship at rest and far away from any other object; inside, objects float freely. Now suppose that the spaceship accelerates in the +*x*-direction in Fig. 3.7. Travellers in the spaceship will now feel pushed in the opposite direction (just as when a bus accelerates from rest). For them, one of the *yz*-planes will seem like a 'floor', the other one like a 'ceiling'. If the acceleration of the spaceship happens to equal g, the acceleration due to gravity on Earth, it will be impossible for them to distinguish their situation from being in a closed room on Earth.

Now suppose that one of the travellers drops two spheres – one of iron, one of wood. Once released, these spheres will obey Newton's First Law and travel at constant speed. They will thus be overtaken by the accelerating 'floor' of the spaceship and hit it *at the same time*. Travellers in the spaceship, though, will be inclined to put an entirely different perspective on the events. It will seem to them that the spheres have fallen under the apparent 'gravity' in the spaceship. The fact that they have hit the floor together will seem to support Galileo, and also to ensure that the two types of mass defined above are identical.

From these fairly general thoughts, Einstein leapt to the idea that the two descriptions are *equivalent*; *all* phenomena must occur in the same way in a gravitational field as under acceleration. This *principle of equivalence* makes the identification of the two masses above obvious. It also points to some important predictions of general relativity. One is that the path of a ray of light should bend under gravity. A second is that, if a clock is moved to a region close to a large mass, it should go slightly slower. For example, a clock on the surface of the Sun should go 0.0001% slower than on Earth. This implies that spectral lines in light emitted from such a body should be shifted towards lower frequencies – the *gravitational red-shift*.

To move beyond these important conclusions, Einstein needed to construct his highly mathematical and elaborate theory, which it is vastly beyond the scope of this book even to sketch. This explained in detail the anomaly in the precession of the planet Mercury (Chapter 2). It gave precise predictions for the gravitational red-shift, and for the curvature of light under gravity, and it was the agreement of the latter prediction with results obtained by Arthur Eddington for the curvature of stellar light passing near the Sun during the solar eclipse of 1919 that led to practically universal acceptance of the general theory of relativity, and hence to Einstein's worldwide celebrity.

Einstein, relativity and quantum theory

In a book of this nature, it is particularly appropriate to discuss what effect the theory of relativity – its content, its general nature, its effect on Einstein's life – had on his attitude towards quantum theory. The first point is technical. A central feature of special relativity was that no object could travel faster than c; slightly more significantly, no *information* could travel faster than c.

This idea is very important in many discussions of the interpretation of quantum theory – such as the famous Einstein–Podolsky–Rosen (EPR) 'paradox' discussed in Chapter 6. Such analysis seems to involve breaking, not the letter, but *perhaps* the spirit, of the prohibition above. It is not surprising that Einstein was prominent among those most adamant in their opposition to such ideas.

More generally, one may contemplate the means by which Einstein had succeeded in replacing Newton's till then magnificently successful laws of motion and gravitation. He had put phenomenal effort into working with exceptionally complicated mathematics based on a very small number of general principles. Until after his great 1905 successes, Einstein had actually been no more than a competent performer of mathematics [1]. His 1905 papers – special relativity, the photoelectric effect, and so on – were brilliant in their physical novelty, but required only basic maths. But after his 1916 triumph, Einstein perhaps became wedded to the idea that solution of fundamental problems in physics automatically required solving extensive sets of complicated equations.

While it could be said that the problem with the orbit of Mercury was the 'anomaly' which Einstein's theory removed, it seemed utterly impossible that one might have worked from Newton's theory and this 'anomaly' towards Einstein's own theory. Rather Einstein had to maintain his aim at abstract general objectives, and create his own theory from first principles. Only then could it become clear that Newton's theory was a good approximation in many circumstances but failed in a few specific points.

So again it is not surprising that Einstein adopted a similar approach to quantum theory. While recognising its many successes, he disliked aspects of it, and felt that there could be no point in modifying it to remove the perceived difficulties. Rather a replacement must be found from first principles and the use of elaborate mathematics. At that stage conventional quantum theory must appear as some sort of approximation for the new theory, and only then could the reasons for its successes and failures be understood. Einstein's approach will be discussed fully in Chapter 6.

4

The slow rise of the quantum

Planck and the genesis of the quantum

During the first quarter-century of quantum theory, it developed by addressing a considerable range of topics – atomic physics, especially atomic spectroscopy; interactions involving fundamental particles – electrons, protons and so on, and also electromagnetic radiation; and the thermal capacities of solids and gases, where deviations from classical physics seem, at least in retrospect, rather straightforward.

Yet the first intimation of quantum ideas, which came to Max Planck in 1900, appeared in rather a recondite area of physics, where there would be no satisfactory *classical* theory for another five years, and where considerations were very much complicated by the fact that statistical methods were required. This was the area of *black-body radiation* or *cavity radiation* [32].

All surfaces emit energy in the form of electromagnetic radiation. This *emission* is of so-called *thermal radiation*, the word 'thermal' emphasising that the amount of energy radiated, and its frequency distribution, depend strongly on temperature. In 1879, Josef Stefan deduced from experiment that the total amount of radiation is proportional to T^4 (T being an absolute temperature, of course), and Boltzmann confirmed this theoretically using thermodynamics.

The nature of the surface is also very influential and, to explain this, it is helpful to start with the *absorption* of energy. Different surfaces absorb different fractions of the energy that is incident on them. In particular, black surfaces absorb more than white ones. (So sports players wear white to minimise the amount of solar energy they absorb; reflectors behind radiators are white so that they *do* reflect rather than absorb the heat radiation.)

An idealised perfect *black body* is defined as one that absorbs *all* the radiation incident on its surface, and an important deduction from thermodynamics by Gustav Kirchhoff in 1859 was that the better a surface is as an *absorber* of electromagnetic radiation, the more strongly it also *emits* radiation – so black bodies must also be the best emitters.

From the time of Kirchhoff's discovery, the study of black-body radiation was a topic of considerable interest. While the Stefan–Boltzmann Law above gave the total amount of energy radiated, physicists sought an understanding of its distribution as a function of frequency – which Boltzmann was able to show was a unique function, independent of the precise nature of the black body.

It should be mentioned that the theoretical model of a black body usually studied was rather different from that explained above. Imagine a cavity with the internal walls at temperature T, made of a material that absorbs at least some of the radiation incident on it. Now imagine that a very small hole is left in one of the walls. Any energy incident from outside on that hole will pass into the cavity, and will undergo a series of reflections at the walls, some of the incident energy being absorbed at each stage. A negligible amount of the incident radiation will emerge from the hole, and so the hole itself has the property of a black body; it absorbs all energy incident on it.

Inside the cavity, the walls will emit as well as absorb radiation; the small fraction of this energy that moves outwards through the hole must be characteristic of a black body at temperature T and, since this is just a sample of the radiation inside the cavity, the *cavity radiation* itself must have a *black-body distribution* or *spectrum*. This cavity idea is extremely useful for theoretical analysis.

Let us first, though, look at the experimental results, shown in Fig. 4.1. These show the spectral distribution of black-body radiation at a number of temperatures from 1000 K to 2000 K. At any frequency the amount of radiation increases as the temperature rises, and of course the total amount, given by the total area under the appropriate curve, increases sharply with T, being proportional to T^4, as the Stefan–Boltzmann Law tells us.

At any temperature, the spectral distribution of energy goes through a maximum at a particular frequency, f_{max}. The most interesting feature of these curves is that, as T increases, f_{max} increases also. In 1893, Wilhelm Wien obtained another result from thermodynamics; f_{max} is proportional to T, and this is known as *Wien's displacement law*.

Since the visible spectrum is roughly between frequencies 4×10^{14} s^{-1} and 8×10^{14} s^{-1}, it is clear that, at all temperatures shown on the graph, most, by far, of the radiation radiated is in the infra-red; it is 'heat'. (When you pay your electricity bill, very little of the cost of having your light-bulbs on has actually gone to the provision of light!)

Fig. 4.1. The spectral distribution of black-body radiation at a number of temperatures. Frequency is given in units of 10^{14} s^{-1}. The visible region extends from 4×10^{14} s^{-1} (red) to 8×10^{14} s^{-1} (violet), so at all these temperatures the bulk of the radiation is in the infra-red. At 1500 K, though, the black body will be 'red-hot', and at 2000 K it will be 'white-hot'. The maximum in the distribution occurs at values of f increasing with T in agreement with Wien's displacement law, and the total area under the curve increases as T^4 in agreement with the Stefan–Boltzmann law.

However, as the temperature increases to about 1500 K, the right-hand end of the curves moves from the infra-red to the red end of the visible; the object glows *red-hot*. At higher temperatures still, say 2000 K, there is radiation throughout the visible region, and the object glows *white-hot* (though, of course, there is still far more energy radiated in the infra-red than in the visible).

As a historical point I would mention that the measurements in the infra-red were, of course, difficult to perform. Reliable values for frequencies lower than about 1.5×10^{14} s^{-1} were obtained (by Otto Lummer and Ernst Pringsheim, and also by Heinrich Rubens and Ferdinand Kurlbaum, both groups working, like Planck, in Berlin) only in the year 1900 itself, immediately stimulating Planck's great contribution.

The laws of Boltzmann and Kirchhoff, and Wien's displacement law, exhaust the 'secure' information available from thermodynamics. After that it was largely a case of guidance 'by the seat of the pants', different physical models giving different results. What eventually did become clear, though – but not till 1905 – is that classical physics *did* give a unique answer for the spectral distribution, though *not* the correct one! The result is called the Rayleigh–Jeans distribution, though, to the extent that it matters, Einstein had an equal claim to

Fig. 4.2. The spectral distribution of black-body radiation at 2000 K according to the distributions of Planck (which agrees with experiment), Rayleigh–Jeans and Wien (P, RJ and W, respectively). Rayleigh–Jeans agrees with experiment for the lowest values of frequency, but is totally unsatisfactory at higher frequencies. Wien's distribution agrees with experiment through the ultra-violet and visible regions, but fails in the infra-red. It was this failure, demonstrated by measurement in 1900, that led to Planck producing his own distribution. (Frequency is in units of 10^{14} s^{-1}.)

its discovery, and particularly to the appreciation that it *was* the unique classical answer.

Most of the background for Rayleigh–Jeans has already been assembled in Chapter 2. Rayleigh was a great expert in the theory of sound, and, by analogy, it was natural for him to start by calculating the density of modes of vibration of electromagnetic radiation in the cavity. In Chapter 2, we discussed the density of modes in the one- and two-dimensional cases, and found that it was independent of frequency f in the first case and proportional to f in the second. The result for the present three-dimensional case, that the density of modes is proportional to f^2, is thus highly reasonable. The constant of proportionality may be calculated in a straightforward way, though Rayleigh actually made a mistake of a factor of eight!

Now equipartition of energy (Chapter 2) tells us that *each* mode will have, on average, energy kT ($kT/2$ for each of two degrees of freedom corresponding to electrical and magnetic energy) – the same for modes of *any* frequency. To calculate the energy density at any frequency, the product of these two quantities, density of modes and energy per mode, was required, and once James Jeans had corrected Rayleigh's error, the result was available for comparison with experiment. It is shown in Fig. 4.2 for one particular temperature. (For other temperatures, each value is increased or decreased in proportion to T.)

The comparison with experiment gives one small piece of good news, and two much bigger pieces of bad news. The good news is that, for very low frequencies, theory and experiment agree. The first piece of bad news is that they don't agree anywhere else. As frequency increases, and the experimental results go through a maximum and return to zero, the f^2 factor in the Rayleigh–Jeans expression means that it just keeps rising – up to infinity!

The other bad news is that the Rayleigh–Jeans results don't really make sense. The *total* radiation emitted at all frequencies corresponds to the *total* area under the curve, and, for Rayleigh–Jeans, turns out to be infinity. The clear, but clearly nonsensical, outcome of classical theory for radiating bodies is that any body, at any temperature whatsoever, is radiating (and absorbing) energy at an infinite rate. (Note that, for Rayleigh–Jeans, the Stefan–Boltzmann Law and Wien's displacement law do not apply; the total energy radiated is infinite, and the curve has no maximum.)

Textbooks now show how the Rayleigh–Jeans calculation may be adjusted – fairly simply in mathematical terms, though with enormous consequences conceptually – to give agreement with experiment. This is useful paedagogically, provided, of course, that we do not delude ourselves that Planck, in 1900, actually achieved his results by manipulation of the 1905 theory of Rayleigh and Jeans.

The problem with Rayleigh–Jeans is actually in its use of *equipartition of energy*. To obtain this result in Chapter 2, we needed the assumption – taken for granted from the classical point of view – that the energy of a mode of electromagnetic radiation of frequency f may take *any* (positive or zero) value. Let us now assume, instead, that its energy may only take the values 0, or hf, or $2hf$, . . ., or we may say, in general, nhf, where n is a positive whole number or zero. Here h is a new constant – chosen for the specific purpose of achieving agreement with experiment. The integer n is the first example of a *quantum number*; we shall meet many more.

The constant h is called *Planck's constant*, and it takes its position, with m and e, the mass and charge of the electron, and c, the speed of light, among the most fundamental constants of physics. The best modern value is 6.63×10^{-34} J s (joule seconds).

Figure 4.3 is an *energy-level diagram* for a mode. Whereas classically *all* energies are allowed, in the *new quantum theory* (or *quantum mechanics*), only *discrete* values are available.

Use of Boltzmann's probability distribution from Fig. 2.22 gives the probability of a mode having any particular value of energy from the *discrete* list above (or, as we may put it, occupying a particular energy-level), and we can thus calculate the *average energy* of a mode of frequency f at temperature T, shown in Fig. 4.4.

Fig. 4.3. The first energy-level diagram, appropriate for a mode of electromagnetic radiation, or a simple harmonic oscillator. Whereas classical theory would say that *any* (positive) value of energy should be allowed, to obtain Planck's distribution one had to restrict energies to nhf, where n may equal 0, 1, 2, 3, . . .

Fig. 4.4. The average energy of a mode of electromagnetic radiation of frequency f at temperature T. Whereas classical theory would say that this should be equal to kT independently of frequency, the limitation of energy to $E = nhf$ implies that the value kT is obtained only for the lowest values of f, and the average energy decreases steadily for higher values.

We see that for the lowest values of frequency the average energy is just kT, as in the classical case; as f increases, it drops rapidly, and tends to zero for high f. Since the black-body spectral distribution is just the product of the density of modes, which is proportional to f^2, and this average energy, it now has exactly the right form to agree with experiment. As f increases, the f^2 factor initially pushes the product up fairly fast; but as Fig. 4.4 comes into play, it first slows down the increase in the product, and then begins to dominate, forcing the distribution function downwards from a maximum towards zero as f increases further – exactly like the experimental curves of Fig. 4.1.

Why do the quantum and classical answers agree for low f? Here the energy-levels of Fig. 4.3 become so close that the difference between classical and quantum predictions becomes negligible. This feeds through to the Rayleigh–Jeans expression agreeing with experiment. We may obtain the same type of result by making T sufficiently large. Also, though, of course, we cannot our-selves vary h, we may in a formal way let it tend to zero in mathematical expres-sions, and again quantum and classical answers will agree. Such occurrence of *classical results* as limiting cases of quantum results played an important part in the further development of quantum theory. Bohr was the chief instigator of such techniques, summed up in his *correspondence principle*, and the same kind of ideas, though from a very different perspective, played a part in his later *principle of complementarity* (Chapter 5).

Now let us return to 1900 and consider, briefly, how Planck, who of course did not have access to the Rayleigh–Jeans analysis, developed the ideas himself. Prior to 1900, in fact, the best-known attempt to obtain a distribution formula had been that of Wien in 1896. Unlike his displacement law above, Wien's dis-tribution law was not based securely on thermodynamics, but used a reason-able model – a radiating heated gas – and a few rather dubious assumptions. The good thing was that, as shown in Fig. 4.2, for the data available at the time agreement with experiment was good. (This was for the *high*-frequency case, rather than the *low*-frequency case, for which it is Rayleigh–Jeans that is good.)

It was only when the new infra-red data became available in 1900 that it became clear that Wien's law failed for these results. From the detailed experi-mental data of October 1900, Planck was able to concoct his new formula, which agreed with experiment, becoming equivalent to that of Wien for high frequen-cies. It included the new constant h, together with Planck's (very good) estimate for it.

This was rather empirical, but Planck, who had already been working for three years on the black-body problem [32], was desperately keen to develop a rigorous theoretical explanation. Unlike Rayleigh and Jeans five years later, he did not consider the modes of electromagnetic radiation in the cavity, but instead ima-gined an array of simple harmonic oscillators present in the cavity, interacting with the radiation and sharing the same energy distribution. Planck used Boltz-mann's statistical techniques to find the most probable distribution of energy among the oscillators; he was then able to relate this to the electromagnetic field density, and hence to the energy of radiation. He discovered that, to obtain agreement with experiment (and his own empirical formula), he had to work in terms of an element of energy equal to hf.

This derivation makes it clear that the *quantisation* applies to simple harmonic oscillators as well as to modes of electromagnetic radiation. The two are mathematically equivalent, and for the rest of the book it is the oscillator case that will usually be regarded as fundamental. I should also mention that, though quantisation *does* apply for macroscopic oscillators – such as a pendulum in a clock, the size of h means that the energy spacings hf are negligible in comparison with the actual energies; it is only for objects of atomic dimension that quantum theory gives results effectively different from classical theory.

There has been disagreement over how abrupt the change was in Planck's methods, and also regarding his appreciation of his results. For Klein [33] and Kangro [34], the change from Planck's earlier total opposition to Boltzmann to a recognition of the centrality of his methods occurred precisely at this time, and these authors take for granted Planck's awareness that his results implied discontinuity of energy.

But, in a well-argued book of 1978, Kuhn [32] suggests that the volte-face over Boltzmann occurred rather earlier. (It is certain that his earlier criticisms of Boltzmann left Planck feeling considerable guilt, particularly after Boltzmann's suicide.) More importantly, Kuhn suggests that Planck, and most other physicists, were unaware of what to us is the central point of his results – energy discontinuity – until at least 1908. (Einstein and Paul Ehrenfest, Kuhn says, were convinced at least three years earlier.) Planck, of course, *was* well aware of the importance of the introduction of h. It was also in this work that he introduced the constant k – 'Boltzmann's constant' for us – and, strangely enough in retrospect, for him this may have seemed the most significant thing of all. The importance of Kuhn's re-assessment, from the point of view of this book, is that it points up even further Einstein's contributions in the first decade of the quantum theory.

Planck fully deserved his Nobel Prize (in 1919, before Einstein and Bohr in 1921 and 1922, respectively) and his towering reputation. His major contribution of 1900 may seem a little incoherent from today's perspective, but since it was just the first step in a passage to a complete quantum theory that would take 25 years, that is scarcely surprising – the first step is the most crucial and usually the most difficult. Planck, though, as distinct from his constant, will not appear much more in this book. He was probably inherently too conservative to take part in the conceptual re-building necessary for further progress. The baton was passed to Einstein and, from 1913, to Bohr.

Einstein and black-body radiation

Einstein is famous, among other things, for the discovery of the photon and his consequent Nobel Prize-winning work on the photoelectric effect. This was only part, though, of a prolonged and widespread programme of study. Kuhn [32] speaks of 'the coherent development of a research program begun in 1902, a program so nearly independent of Planck's that it would almost certainly have led to the black body law even if Planck had never lived' and states, of a 1906 paper of Einstein, that '[in] a sense, it announces the birth of the quantum theory'.

The programme began with Einstein's 1902–3 papers on *statistical mechanics*, in which he built up the foundations of the subject in such a way that it could freely be applied to general physical systems, rather than being largely restricted to gases as the work of Maxwell and Boltzmann had been. Unfortunately, Einstein's results, including those on *fluctuations*, only duplicated those published by Gibbs in 1902, though the actual approaches were rather different.

On the basis of this general work, it was comparatively easy for Einstein to consider Planck's oscillators interacting with gas molecules, from which classical theory would demand that they *must* acquire average energy kT. From the connection already obtained by Planck between oscillator energy and density of radiation, Einstein produced independently what we have called the Rayleigh–Jeans formula (and insisted that it was the unique classical answer).

The main concern of the paper in which he announced this result, though, was much more startling – Einstein's only contribution to physics that he himself was prepared to describe as 'revolutionary'. He assumed Wien's distribution to be correct, and used it to calculate the entropy of the radiation. Next he examined how this entropy varied with volume, and found, to his surprise, that the variation was precisely that of a gas of particles each of energy hf.

These 'particles' were what we now call *photons*, essentially *quanta* (or chunks) of light, and are now fundamental in all branches of physics and beyond. (The term 'photon' was not introduced, by the chemist Gilbert Lewis, till 1916, but for convenience I shall use the word in describing events from 1905 on.) In 1905, though, any motivation for photons seemed dubious, and their existence appeared to be quite foreign to the great wealth of evidence for the *wavelike* nature of electromagnetic radiation built up over the better part of a century. (This is certainly a major issue, which will be discussed much more in the rest of the book.) The idea of photons was taken seriously only by Einstein and one or two others until the early 1920s.

Indeed, in 1914 a number of the most influential Prussian physicists, including Planck, gave an extremely generous recommendation for Einstein's membership of the Prussian Academy. They thought it best to comment on the photon idea as follows: 'That he may sometimes have missed the target in his speculations, as, for example, in his hypothesis of light-quanta, cannot really be held too much against him, for it is not possible to introduce really new ideas even in the most exact sciences without sometimes taking a risk.'

In his 1905 paper, Einstein made an important experimental prediction from his idea of light-quanta. It concerned the *photoelectric effect*. The effect itself had been discovered by Heinrich Hertz in 1887, though it was not till after the identification of the electron that it could be described as follows: when light is incident on a metal surface, electrons are emitted.

Successful quantitative experiments on the effect were extremely difficult to perform. The effect occurs at the surface of the metal, but, unless a very good vacuum is used, a layer of gas molecules is speedily deposited on the surface and the results are completely meaningless. By 1905, the best investigation had been that of Philipp Lenard, who studied the energy of the photoelectrons. He reported that this was independent of the intensity of the light, but increased with its frequency. Neither of these observations seemed easy to understand on the basis of the wave model of light, and it is, in any case, almost certain that Lenard's work did not influence Einstein.

From Einstein's theory of light-quanta, it seemed reasonable that an electron would gain the energy from a whole quantum (or none at all). Thus its energy *inside* the metal would be hf, just the energy of the quantum. For an electron it is energetically favourable to be *inside* the metal; in other words, when it passes through the surface it *loses* an amount of energy which is characteristic of the metal, and called the *work-function*, ϕ. So when the electron is detected outside the metal, it should have an energy less than hf by ϕ. (Actually this is a *maximum* energy; some electrons will have less for a variety of reasons including energy losses in collisions on their way out of the metal.)

Einstein's prediction agrees with Lenard's result, but is much more specific. His theory would suggest that the intensity of the light should not affect the electron energy; increasing the intensity means *more* photons, but each photon has only the same energy, hf. (Of course it may occur to the reader that an electron might obtain the energy from two or more photons; this *can* happen, but is exceptionally unlikely at normal intensities.)

Over the next decade, several experiments gave results in general support of Einstein's proposed law, but it was not until the work of Robert Millikan between 1914 and 1916 that all doubt was removed. The general form of his results clearly followed Einstein's suggestion, and Millikan obtained a value of h in very good agreement with that of Planck. This was a triumph for Einstein; his 1905 paper had been the first application of quantum theory to any system other than radiation. Because this work had practical uses (the industrial photocell), the Royal Swedish Academy of Sciences found it easier to give him the Nobel Prize for this work than for relativity, and Millikan received the 1923 prize himself for his own work, together with his accurate determination of the electronic charge (Chapter 2).

Yet one must still stress the point that, till 1922, it was the *law* that was accepted, *not* the photon. Millikan himself considered the light-quantum hypothesis 'wholly untenable' in 1915, and 'bold, not to say . . . reckless' in 1916. More than anything, this confirms how far Einstein was in advance of his contemporaries.

It is a very great pity that virtually all textbooks and elementary accounts of quantum theory choose to rewrite history, presenting Einstein *reacting* to very complete experimental evidence concerning the photoelectric effect when he produced his idea of the photon. This certainly produces a clear story-line, of course, but at the expense of making one of the greatest scientific feats of all time appear practically mundane.

By 1906, Einstein was able to produce the first coherent account of the status of quantum theory with respect to black-body radiation. His statistical mechanics gave *either* the Rayleigh–Jeans *or* the Planck law, depending on whether the energy-levels for the oscillator were continuous or discrete, and it was this analysis that Kuhn [32] says could be regarded as the birth of quantum theory. (He acknowledges that Ehrenfest provided similar arguments, although not the same simple conclusion.)

For the Planck case, there was one additional and very heavy price to pay; because energy is emitted and absorbed by oscillators discontinuously, Maxwell's Laws themselves can apply only to the *average* energy of oscillators. Again one sees how the wave theory of light, intimately associated with Maxwell's Laws, becomes inconsistent with the new quantum ideas.

Einstein emphasised this in 1909 when he calculated the (statistical) fluctuation in the energy of black-body radiation, now using Planck's Law. He obtained two terms, one corresponding to particles, one to waves. (In contrast, Pais [1] points out that Wien's Law gives only the first term, Rayleigh–Jeans only the second.) Einstein had fully succeeded in mixing wave and particle together; he would spend much of the rest of his life trying to unscramble them – never to his own satisfaction.

Einstein and thermal capacities

Till 1907, the quantum literature was restricted to studies of black-body radiation, together with Einstein's proposed extension to the photoelectric effect.

In that year, Einstein introduced a new topic – the thermal capacity of solids. In Chapter 3 we saw that classical statistical mechanics led to the equipartition of energy – each degree of freedom would have, on average, an energy $kT/2$, where k is Boltzmann's constant and T the temperature in Kelvin. In a solid, each atom vibrates in three dimensions, each dimension contributing two degrees of freedom (corresponding to kinetic and potential energy) for each atom. So at temperature T, according to classical physics, the total amount of energy should be the product of N, the number of atoms, and $3kT$.

The thermal capacity of the solid is the amount by which the energy increases when the temperature goes up one degree – this should be just $3Nk$, independently of temperature, corresponding to the rule of Dulong and Petit mentioned in Chapter 2, which worked well for most substances. There *were* exceptions, though; in particular, at normal temperatures the thermal capacities of carbon (diamond), boron and silicon were much too low.

Einstein predicted that, for all solid elements, the Dulong and Petit result should hold only above a certain temperature (with different temperatures for different elements); as the temperature decreased, the thermal capacity too should fall, and it should tend to zero at the lowest temperatures.

His argument was based on Planck's prescription, and our resulting Fig. 4.4 for the energy of each vibrator. Einstein made the assumption, which was simple and reasonable, though not exactly correct, that one could consider each atom vibrating independently at the same frequency f, and simple mathematics then led to Fig. 4.5 for the thermal capacity. This is a *universal curve* – the same *shape* applies for each element, but the temperature scale is different for each.

Einstein worked with diamond, for which θ, the value of temperature marked on the graph, that at which the thermal capacity is reduced to five-sixths its classical Dulong–Petit value, is about 900 K. He obtained good agreement with experiment. Diamond has a high θ because it has a high value of f, which in turn is because it is a hard substance. Much more typical is aluminium, for which θ corresponds to about 200 K, well below normal temperatures; this explains why, for most elements, the Dulong–Petit value is reasonably accurate at such temperatures.

The assumption that each atom vibrates independently was eliminated in the more rigorous theories of Peter Debye, and of Max Born and Theodore von Kármán, in 1912. These authors considered coupling of the vibrations of different atoms, which led to a range of frequencies of vibration, and they obtained improved agreement with experiment.

Study of the thermal capacities of solids and their temperature dependences had received a great boost in 1906 when Nernst's work, which was moving

Fig. 4.5. Einstein's calculation of how the thermal capacity, C, of an element should vary with temperature, approaching the classical Dulong–Petit value only above a certain temperature. The curve is the same shape for any element, but the temperature scale is different. θ denotes the temperature at which C reaches five-sixths its classical value; this is 900 K for diamond, but only 200 K for aluminium.

towards the Third Law of Thermodynamics mentioned in Chapter 2, required such knowledge, particularly at low temperatures. Nernst [35] was an extremely powerful scientist, and a mass of data was soon collected; it was in good agreement with Einstein's theory, and in particular the results tended to zero at low temperatures.

Nernst thus became interested in quantum ideas, and he was responsible for the first Solvay Congress in 1911 when the Belgian industrialist, Ernest Solvay, was persuaded to support a conference to discuss problems associated with the quantum theory. It was interesting that the theory was moving away from areas of physics known to be problematic – cavity radiation, electron theory – to areas such as atomic vibrations, which had previously been considered to be well understood by classical theory.

I shall conclude this section by briefly referring to thermal capacities of gases with molecules containing more than one atom. In Chapter 2 we saw that the translational motion of the molecule should give a thermal capacity of $\frac{3}{2}k$. Rotational and vibrational motion of the molecule should add to this, but these degrees of freedom are 'frozen out' at lower temperatures in rather a similar way to that just discussed for vibration in solids. (We have not yet considered quantisation of rotational motion, but the techniques to handle this were developed by Niels Bjerrum in 1912.)

The thermal capacity of hydrogen is sketched in Fig. 4.6 (though with a little licence, as explained in the caption) [36]. Below about 60 K it takes the value $\frac{3}{2}k$

Fig. 4.6. A sketch of the dependence of the thermal capacity of molecular hydrogen (H_2) on temperature. Below roughly 60 K, the value corresponds to translational motion only. At higher temperatures there is a gradual increase due to rotation, and above roughly 800 K a further increase due to vibration. (The section of the graph for the highest temperatures is calculated, since the hydrogen molecule dissociates into atoms at such temperatures; this section of the curve can be investigated experimentally for other gases.)

per molecule corresponding to translational motion only. Over the range from 60 K to 300 K, it increases to $\frac{5}{2}k$ per molecule, the increase being due to two degrees of rotational freedom. Finally, above about 800 K, there is a further rise to $\frac{7}{2}k$ per molecule due to two degrees of vibrational freedom. (It should be mentioned that our thermal capacities are all at *constant volume*; also important are thermal capacities at *constant pressure*, which are higher because of work done by the gas or solid in expanding.)

Bragg and X-rays

In Chapter 2, I described how the 'classical' understanding of X-rays was built up. As Wheaton [29] in particular has shown, at the very same time what may be called a 'quantum' understanding was being championed, particularly by W. H. Bragg.

From the time of their discovery in 1895, there was considerable controversy over the nature of X-rays (and, once they had been discovered, γ-rays came into

Fig. 4.7. The electromagnetic pulse (or impulse) model of an X-ray. Two possibilities were a single localised disturbance and a double disturbance of average amplitude zero. The horizontal axis may represent either distance or time.

the same category). Röntgen himself, and others, initially suspected that they might be *longitudinal* electromagnetic waves (though Maxwell's theory predicted only transverse waves). This belief was short-lived. Another view was that X-rays were a wave of just the same form as light, but much higher frequency. A third was that they were *electromagnetic pulses* (or impulses).

An impulse is an electromagnetic disturbance very different from a long wavetrain. It consists of a single localised disturbance, or alternatively a double disturbance of average amplitude zero, as shown in Fig. 4.7. Such a model appeared able to explain the absence of reflection, refraction and polarisation in the initial experiments. It is not *mathematically* distinct from the usual waveforms; a number of infinitely long wave-trains of different wavelengths may be superimposed mathematically to give *any* wave-form, including the impulse. (Mathematically this is called Fourier analysis.) But the impulse idea seemed able to explain *physical* problems, and it was developed extensively by G. G. Stokes, Wiechert, J. J. Thomson and Arnold Sommerfeld. Charles Barkla, an experimentalist, was its chief advocate.

His opponent was Bragg, who argued persistently that X-rays were particles. (His ideas on the precise nature of the particles varied from time to time, and are not important here.) Much of the argument between Barkla and Bragg was over experimental and theoretical details, the nature of X-ray scattering being particularly controversial. However, Bragg's central argument concerned the interaction of an X-ray with a gas containing many atoms. It was found that an X-ray would ionise just one particular atom; moreover the electron produced by the ionisation would have nearly the full energy of the X-ray. (These are Wheaton's *paradoxes of quantity and quality* [29].)

These observations seem impossible to explain if X-rays are extended wavetrains, and the energy should be spread evenly over a large volume; they are explicable with an impulse theory only if it is stretched to its absolute limit, but they are readily comprehensible – indeed obvious and expected – if an X-ray is a particle, and hence its energy is localised.

For us, of course, this just reads like an acknowledgement of the truth of Einstein's 1905 theory of light-particles. But, although the main Barkla–Bragg controversy took place in 1908, Bragg had not at first heard of Einstein's idea. Even when told of it, told in fact of a conflict between Johannes Stark, practically Einstein's only supporter at the time, and Sommerfeld supporting impulses, Bragg was unimpressed. For Einstein and Stark, light and X-rays were of the same type, and both exhibited particle-like *as well as* wavelike behaviour. For Bragg, light was a wave; X-rays were totally distinct from light, and were particles pure and simple.

As described in Chapter 2, the discovery of X-ray diffraction radically changed the situation. It seemed that one was forced back to the idea that X-rays were precisely analogous to light, and that at least the characteristic X-rays of Chapter 2 existed as long wave-trains. But with the acceptance of Einstein's photon concept in the early 1920s, the Barkla–Bragg struggle could be seen in retrospect as a conflict between classical and quantum ideas of X-rays, obscured at the time by the lack of basic knowledge concerning X-rays and the availability of the impulse idea, which was tempting, but eventually found to be rather misleading.

Quantum theory – the first 13 years

In this section I shall first mention a few other applications of quantum theory between 1905 and 1912. In his light-particle paper of 1905, Einstein explained *Stokes' rule* in *fluorescence*, according to which the frequency of the fluorescent radiation is less than or equal to that of the radiation that excites the fluorescing atom. In terms of photons, the energy of the outgoing photon cannot be greater than that of the photon causing the initial excitation.

Einstein also applied his idea to photoionisation; when light frees an electron from an atom, the electron must emerge with energy less than hf, where f is the frequency of the light. Later he discussed the reverse photoeffect, whereby electrons incident on a surface excite photons; the generation of secondary electrons by X-rays; and the high-frequency limit of an X-ray tube. He also commented that his light-particle concept would play an important part in photochemistry, where light stimulates chemical reactions.

He was not the first to recognise all these possibilities. Wien and Stark made several similar suggestions at roughly the same time. As Kuhn [32] points out, such work often required little more than identifying a related frequency and energy, and putting E equal to hf. A good example is the relationship between f, the maximum frequency of an X-ray tube, and its potential, V; hf should be equal to eV, since the energy of the electron responsible for exciting the X-ray must be eV. This relationship was confirmed experimentally by William Duane and Franklin Hunt in 1915, and is known as the Duane–Hunt limit.

Stark was at first a particularly keen advocate of the quantum idea, an enthusiastic supporter of Einstein, and, as mentioned above, practically the only other champion of the photon. However, by 1913, the two were embroiled in a priority dispute over photochemistry, and by the 1920s he and Lenard were the leading physicists in the Nazi party, and were vilifiers of Einstein as the representative of 'Jewish physics'.

Let us now summarise the years 1900–1912 for the quantum theory. It had spread from its roots in black-body radiation to tackle a growing range of problems. Despite the courage, determination, flexibility and professionalism of Planck, it is Einstein who stands supreme. Over and above the solving of specific problems, his deep understanding of the fundamental theories of classical physics, in particular statistical mechanics, enabled him to appreciate more fully than anyone else the full *implications* of the quantum results.

In the period from 1913 on, Einstein was still prominent in quantum matters, as much through his influence on others as through his own ideas. (Also one should not forget the little matter of general relativity!) But rather more prominent was a newcomer. Even before 1912, a few physicists – Arthur Haas, John Nicholson and others – had made interesting, but not totally convincing, attempts to apply quantum theory to atoms. In the hands of the newcomer, Niels Bohr, the emphasis of the still-developing theory was to move in this direction.

The basis of atomic spectroscopy

Bohr's great step in 1913 was the application of quantum ideas to *atomic spectroscopy*, an area that was from then on central in the development of the quantum theory. I shall explain the basic facts of this topic, and then describe some of the information available to Bohr.

To produce an atomic spectrum, a hot gas containing atoms of the substance is required. Today an electric discharge is passed through it, though in the pioneering work of Kirchhoff and Robert Bunsen last century, the famous *Bunsen burner* was used. It is found that the light emitted by the atoms is not continuous through the spectrum, but occurs at a number of discrete wavelengths. In an experiment, the different wavelengths are separated by a *diffraction grating* (explained in Chapter 2), and produce *lines* on a photographic plate, each line corresponding to a particular wavelength in the atomic spectrum.

A good example is atomic sodium, for which there are two strong lines very close together in the yellow part of the visible region called the D-lines. (There are also some other wavelengths, other lines, in the spectrum, but they are very much weaker.) Everybody who has studied chemistry knows that sodium glows yellow in a bunsen flame, and sodium street-lights work on the same principle.

Atoms of other elements often have very many prominent lines, and *spectroscopists* in the nineteenth century went to great efforts to measure them. Considerable accuracy could be achieved, easily better than one in a thousand, because there were not moving parts or other awkward sources of error in the experiment. The initial importance of this work was that the list of wavelengths became a 'finger-print' for the element, of inestimable worth in analytical work, since very small quantities of the substance were required.

(I might mention that atomic spectroscopy is just the beginning; since then, literally scores of analogous techniques have been invented in which atoms, or their nuclei or electrons, are stimulated in one of a myriad of different ways, and a spectral response is obtained. Such work can give an enormous amount of information, and is useful right across the sciences.)

The discussion so far has been of the *emission spectrum*. An *absorption spectrum* may be obtained by shining white light (containing all wavelengths) through the (cold) substance under investigation. Certain wavelengths are found to be missing from the emergent light, because they have been absorbed by the atoms. Kirchhoff's Law, explained earlier in this chapter, leads us to expect that absorption and emission spectra will be related. They are not, though, identical (because, in fact, the gases are at different temperatures in the two cases). The absorption spectrum contains only some of the lines in the emission spectrum, for reasons that will emerge in the following section.

A particularly interesting application of absorption spectroscopy is to the solar spectrum. In the early nineteenth century, William Wollaston and Joseph von Fraunhofer observed dark lines in this spectrum, including two at the wavelengths of the sodium D-lines. In today's terms, this tells us that there are atoms of sodium in the (gaseous but relatively cold) outer layers of the Sun, and these absorb particular wavelengths in the light emerging from the hotter interior.

New elements were found in this way by the discovery of new sets of spectral lines – in particular, *helium* discovered in the solar absorption spectrum in 1869. Whenever new elements were discovered on Earth, the key point of the proof became the demonstration of a fresh set of spectral lines.

Atomic spectroscopy is all that will be required in this book, but I shall briefly discuss molecular spectra (the molecules having more than one atom). Here the key new feature is the rotation and vibration of the molecule. This leads to a *band* spectrum, consisting of very many lines close together, resulting from different

Fig. 4.8. The frequencies of radiation in the spectrum of atomic hydrogen (in units of 10^{15} s^{-1}). The only lines in the visible region of the spectrum are those of the Balmer series, marked B and tending to the series limit B_L. The Lyman series, marked L and tending to limit L_L, is in the ultra-violet. The Paschen series, marked P and tending to limit P_L, is in the infra-red. Further series, Brackett, Pfund and so on, occur at lower values of frequency than the Paschen series.

rotational and vibrational states. These usually lie in the infra-red region of the spectrum, whereas the atomic spectra discussed so far have bright lines in the visible and ultra-violet.

Lastly I mention a solid. The solid, of course, consists of atoms that are far closer together than in a gas, and hence interact strongly. The emission of energy now really relates to the solid as a whole rather than to the individual atoms, and has an effectively continuous spectrum (as in any source of electric light where a solid, such as a filament, not a gas, is heated).

So far we have seen that spectroscopy must be of immense use in analytical science, but it may have occurred to the reader that the vast amount of information obtained should also help us to understand the atom. To obtain such understanding, scientists had to search for regularities in the set of frequencies of spectral lines. (Till now, I have been talking in terms of wavelengths, which is the long-standing experimental tradition, but from now on I shall discuss frequencies, which are more useful in theoretical analysis. Remember that wavelength λ and frequency f are inversely proportional because $\lambda f = c$, where c is the speed of light.)

The simplest spectrum is that of atomic hydrogen. (Note that, in obtaining such a spectrum, the electric discharge must cause at least some of the hydrogen molecules, H_2, to dissociate into single atoms.) In the visible region, hydrogen has four clear lines, one in the red, one in the blue and two in the violet, though there are many further lines, more easily obtained from astronomical data, moving into the ultra-violet, becoming closer and closer together, and tending to a *series limit*, as shown in Fig. 4.8.

It was natural for physicists to search for a formula that would give all these frequencies, but it was also very natural for them to think in terms of the fundamental–overtone scheme explained in Chapter 2, where all the frequencies would be multiples of a fundamental. Yet such schemes, however flexibly applied, were doomed to failure.

A completely different approach brought success in 1885 to Johann Balmer, a Swiss schoolteacher of mathematics and a great experimenter with numbers. He initially knew about only the four lines first mentioned, and he came up with a formula that, in our terms, gives the frequencies of these lines as simple multiples of Rc, where R is a new and important constant, the Rydberg constant, named after Johannes Rydberg. (The speed of light, c, is there for historical reasons.)

The multiples of Rc for the four lines are $(1/2^2 - 1/3^2)$, $(1/2^2 - 1/4^2)$, $(1/2^2 - 1/5^2)$ and $(1/2^2 - 1/6^2)$. With a suitable value of R, Balmer found excellent agreement with experiment. More than that, when he was told of the other lines, he continued in the same fashion, using the numbers $(1/2^2 - 1/n^2)$, with n equal to 7, 8, 9 and so on, and again his calculations agreed with experiment. It is easy to see how the series limit arose from his expression, because when n is very large, the frequencies become closer and closer to $Rc/4$, which is the limit.

One is inclined to suspect that there might be other such series in the spectrum with frequency equal to Rc times $(1/m^2 - 1/n^2)$, where m would be 1, 3, 4 and so on, and n would take integer values greater than m. (These frequencies would not lie in the visible region.) This turned out to be the case. In 1908, Friedrich Paschen discovered the series with $m = 3$; it is in the infra-red. Then (after Bohr's own work, in fact), in 1914 Theodore Lyman found the series with $m = 1$ in the ultra-violet; and the series with $m = 4$ and 5 were found by Brackett and Pfund in 1922 and 1924, respectively.

The spectra of other elements could be represented by similar formulae, though never quite such simple ones and never so exactly. The simplest case apart from hydrogen is that of the alkali metals (lithium, sodium, potassium and so on). For these spectra, m and n above must be replaced by $m - a$ and $n - b$, where a and b are small and (nearly) constant correction factors. Then the same type of formula ensues; the appropriate value of R is nearly but not quite the same as for hydrogen.

Such schemes were built up principally by Rydberg; in the first decade of the present century, he and Walther Ritz pointed out that each frequency could be expressed as the difference between two *terms* (or *spectral terms*). It seemed that these terms were more fundamental than the actual frequencies, but, till Bohr, nobody knew why.

The Bohr atom

Bohr began work on atomic models in 1912 in Rutherford's laboratory, and fairly naturally he attempted to develop the basic Rutherford atom described in Chapter 2. He did not initially work with the spectrum as described in the previous section for the very good reason that he did not know of the simple Balmer formula, and in any case did not imagine that spectra could be fundamental. The colours on a butterfly's wing were very interesting, he said much later, but you wouldn't use them as a basis for studying the fundamentals of biology [37].

Instead he attempted to adapt the kind of atomic model used by J. J. Thomson and mentioned in Chapter 2 to the Rutherford case [38]. This involved building up rings of electrons, and analysing electron–electron and electron–nucleus interactions. However, he found two major problems. First, there could be no *mechanically* stable structure; the system would break up rapidly. Secondly, it is also *radiatively* unstable. To understand this, we must remember, from Chapter 2, Maxwell showing that any accelerated charged object would emit radiation. In the Rutherford atom the electrons presumably travelled in circular orbits, and this implies that they are undergoing acceleration towards the centre of the orbit, so one expects them to radiate their energy away, and spiral into the nucleus.

Bohr was not too distressed by these problems because he already believed that atoms would not be explicable on the basis of classical physics. After some time working on this problem, he became aware, through Hans Hansen, of the Balmer formula, and within weeks he had produced his great work on hydrogen.

The ideas involved may be described fairly simply; I shall make no attempt to follow Bohr's own arguments, which were rather more convoluted. For his model of hydrogen, Bohr took one (singly negatively charged) electron travelling in a circular orbit round the (singly positively charged) nucleus – just a proton. The radius of the orbit and the speed of the electron in the orbit may be written as r and v – both unknown so far, but both to be determined during the course of the analysis. As I mentioned in Chapter 2, the problem so far is analogous to that of the Earth going round the Sun, or the Moon round the Earth. Circular motion requires a force towards the centre, which is provided by the Coulomb attraction between electron and proton (in analogy with the gravitational attraction between Earth and Sun). Mathematically this gives us one equation involving v and r.

So far everything has been totally classical, but now Bohr brought in a completely new and quantum idea involving Planck's constant h. It is that the magnitude of the angular momentum of the electron in its *Bohr orbit* (given by m

times v times r, from Chapter 2) can take only certain specified values. It must be equal to an integer n times h divided by 2π. The integer n may take the values 1, 2, 3, . . ., and since $h/(2\pi)$ comes up a great deal, much more than h on its own in fact, it is convenient to give it a special symbol – $\hbar$ (which is read as h-bar). So, according to Bohr, the magnitude of the angular momentum is given by $n\hbar$. An electron in such an orbit is said to be in a *stationary state* of the system.

The *precise* form of this (the 2π factor in particular) is frankly to provide agreement with the Balmer formula in the end. But the expression is exceptionally simple and broadly along the lines of Planck's or Einstein's ideas, though of course Bohr is quantising angular momentum, whereas Planck and Einstein were quantising energy. What Bohr's *quantisation postulate* means is that only certain orbits are allowed; of course, from a classical point of view, as with the ideas of Planck and Einstein, one cannot hope to justify such a restriction.

Mathematically this postulate gives us a second equation in v and r, and, with now two equations in two unknowns, it is possible to solve for v and r in terms of the important and fundamental constants $\hbar$, and e and m, the charge and mass of the electron; also involved is *quantum number n*.

Let us consider r, and we shall first put n equal to its lowest value, 1. We shall shortly learn that this value of n gives us the lowest value of the energy, for which the state of the atom is called its *ground-state*; in many, though not all, cases, a system nearly always sits in its ground-state at normal temperatures, and hydrogen is a good example of this. The radius of the orbit for this ground-state is a function of the values of m, e and $\hbar$; in numerical terms it is about $5, \times 10^{-11}$ m. This is a very acceptable number, since it gives us a diameter close to 10^{-10} m, which is in the range of atomic diameters estimated from X-ray crystallography, and at the low end of the range because the hydrogen atom is the simplest. If we now vary quantum number n, we find that the radius of the orbit increases as n^2.

We may also calculate v. In the ground-state, this is about 2.2×10^6 m s^{-1}. While we have nothing to compare this with very directly, we may at least be pleased that v is much less than c, the speed of light. If this had *not* been the case, our method, which uses non-relativistic expressions, would have been inconsistent. As n increases, v decreases proportionally to $1/n$.

From v we can calculate the kinetic energy of the electron, from r we can calculate the electrostatic energy between electron and proton, and we can add these together to give the total energy of the atom – as a function of n, of course.

Before thinking about magnitudes, let us consider the *signs* of these quantities. Kinetic energy is always positive; the electrostatic energy between the opposite charges of the proton and electron is negative (as shown in Chapter 2). There is another condition, though; the electron is *bound* to the nucleus; it cannot escape to infinity. In other words, it does not have enough *energy* to escape to infinity, where its energy would be zero, so its total energy must be negative; this is a criterion for a *bound state*. So the magnitude of the electrostatic energy must be greater than that of the kinetic, in order that the total can be negative. In fact, for each state, that is to say, for each value of n, if the kinetic energy is equal to E_k (positive), the potential energy is equal to $-2E_k$, and the total is $-E_k$.

Now to values. It turns out that, for the ground-state with $n = 1$, the energy of the system is -2.2×10^{-18} J (J for joules). The smallness of this quantity tells us that the joule is much too large a unit for atomic systems, and a far more suitable unit is the election volt (eV), the amount of energy required to increase the voltage of a charge e by one volt. 1 eV is equal to 1.6×10^{-19} J, so the ground-state energy of hydrogen becomes -13.6 eV.

States with n greater than 1 are called *excited states*. For these, the ground-state energy is divided by n^2, so the energy for $n = 2$ is -13.6 eV/4, or just -3.4 eV, and so on. An energy-level diagram for hydrogen, analogous to that shown earlier for the simple harmonic oscillator, is given in Fig. 4.9. It will be noticed that, as n increases, the energy-levels are packed closer and closer together; they tend to zero as n tends to infinity.

When we move on from the Bohr model to modern quantum theory, these energy-levels are unchanged. In addition, however, there are now *unbound* states with positive energy-levels – with *any* positive value of energy, in fact. These correspond to situations in which the electron has freedom to move right away from the nucleus. It is a general rule that, while *bound* states have *discrete* energy-levels, *unbound* states have *continuous* levels – a *continuum*.

One last important quantity for the hydrogen atom is its *ionisation energy*, the amount of energy required to remove the electron from the nucleus altogether. Since the atom is normally in its ground-state where it has energy -13.6 eV and, for ionisation, must be raised to the very bottom of the continuum, where it has energy zero, the ionisation energy is just 13.6 eV.

The Bohr calculation is pleasant and simple, but the reader may think 'Does it not defy what you said earlier about the lack of stability of such a system?' Classically the electron should spiral into the nucleus and, in doing so, it should radiate energy at *all* frequencies. The reader would be quite correct, but Bohr was willing to suspend this element of classical theory. How then *does* the atom emit or absorb energy? Bohr said (again quite contrary to classical physics) that the electron may jump, at random times and instantaneously, from one orbit to another of lower energy; the energy that it has lost is emitted as electromagnetic radiation at a specific frequency, f.

If, for example, the electron drops down from the $n = 2$ level to $n = 1$, the amount of energy emitted will be $E_2 - E_1$, the difference in energies of the two

Fig. 4.9. The energy-levels of atomic hydrogen, with values of quantum number n, and energy in eV. Transitions between levels ending on $n = 1$ give rise to the Lyman series, those ending on $n = 2$ give the Balmer series, and so on. Not shown is the continuum for positive values of energy; here *any* value of energy is allowed.

levels, or just 10.2 eV. During the process, a photon is emitted with frequency such that hf is equal to 10.2 eV. Similarly, an electron with $n = 1$ may absorb a photon of the correct frequency and jump to the $n = 2$ level.

In this way one may build up *all* the spectral series. Downwards transitions ending on $n = 2$ give the Balmer series, those ending on $n = 1$, the Lyman series, and so on; they are shown on Fig. 4.9. The $(1/m^2 - 1/n^2)$ factor appears exactly as in the Balmer type of formula, and, the jewel in the crown, Bohr was able to predict the value of R, the Rydberg constant, to an accuracy of roughly one part in 2000.

Actually he shortly showed that he could do rather better. Bohr's theory was limited to the case of a one-electron atom. Now neutral helium atoms have atomic number Z equal to 2, so they have two protons in the nucleus, and two electrons. A *singly ionised* helium atom, on the other hand, which is an atom with one electron removed, *is* a one-electron atom, and the Bohr calculation may be applied; it turns out that the presence of the *two* protons in the nucleus results in all the energy-levels being 2^2 times the corresponding ones for hydrogen. It is easy to see that all the frequencies are multiplied by the same factor. (Similarly, since lithium, Li, has $Z = 3$, a *doubly* ionised lithium atom, Li^{++}, has energy-levels and frequencies 3^2 times those of hydrogen, and so on.)

Now spectral lines close to the positions predicted for He^+ had been found by Charles Pickering in 1896 in starlight, and again by Alfred Fowler in 1912 in the laboratory. Bohr's triumphant identification was, however, hotly disputed by Fowler; the lines were close to the predictions, he said, but not close enough. Bohr responded with his *coup de grâce*. Till now we have been assuming that the nucleus is fixed, and the electron circles it; strictly this would be the case only if the nucleus were infinitely massive. (There is a direct analogy with the Sun–Earth case mentioned in Chapter 2.) It is a good approximation, because the nucleus is roughly 2000 times more massive than the electron, but, to be precise, both nucleus and electron rotate around a point *very* close to the nucleus.

Bohr's calculations may readily be adapted to this case, and the change causes a small difference in all the energy-levels, and thus in the frequencies of the radiation – precisely the correction demanded by Fowler. It was game, set and match to Bohr. (Elsewhere this correction makes R slightly dependent on Z, as experiment demands, and the predicted hydrogen frequencies now become equal to the experimental values to within the uncertainties of the various physical constants.) It was only at this point that Einstein concluded that 'This is an enormous achievement. The theory of Bohr must be correct' [3].

I shall now explain the differences between the emission and absorption spectra. In emission, electrons are flung to all energy-levels by the electric discharge, and one merely observes the photons emitted as they drop down. All spectral series will be present. For absorption, however, there is no discharge; the atoms sit at their normal temperature, and, as already stated, atoms are then virtually always in their ground-state. (This is because, from our statistical mechanics of Chapter 2, levels will be appreciably occupied only if they are within kT of the ground-state, where k is Boltzmann's constant; at room temperature, kT is roughly (1/40) eV, while, for hydrogen, $E_2 - E_1$ is 10.2 eV.) So in absorption only transitions starting at the ground-state will be observed – just the Lyman series.

In stellar atmospheres, of course, temperatures may be very much higher, kT may be in the eV range, and *all* the spectral lines may be seen in absorption.

I shall make a few last comments on Bohr's theory. First, I have used the idea of photons freely, but it should be remembered that Bohr did not have this idea to assist him; as I have said, very few physicists took photons seriously until the 1920s, and Bohr was one of the last to do so. For him, the radiation came off as a train of waves at a particular frequency. Rutherford was always a strong supporter of Bohr, but he did point out a difficulty. As the electron 'moved from' the $n = 3$ level, say, it must immediately commence radiating at the correct frequency, before it reaches the $n = 2$ or $n = 1$ level and therefore 'knows where it is going' and therefore what frequency it should be radiating at. What actually happens during the transition? Such questions plagued the theory. In modern quantum theory we learn not so much how to answer them as not to ask them.

Another important point is that, in Bohr's theory, the transitions were inherently probabilistic, just like those in radioactivity, and very much anticipating how things would emerge in the modern theory, at least according to orthodox interpretations. (See Chapter 5.)

I would also stress that Bohr's theory broke the well-established rule that an oscillating system radiates energy at the frequency of its oscillation. For Bohr, the frequency depends on *two* levels; each level has a frequency of rotation, but there was no direct connection between either of these frequencies and the frequency of radiation (at least in the general case – see the discussion of the correspondence principle below).

Indeed, we should recognise the limitations, as well as the strengths, of Bohr's ideas. They applied only to periodic one-electron systems, and were rather an unsatisfactory blend of classical and quantum notions. These defects would not be removed from quantum theory for more than a decade.

The heyday of the old quantum theory

The term 'old quantum theory' is used for the period between Planck's work in 1900 and the emergence of modern quantum theory in 1925. In retrospect it may be seen as a time when many correct results were obtained – but often only by use of half-understood, sometimes frankly ad-hoc procedures. A more positive characterisation might be that much was learned, sucessful ploys were gradually sifted from unsuccessful ones, and the ground was prepared for the creation of the new theory by Werner Heisenberg and Erwin Schrödinger.

There was a period following Bohr's 1913 theory, lasting most of that decade, when it seemed that the old quantum theory might perhaps itself form the basis of a new and complete theory. Even this statement, though, would have meant different things to different people. For Arnold Sommerfeld, who became the most active and successful proponent of the old quantum theory, the Bohr type of model could be taken fairly much at face value, and all that was required was to develop such models in closer and closer agreement with experiment.

Bohr, on the other hand, always realised that, even at their most numerically successful, such models were a conceptually incoherent mixture of quantum and classical. He himself had considerable success in bridging the divide with his *correspondence principle* described below, but it gradually became clear that a more fundamental break with tradition was required, and that the resulting theory might well be more 'abstract', less physically transparent, than the 1913 model of Bohr, indeed than virtually all physics up to that time.

Let us return to some immediate successes of the Bohr theory. I mentioned Moseley's results on characteristic X-rays in Chapter 2. These could not be rigorously understood on the basis of the Bohr model, because that model is applicable only to atoms with one electron, but a rough approach, in which those electrons not directly involved in transitions are treated in an approximate way, was very successful.

On the experimental side, in 1914 James Franck and Gustav Hertz published a paper providing clear support for Bohr's ideas. In their experiment, electrons were passed through mercury vapour. For electron energies up to 4.9 eV, nothing of interest happens, but when energies are above 4.9 eV, the electrons are found to *lose* that amount of energy, and the mercury emits ultra-violet light. In terms of the Bohr theory, one may interpret 4.9 eV as the energy difference between the ground-state of the mercury atom and its first excited state. If the incoming electron has energy somewhat greater than 4.9 eV, 4.9 eV is used to raise the atom to its first excited state; subsequently it returns to its ground-state, emitting radiation at the appropriate frequency.

I now move on to Sommerfeld's most famous work – his analysis of the fine-structure of the hydrogen atom. It had been shown by Michelson in 1892 that the Balmer 'lines' were not actually single lines, the $n = 2$ to $n = 1$ transition, for example, consisting of two lines. This *fine-structure*, though, is aptly named, at least for hydrogen, where the difference in frequencies is typically of order 10^{-5} times one of the actual frequencies. (For more massive atoms, the fine-structure becomes rather less fine.)

Sommerfeld wished to explain theoretically why the hydrogen fine-structure occurred. To do this, he needed to consider the hydrogen atom as a genuinely three-dimensional system, and this required a generalisation of the quantum conditions – omitted here because it is not really of long-term significance. Applied to a three-dimensional system, his generalised conditions give three

Fig. 4.10. Sommerfeld orbits for the first three Bohr energy-levels. For each value of n there are n distinct orbits, which we distinguish by quantum number l equal to 0, 1, ..., $n - 1$. The semi-major axis of the orbit is equal to $n^2 a_0$, where a_0 is the Bohr radius, and the semi-minor axis is equal to $n(l + 1)a_0$.

quantum numbers, rather than one as in Bohr's simple approach, or as for the (one-dimensional) simple harmonic oscillator. The first, n, corresponds closely to Bohr's original quantum number, but discussion of the orbit is more complicated for Sommerfeld. For a given n, he has n different orbital paths – ellipses, in fact, provided we remember that the circle may be regarded as a special case of an ellipse. (The occurrence of the ellipse should not surprise us; we remember from the planets that the *general* orbit under an inverse-square law is an ellipse.)

To describe the ellipses, we remember that Bohr's circular orbits had radii $n^2 a_0$, where a_0 is the so-called *Bohr radius*, the radius of the orbit for the ground-state. So for $n = 1$, Bohr gives a circle with radius a_0; Sommerfeld gives the same. For $n = 2$, Bohr gives a circle with radius $4a_0$. Sommerfeld gives two orbits – one the same as Bohr, the other an ellipse of semi-major axis (defined in Chapter 2) $4a_0$ and semi-minor axis $2a_0$. For $n = 3$, Bohr gives a circle of radius $9a_0$. Sommerfeld gives a circle of radius $9a_0$ and two ellipses, each with semi-major axis $9a_0$ and with semi-minor axes $6a_0$ and $3a_0$. And so on, as shown in Fig. 4.10. In all elliptical cases, the nucleus lies at one of the foci of the ellipse.

We may discuss this in terms of two quantum numbers, the first, as usual, n; for the second I shall cheat a little by giving the *modern* quantum number which plays the role of Sommerfeld's. I introduce quantum number l, which, for a given value of n, may take any integrally spaced value between 0 and $n - 1$. So, for $n = 1$, we must have $l = 0$; for $n = 2$, l can be 0 or 1; for $n = 3$, l can be 0, 1 or 2; and so on. For given n, the semi-minor axes of the n possible ellipses take values $n(l + 1)a_0$, for each possible value of l.

Having got this far, one can imagine Sommerfeld calculating feverishly to check whether states with the same value of n but different values of l would have different energies, which could then feed through to differences in transition frequencies and to the experimental fine-structure. At first he restricted himself to non-relativistic expressions, and the disappointing answer was no! States with the same n but different l had the *same* energies. But when he repeated the calculations using the expressions from special relativity, the ellipses precessed (like planetary orbits), energy differences emerged, and, triumphantly, agreement with experiment was found for the fine-structure.

I have mentioned that there is a third quantum number, which is related not to the shape of the orbit, but to its orientation. For a given value of l, there are $2l + 1$ values of this third quantum number. I shall again cheat by using specific numbers from modern quantum theory, which are very similar to, but different in detail from, those of Sommerfeld. I shall call this quantum number m_l, and say that, for given l, it may take integrally spaced values between $-l$ and $+l$. So for l equal to zero, there is only one possible value of m_l; this is zero. For l equal to 1, m_l may take the values -1, 0 or 1, and so on.

We may explain the meaning of the quantum number m_l by remembering from Chapter 2 that an electron in an orbit has an angular momentum, that angular momentum is a vector, and that its direction is perpendicular to the plane of the orbit. While l determines the length of the vector, which is $l\hbar$, m_l determines its orientation. If m_l takes its highest value of $+l$, the direction of the vector is along the z-axis, so the orbit is in the xy-plane. When m_l is equal to $-l$, the direction of the vector is *opposed* to the z-axis, so the orbit is again in the xy-plane, but is traversed in the opposite direction. For intermediate values of m_l, the vector is at specific angles to the z-axis, so its z-component is equal to $m_l\hbar$, and the plane of the orbit is always perpendicular to the direction of the vector. (We shall see later that this result of the old quantum theory is not *quite* true in the modern theory.)

The energy cannot depend on the value of m_l in the absence of a magnetic or electric field that could single out a particular direction as special. Effects when such fields were present had been known for some time. In 1897, Pieter Zeeman discovered that, in a magnetic field, a spectral line splits into a number of separate lines or *components* – the *Zeeman effect*. Lorentz immediately produced a theory based on a classical model of the electron, showing that the pattern should be of three equally spaced lines, symmetrically disposed with respect to the position of the single line when there is no field. Some lines *did* split in this way – the so-called *normal* Zeeman effect. Many others did not; far more complicated patterns were found, with many different numbers of components. This is the so-called *anomalous* Zeeman effect, in reality the rule rather than the exception! It was to give theoretical physicists a lot of trouble.

The normal effect could be dealt with fairly easily on the basis of the old quantum theory, as Sommerfeld and Debye showed. Indeed, we may gain some understanding of this from the ideas of Chapter 2; a rotating electric charge, such as the electron in the Bohr atom, is equivalent to a current in a loop, which, according to Ampère, constitutes a magnetic dipole with magnetic moment perpendicular to the plane of the orbit. To calculate the energy, we need to know the direction of the magnetic moment relative to that of the magnetic field, which is taken to be along the z-axis. It is easy to see how central the value of m_l is in this, and so this quantum number is called the *magnetic quantum number*. To go further and obtain the spectrum, we need a selection-rule, for which we require the idea of the correspondence principle, to be discussed shortly.

I shall mention here, though, that, just as we may regard $\hbar$, the angular momentum of the electron in the ground-state of the Bohr atom, as a basic unit of (electronic) angular momentum, so the magnetic moment of the electron in this orbit may be regarded as a basic unit of (electronic) magnetic moment – the *Bohr magneton*. To call these 'basic units' does not mean (as might have been hoped at first) that *all* electronic angular momenta and magnetic moments are integral multiples of such units; they do, however, set a suitable scale for these quantities. We are reminded by these considerations of what we learnt in Chapter 2 – there is an intimate connection between angular momenta (of charged particles) and magnetism.

I briefly mention the splitting of spectral lines in *electric fields* – the *Stark effect* (discovered by Stark in 1913). In fact I concentrate on the *linear* effect, in which splitting is proportional to the electric field; Stark demonstrated this for the excited states of hydrogen. Three years later Karl Schwarzschild and Paul Epstein independently explained this effect on the basis of the old quantum theory. For all other atoms, a *quadratic Stark effect* is exhibited, the splitting being proportional to the *square* of the electric field; understanding of this effect had to wait for modern quantum theory.

Einstein: spontaneous and stimulated emission

Before discussing the correspondence principle, I want to give an account of an important contribution to quantum theory of Einstein, made in 1916, in the period when he was more concerned with general relativity. It was highly significant in its own right, and Bohr also made use of it in the correspondence principle.

Einstein considered entities with two energy-levels in thermal equilibrium with electromagnetic radiation. He was able to derive Planck's Law for the radiation by assuming transitions occurring at random times between the energy-levels. To do so, he required transitions of *three* types. The first occurs from lower level E_1 to upper level E_2, and the probability of its occurrence is proportional to the energy density of the radiation; this is just *absorption* of energy as we have discussed it so far. The other two types of transition occur from E_2 to E_1, and hence they are *emission* processes. One occurs independently of the energy density, and is the emission process discussed so far; here we call it *spontaneous emission*. The rate of the other is proportional to the energy density. It is called *induced* or *stimulated emission*, and it was here discussed for the first time in quantum theory, though the corresponding process does occur classically.

For Einstein perhaps the most important result of his calculation was that it provided a direct connection between Planck's Law and Bohr's quantum hypothesis – a transition between E_1 and E_2 accompanied by emission or absorption of a packet of radiation of frequency f, with hf given by $E_2 - E_1$.

For us the most interesting feature may be the stimulated emission. It lies at the heart of the 1950s invention of the *laser* (light amplification by stimulated emission of radiation), a tool of immense power in industry and surgery.

There are two more important aspects of Einstein's work. First, it may be felt that, by following radioactivity and Bohr's theory in assuming transitions occurring apparently randomly in time, Einstein was abandoning determinism. Certainly he was much concerned that this *might* be necessary. Jammer [39], however, suggests that this was not his position, or at least not necessarily. It was possible for all these cases to believe that there was an underlying causal mechanism giving rise to transitions that only *seemed* to be spontaneous. It could be said that, broadly speaking, this remained Einstein's position, while the great majority of other physicists, including Bohr, came to embrace non-deterministic notions fully.

The second point concerns the introduction by Einstein as part of this work of the idea that the photon possesses momentum as well as energy. Classically electromagnetic radiation does transport momentum, so this new point should not be a real surprise. To obtain a value for p, the magnitude of this momentum, one cannot use the expression given in Chapter 3, γmv, since the photon has mass zero, and also, since it travels at the speed of light, γ must be infinite. But it is possible to deduce that, for the photon, p must be equal to E/c. Since E is hf, p must be given by hf/c, or just h/λ, where λ is the wavelength. This concept, and the explicit expression, were to be important both for Compton and for de Broglie.

The correspondence principle

Sommerfeld showed that for hydrogen the nature of the orbit depended on two quantum numbers, n and what we call l, following the modern notation. Its orientation depended on a third quantum number, which we call m_l. For hydrogen the energy depended only on n in the non-relativistic approximation and in the absence of a magnetic or electric field. When relativity is taken into account, different values of l give *very* small differences in energy, and when a magnetic field is applied, the energy depends on m_l as well.

When one turns to any other atom, the main difference is that, even non-relativistically, the energy depends very much on l as well as on n. This fact raises a new question. Not every pair of states may be connected via a transition. We did not have to consider this point in the original Bohr case, because in fact there is *no* restriction on transitions between states with different values of n. Experimentally there *is* such a restriction on states with different values of l; it is called a *selection-rule* and says that transitions will take place *only* between states with values of l differing by unity. Similarly, in the presence of a magnetic field, there is a selection-rule for m_l; in a transition it must change by unity *or* remain unchanged. (This m_l rule enables the deduction of the normal Zeeman effect, mentioned above, to be completed.)

A complete theory should give us all this information. It should also tell us about the *polarisation* of the various lines in the case of the Zeeman effect – there is much important information here – and also give the *intensities* of the various lines with or without a magnetic field present. The old quantum theory had no direct means to give these results, so, in that sense, it was not a complete theory; rather Bohr aimed to complete it with his *correspondence principle*.

The principle used the idea of a *classical limit*. We have met this idea in connection with the Planck distribution. At low frequencies, or, in fact, at high temperatures, this became equivalent to the classical Rayleigh–Jeans distribution, so low f or high T constitutes a 'classical region'. For the hydrogen atom, Bohr realised that the energy-levels became closer and closer together as n increases, and took this to mean that high values of n constituted a classical region. He attempted to show that this was true in general, with considerable success, though the arguments could never be made totally rigorous.

An interesting example here is to compare the frequency of rotation f_R of the electron in a Bohr orbit with the frequency of radiation in a transition. From the classical point of view, one would expect the system to radiate at f_R. In the general Bohr case there is no clear connection; in fact a particular frequency of radiation relates to *two* orbits, not one. But for large n, the frequencies of radiation for the transitions between states with quantum numbers $n - 1$ and n, or n and $n + 1$ become indistinguishable from each other, and also from the values of f_R for the states with quantum number n or $n - 1$ or $n + 1$. For a transition from (large) n to $n + 2$, say, the frequency is doubled, giving us a fundamental frequency and overtones, as one might expect classically.

Here the connection between classical and quantum regions emerges from the final result. However, it is possible, and more in the spirit of the correspondence principle, to reverse the logic somewhat, and this is what Bohr actually did in his original paper of 1913. He did not *postulate* his quantum restriction of angular momentum (as I implied above), but was able to deduce it, by making explicit use of the correspondence-principle argument of the previous paragraph.

The correspondence principle was not used systematically until 1918 (and not named until 1920). In its first formal development, Bohr considered electrons emitting very-low-frequency radiation – the classical, high-n case. He analysed the radiation classically, determining which frequencies were present, and then translated this into quantum terms, deducing which selection-rules would allow these frequencies and no others. All this was assured only for the large-n case, but Bohr boldly predicted that these rules applied for transitions between states with *any* values of n. He hence deduced the selection-rules for l and m_l given above. He was also able to show that, if m_l does not change, the radiation emitted is polarised parallel to the direction of the magnetic field, whereas if it changes by one, the polarisation is perpendicular to that direction. This was again in good agreement with experiment.

With the use of Einstein's work on emission described in the previous section, Bohr's associate, Hendrik Kramers, was able to calculate the relative intensities of the components of the fine structure, and also those present in the Stark effect, for a range of cases and with a good level of success.

Not everybody appreciated the correspondence principle. Sommerfeld had the most to lose, since he saw his hard-earned results from 'proper' physics being overtaken by facts seemingly pulled out of a hat. He described [40] the principle as 'a magic wand'; he seems to have been particularly frustrated by the fact that, even in the large-n case, there was no suggestion that classical and quantum *physics* are the same – just the experimental predictions of the theories, but he was open-minded enough to admit that 'in view of the achievement, the question as to whether Bohr's procedure is . . . satisfactory logically . . . becomes only of secondary importance'.

The correspondence principle was certainly successful in its own terms – a vast amount of detailed work went towards making the old quantum theory complete in its predictive ability. Logically it seemed less satisfactory to have to use classical results explicity to complete the quantum theory. However, even when the old quantum theory was replaced by the modern theory, the correspondence principle still had two important roles.

First, it assisted the development itself. Jammer [39] writes that 'there was rarely in the history of physics a comprehensive theory which owed as much to one principle as [modern] quantum mechanics owed to Bohr's correspondence principle', van der Waerden [41], that modern quantum theory may be described as 'systematic guessing, guided by the Principle of Correspondence'. Secondly, even in the modern era, Bohr's principle of complementarity, though very different from the correspondence principle, retains the same idea of classical physics having a status as more than *just* a limiting case of quantum physics.

The Compton effect

In the early 1920s, a number of new experimental results and theoretical ideas made it seem less and less likely that the old quantum theory, even buttressed by the correspondence principle, had the resources to describe Nature adequately. The first of these was the 1923 work of Arthur Compton. The *Compton effect* is very easy to explain, but Stuewer [42] has described comprehensively the experimental ingenuity required and also the conceptual struggle before the 'simple' explanation emerged.

The experiment involved the scattering of X-rays by a metal foil. As well as a beam scattered without change in wavelength, Compton identified a beam at a wavelength higher than the initial one; the *change* in wavelength depended only on the angle of scattering, not on the initial wavelength or the particular metal.

Compton interpreted his results as describing a collision between an initial X-ray photon and an effectively free electron in the scattering foil. The collision could be analysed in exactly the way sketched in Chapter 2, provided that the X-ray photon was assumed to have energy hf and momentum h/λ. In the collision, the X-ray photon would lose some of its initial energy to the electron, so its frequency would be decreased and its wavelength increased. The experimental results agreed in detail with calculations performed by Compton (and independently by Debye).

I shall note two further points. First, it is necessary to use *relativistic* expressions for the energy and momentum of the electron. Secondly, it might be queried why the electron may be treated as initially free, when it must really be bound to an atom in the scattering foil. The reason is that because of its high frequency the X-ray quantum has far more energy than, say, an optical or ultra-violet quantum involved in the photoelectric effect, and compared with this energy the initial binding energy of the electron to its parent atom may be ignored.

With the Compton effect, the photon concept had definitely arrived. There now seemed little alternative to accepting that electromagnetic radiation was here behaving as particles with energy and momentum. Other recent relevant experimental evidence was the detailed extension of Millikan's photoelectric work to the X-ray region by Maurice de Broglie and also by Charles Ellis in 1921. De Broglie had been joint secretary and editor of the first Solvay Congress mentioned above, and had developed an intense interest in quantum matters [37]. He had, however, remained rather an outsider from the main centres and had maintained independent views; for some time he and his brother Louis had believed in the corpuscular aspect of light. We shall hear much more of Louis shortly.

As far as the photon was concerned, another way in which the early 1920s perhaps differed from, say, 1914, was Einstein's new stature; by 1921 what had been a (pardonable) idiocy seven years earlier demanded to be taken seriously.

I said that there was little alternative to the photon but Bohr, still an opponent, exploited what little there was. His alternative involved making conservation of energy statistical. As Charles Darwin (great-grandson of the famous naturalist) in particular had stressed, any theory which held that atoms emitted and absorbed energy *discontinuously*, but retained *continuity* of the electromagnetic field (through ruling out photons and sticking totally to Maxwell's Equations), must *inevitably* hold that energy is not conserved at the level of individual systems, only statistically.

In 1924, Bohr seized on an idea of John Slater, an American visitor to Copenhagen. Slater had suggested that each atom is the source of a 'field' interacting with all the other atoms and oscillating with the frequencies of the various quantum transitions. The 'field' would then guide the photons and specify the probabilities that they take particular paths between the atoms. Slater had blended together wave and particle.

Bohr first got rid of the photons. The 'field' now acted directly on the atoms undergoing transitions. These transitions were discontinuous, and the connection with the continuous electromagnetic field had to be statistical. Determinism disappeared, and conservation of momentum and energy also became statistical. Such was the content of the 1924 paper by Bohr, Kramers and Slater (BKS).

Einstein hated this abandonment of determinism. Schrödinger, however, who had for a long time been interested in the possibility that the conservation laws were statistical, was strongly positive. (In this book, where I often imply that Einstein and Schrödinger were in adjacent camps, if not the same one, it is salutary to remember that different issues divide physicists in different ways.)

For the Compton experiment, BKS implied that there would not usually be a coincidence between recoil of an electron following the collision and scattering of an X-ray quantum. Almost immediately, however, experiments of Walther Bothe and Hans Geiger demonstrated that there was *always* such coincidence, while Compton himself and Alfred Simon checked from the angles of emergence of electron and quantum that the conservation laws were obeyed for *each* event.

So Bohr realised that BKS had to be given an 'honourable funeral', while Slater [43] developed a lifelong antipathy towards him for the fruitless dismemberment of his own ideas. For Werner Heisenberg [37], it became clear that there were to be 'no cheap solutions'.

In a book on Einstein and Bohr, is it necessary to pronounce the whole photon saga a triumph for Einstein, a failure on Bohr's part? Certainly the first part must be true. Einstein's initial conception of the photon was no less than an act of genius, and his perseverance with it, despite the negative response and his own misgivings over the relation between wave and particle concepts, showed great determination and courage.

It is easy to understand why Bohr, approaching quantum theory more from the particle than the radiation side, may have initially felt that the problems of absorption and emission of electromagnetic radiation could be dealt with via the correspondence principle. His final advocacy of BKS meant that every possibility of avoiding wave/particle duality would have been tried. There could be no turning back from the conceptual turmoil to come.

Bohr, long-term arch-enemy of the photon concept, was to feel comfortable within a very few years in explaining wave/particle duality as an important component of his framework of complementarity, as described in the next chapter. Ironically it was Einstein, the progenitor and for so long nearly the only supporter, of the photon concept, who never came to terms with wave/particle duality. In 1951, he wrote to his long-standing friend Michele Besso 'All these fifty years of brooding have brought me no nearer to the answer to the question "What are light quanta?" Nowadays every Tom, Dick and Harry thinks he knows it, but he is mistaken.'

De Broglie, Bose and Einstein

I have already mentioned Maurice de Broglie and his early acknowledgement that radiation exhibited wave/particle duality. His brother Louis, younger by 17 years, became absorbed by these issues, and was inclined to think of the light quantum as possessing a mass that was very small but not zero. In 1923, he conceived of an intrinsic frequency, f_i, where hf_i is equal to mc^2, the mass-energy of the photon (as he perceived it). But this led to a contradiction. On the one hand, special relativity would say that, for an observer moving with respect to the photon, its energy, and hence its frequency, should be *increased*. But time-dilation, another component of special relativity, would say it should be *decreased*.

De Broglie attempted to solve the problem by imagining a wave associated with the photon. He was able to show that this wave and the intrinsic oscillation of the photon remained in phase at the position of the photon, but not elsewhere. What is most important came next. Since his photons had a non-zero mass, it seemed reasonable to extend the ideas to what had till then been thought of as particles pure and simple – electrons and so on. (When I say 'reasonable', I mean in the context of the flow of ideas; his suggestion was contrary to all the evidence at the time and also against what would have been called common sense.) De Broglie suggested that such 'particles' should also have a wavelike nature, which could be manifested in diffraction experiments using gratings of suitable dimension. This was his great and crucial contribution to the development of quantum theory.

The analysis was rather subtle, and not all of it would be considered correct today, but one point may be stated clearly: the formula for the wavelength of the associated wave could be obtained from the expression for photon momentum. For the latter we put p equal to h/λ; here we invert the same expression to obtain λ, the *de Broglie wavelength*, equal to h/p. This is at least reasonable in the sense that, for objects of macroscopic dimension – buses and so on, and usual speeds, λ will be extremely small, so the proposed wavelike properties would not be observable.

One nice aspect of the de Broglie idea related to the Bohr theory. For one could put together Bohr's quantisation rule, according to which angular momentum, the product of momentum p and radius r of the orbit, is equal to $n\hbar$, and de Broglie's expression for λ. It is easy to obtain the result that the circumference of the allowed Bohr orbits must be equal to $n\lambda$, where n is a whole number. Thus the allowed orbits are those for which the de Broglie wave interferes *constructively* as the electron orbits the nucleus. This seems good physical sense.

It might be mentioned that de Broglie's vision was analogous to hydrodynamics. Here the water certainly consists of molecules, but one is able to describe the physical behaviour in terms of waves moving through the water. In much the same way, de Broglie perceived his waves as 'pilot waves' directing the electrons or photons around. De Broglie's models – always very specific – will be discussed further in Chapter 7.

Back in the 1920s, his ideas did not immediately find favour. In 1924 they formed his doctoral thesis, for which his principal examiner was Paul Langevin, an extremely talented physicist famed for his work on magnetism. Langevin was impressed by the boldness and ingenuity of de Broglie's ideas, but bemused as to whether they really contributed to an understanding of physics. He therefore sent a copy of the thesis to Einstein for his views. Einstein was impressed, feeling that the proposal was 'a first feeble light on this worst of physics enigmas' [1]. So M. de Broglie became Dr de Broglie, while Einstein continued to cogitate

Also in 1924 Einstein received a letter from the unknown Satyendra Bose of Dacca, then in India. Bose had a new derivation of Planck's Law. Since 1900, quite a few of these had been produced. Some were correct almost fortuitously, others more rigorously, but none had provided what could now be seen as the real requirement – a statistical mechanics of photons; such must be the most direct route to a study of thermal radiation.

The reason why Planck had had to struggle somewhat to obtain agreement with experiment using Boltzmann statistics is simply that photons do *not* obey Boltzmann statistics! Bose and then Einstein took considerable steps towards the new form of statistics that they do, in fact, obey, now called Bose–Einstein statistics, though the full task could not be completed rigorously till the modern quantum theory was created the following year.

Bose introduced several assumptions that were contrary to Boltzmann, though he almost certainly did not recognise this at the time [1]. First, the total number of photons was not held constant; photons were not conserved. Secondly, photons did not behave independently, as Boltzmann decreed. Thirdly, and I think most significantly, Bose effectively treated photons as indistinguishable. I shall come back to these matters towards the end of this chapter in the context of the modern theory.

Bose used his assumptions to obtain Planck's Law, but the novelty of his approach not surprisingly meant that, before he sent his paper to Einstein, it had already been rejected by one journal. Einstein recognised its importance, translated it into German himself, and had it published; its fundamental significance was soon appreciated. Sixty years later, Pais [1] called Bose's work the 'fourth and last of the revolutionary papers of the old quantum theory', putting it alongside the seminal works of Planck, Einstein and Bohr. (That makes one ponder on Bose's personal lack of international recognition. He had little interaction with the leaders of quantum theory, even Einstein, and what he had was not particularly profitable. He does not seem to have come near a Nobel Prize. In India, of course, more particularly in Bengal, he is justly famed.)

Einstein then adapted the Bose procedure to the statistical mechanics of a gas of non-relativistic particles. In contrast to Bose, he needed to consider particles of non-zero mass, and retained conservation of particles. This work was of general great importance, but one aspect is crucial here. In 1909 Einstein had calculated the energy fluctuation of electromagnetic radiation, and found an unexpected particle-like term, as well as the expected wavelike one. Now he repeated the calculation for his gas. He obtained two terms, the expected particle-like one, but also a wavelike one, which must have seemed totally unexpected – except, I suppose, to de Broglie and Einstein himself.

Einstein was in a unique position in 1924 and early 1925 to assimilate the unorthodox views of de Broglie and Bose, to relate them, to extend them, and to make them plausible for the physics community. Thus he played the main role in stimulating Schrödinger's formulation of modern quantum theory, which is often called *wave mechanics*. We shall meet this shortly.

Lastly in this section I come to the experimental evidence for a wave nature of electrons. It is not last in importance, but appreciation of the experiments *did* largely follow the theoretical speculation, even though some of the actual experiments were performed earlier.

From 1921, Clinton Davisson and Charles Kunsman had performed experiments on the scattering of low-energy electrons by metal targets; they obtained unexpected peaks. Also Carl Ramsauer, and independently John Townsend, had shown that, when low-energy electrons were scattered by gas atoms, the scattering cross-section went through maxima and minima as the energy of the electrons was altered, and indeed went to zero for certain energies (the Ramsauer–Townsend effect).

Walter Elsasser, together with James Franck, interpreted these results as diffraction of waves. The results of Davisson and Kunsman were analogous to those from X-ray diffraction and indeed one could calculate the de Broglie wavelength corresponding to the electron energy of 54 eV; it was about 2×10^{-10} m, very typical of inter-atomic distances in crystals. The total transparency of the Ramsauer–Townsend effect may be explained if the diameter of the scattering atom is closely related to the de Broglie wavelength, so as to provide constructive interference between incident and scattered waves in the original direction of motion of the electron.

Elsasser's paper on these matters was published in 1925 – but again only after the editor had consulted Einstein [39]. The conclusive experimental results had to wait till 1927 when Davisson and Lester Germer produced clear diffraction patterns for electron scattering from nickel, together with confirmation of the value of the de Broglie wavelength from the angle of the diffraction peak; and George Paget Thomson obtained direct evidence of electron diffraction from thin films. This work was too late to influence Schrödinger, whose great breakthrough had been made at the end of 1925.

Davisson and Thomson shared the Nobel Prize for Physics in 1937. G. P. Thomson was the son of J. J. Thomson, who had been awarded the 1906 Prize for showing that the electron was a particle; G. P., it might be said, a little loosely, was awarded the Prize for showing that it was a wave!

Spin, the exclusion principle and the Periodic Table

The chief problems that the old quantum theory never seemed likely to solve were the anomalous Zeeman effect and the spectrum of helium. The latter was perhaps particularly disappointing. Helium was the second smallest atom, and, while it was not clear how the Bohr approach should be extended to the two electrons of helium, the sad fact was that *any* sensible method that was tried failed dismally to agree with experiment.

The anomalous Zeeman effect was closely linked to the so-called *multiplet structure* of energy levels. I have already said that, once one leaves hydrogen, the energy of a particular level depends fairly strongly on quantum number l (again, for convenience, using modern nomenclature), as well as on n. In addition, though, most levels of given n and l are further split into a multiplet – doublet, triplet or higher.

Other important and relevant experimental evidence came from the *Stern–Gerlach effect*. From 1921, Otto Stern and Walther Gerlach had sent beams of atoms through inhomogeneous magnetic fields onto a screen where the atoms are deposited. Different values of quantum number m_l (in our terms) correspond to differently oriented angular momenta and hence to differently oriented magnetic moments. These should travel along different paths through an *inhomogeneous* magnetic field. (They would not do so through a *homogeneous* field.) The result is that one should see piles of atoms being deposited at specific points on the screen – one for each value of m_l, or, in fact, $2l + 1$ piles for a particular value of l for the ground-state of the atom.

Since l is an integer, this would suggest that there should always be an *odd* number of piles of atoms, corresponding to an odd number of paths through the field. However, for silver atoms it turned out that there were *two* piles, and hydrogen was later to give the same result. (I would mention that, though the number of piles was difficult to explain, the distances between the various piles fitted in reasonably well with the value of the Bohr magneton mentioned above.)

A last problem area was the failure to understand the Periodic Table. Bohr actually made good progress with this in the early 1920s, and indeed had successfully predicted the properties of a new element with $Z = 72$, which was subsequently identified at Copenhagen in 1922 by Georg Hevesy and Dirk Coster. But Bohr's ideas were a fruitful combination of genius and guesswork, not correct in detail, and certainly lacking a firm theoretical basis.

Such was the range of difficulties faced by the old quantum theory in the early 1920s. There was, of course, a huge amount of information available, particularly from atomic spectra – the problem was how to make sense of it! The principal performers here were Sommerfeld and Alfred Landé, and they recognised that many of the problems mentioned above (not helium!) appeared tractable given a further quantum number. Since it seemed clear that the electrons in any atom could be divided into two categories – a small number of outer *valence electrons*, which participated in atomic transitions, and the remaining *core electrons* which did not – it seemed reasonable to allocate a new quantum number, together with angular momentum and magnetic moment, to the *core*.

Such a manoeuvre led to many of the Stern–Gerlach results becoming explicable. With judicious choice of quantum numbers, including introduction of some half-integral ones, and replacing a quantum number j, related to *total* angular momentum of valence and core electrons, by $\sqrt{j(j+1)}$ or $\sqrt{(j+1/2)(j-1/2)}$, Sommerfeld and Landé (and also the young Heisenberg) were able to correlate much of the data. They could explain many of the strange features of the anomalous Zeeman effect, and produced empirical expressions for energy-splitting in multiplets (such as the famous Landé interval rule, still very useful today). They did not, however, add much to genuine understanding!

At this stage, Wolfgang Pauli entered the fray. Already, in 1921 and at the age of 21, he had published an important account of relativity, which had been highly praised by Einstein, and remains just about the most erudite review of the subject. Much later [1], Einstein was to consider Pauli, in effect, his successor. But in the years from 1922 he was very closely associated with Bohr, though always fiercely independent in his outlook.

In 1924 Pauli showed that the attribution of a quantum number and angular momentum to the core could *not* be correct. (He showed that there should be a relativistic correction, which was never seen in the experimental data.) Instead he postulated a new non-classical quantum number, which took two values only, and was associated with the valence electron itself.

Later in this year he published his famous exclusion principle. First, he allocated four quantum numbers, n, l, m_l and now his new non-classical one, to each electron, and re-assessed the evidence for the values taken by each quantum number. In several points he disagreed with Bohr, following instead recent work by Edmund Stoner, itself based in part on the results of X-ray-absorption experiments of Louis de Broglie and Alexandre Dauvillier [39]. Pauli's crucial postulate then (the exclusion principle itself) was that each electron in an atom must have a *different* set of quantum numbers. Before we examine how the Periodic Table may be built up using this principle, let us follow the story of the fourth quantum number into 1925.

In that year, Samuel Goudsmit and George Uhlenbeck decided that the fourth quantum number must be related to an intrinsic *spin* of the electron. (This is very much analogous to the Earth *orbiting* the Sun, and at the same time *spinning* about its own axis.) The spin angular momentum must have magnitude $\frac{1}{2}\hbar$, and it has two orientations in a magnetic field, parallel or anti-parallel to the field. In analogy with our modern terminology for the *orbital* angular momentum, where l is an integer, and m_l may take integrally spaced values between $-l$ and $+l$, we may use the quantum numbers s and m_s for the *spin* angular momentum; s must always be $\frac{1}{2}$, and m_s may equal either $-\frac{1}{2}$ or $+\frac{1}{2}$. It is m_s that is Pauli's fourth quantum number. (Since s does not vary, it does not need to be included.)

Our previous ideas would suggest that, since spin angular momentum has magnitude $\hbar/2$, its magnetic moment should have a magnitude of *half* a Bohr magneton; actually a factor of two, which was unexplained at the time, emerges to make it a *whole* Bohr magneton. This enables the anomalous Zeeman effect to be explained. (Without this factor of two, we would always have the *normal* effect.) The Stern–Gerlach results were readily explicable, and some at least of the gyrations of Landé and Sommerfeld made more sense.

The idea of spin also helped with the hydrogen fine-structure. By 1925, a few minor discrepancies had emerged between Sommerfeld's 1916 calculation and experiment. Goudsmit and Uhlenbeck hoped to show that inserting a coupling between spin and orbital angular momentum would cause a further precession of the electron orbit, and remove these problems. However, for a considerable time the theoretical prediction for the total precession remained stubbornly too high by a factor of two. It was not till 1926 that Llewellyn Thomas pointed out that there should be a relativistic correction factor of $\frac{1}{2}$ between the rest-frame of the electron and the laboratory frame; this made everything perfect. Relativity has not come into this chapter much, but occasionally it *has* made a decisive contribution!

Despite the empirical success of the idea of spin, it still appears a concept most difficult to justify rigorously. From the point of view of formal electro-dynamics, if the electron were indeed spinning considerations of mass-energy meant that the electron must have either a ridiculously large mass or a ridiculously large volume [39], and for this reason Lorentz advised Goudsmit and Uhlenbeck not to publish their original paper; Ehrenfest had, though, already arranged publication! More unfortunately, Ralph Kronig had thought of roughly the same ideas a little before Goudsmit and Uhlenbeck but was ridiculed by Pauli, and so did not publish them. Really a full appreciation of what was involved in the spin concept had to wait for Dirac's 1928 relativistic quantum theory of the hydrogen atom, which is described very briefly later in this chapter.

As one last point on spin, and in the context of the present book, I mention that Bohr, almost from the beginning, was very supportive of the idea of electron spin. Einstein too had made a telling point at a crucial moment. The spin can couple only to a magnetic field, not to an electric one, but in an atom it seems that there is only an electric field; this appears to rule out the precession mentioned above. The problem was solved by Einstein, who pointed out that, according to relativity, a charged object moving through an electric field experiences an effective magnetic field.

Lastly in this section, I shall explain, in totally modern terms, how, using Bohr's general idea of 'building up' the atom, Stoner and Pauli's identification of the correct ranges of quantum numbers, and Pauli's exclusion principle, the Periodic Table may be understood.

Each electron has to have its own set of quantum numbers. As one builds up the atom, n takes successively the values 1, 2, 3 for individual electrons. For any particular value of n, l may take integer values from 0 to $n - 1$, so, for n equal to 1, l must be 0; for n equal to 2, l can be 0 or 1; and so on. For any value of l, m_l may take values from $-l$ to $+l$. Finally, we do not need to take into account s, since this is always $\frac{1}{2}$, but m_s may be $\frac{1}{2}$ or $-\frac{1}{2}$.

So, for $n = 1$, the set of quantum numbers (n, l, m_l, m_s) may only be $(1, 0, 0, \frac{1}{2})$ or $(1, 0, 0, -\frac{1}{2})$. For $n = 2$, we may have $(2, 0, 0, \frac{1}{2})$, $(2, 0, 0, -\frac{1}{2})$, $(2, 1, 1, \frac{1}{2})$, $(2, 1, 1, -\frac{1}{2})$, $(2, 1, 0, \frac{1}{2})$, $(2, 1, 0, -\frac{1}{2})$, $(2, 1, -1, \frac{1}{2})$ or $(2, 1, -1, -\frac{1}{2})$ or eight sets in all. In fact a not-too-difficult sum tells us that the number of sets for given n is just $2n^2$, so that is 2, 8, 18, 32, . . . for $n = 1, 2, 3, 4, \ldots$

As we commence to build up the Periodic Table, the first two electrons will attain their lowest possible energy by having n equal to 1, or, as it is said, being in the *first shell*. The first electron gives us hydrogen, and when we add the second electron we have helium. The third electron must go into a new shell, which, to minimise energy, will have $n = 2$, the *second shell*. Because this entails a much higher energy for this electron than for an electron in the first shell, the electrons in helium are strongly bound, so helium is extremely unreactive chemically. It is a *noble gas* with a *closed-shell* electronic structure.

The third electron gives us lithium. This electron has $n = 2$ and $l = 0$, or we say it is in the [2,0] sub-shell. (It chooses $l = 0$ because electrons in sub-shells of given n but different l have lower energies the lower the value of l.) Lithium is an *alkali metal*, keen to react chemically to lose an electron and so achieve the energetically favourable closed-shell electronic structure. We may put eight electrons altogether in the second shell, two in the [2,0] sub-shell and six in the [2,1] sub-shell, as our listing above shows. After lithium come beryllium, boron, carbon, nitrogen, oxygen, fluorine and neon. Neon is a noble gas analogous to helium. Fluorine is a *halogen* – reactive because strongly desirous of gaining an electron to reach the closed-shell structure.

This is beginning to look easy! The next eight electrons go into the [3,0] and [3,1] sub-shells – sodium, magnesium, aluminium, silicon, phosphorus, sulphur, chlorine, argon. There is a strong chemical resemblance between these eight elements and the previous eight – the second and third periods of the Periodic

Table – sodium is an alkali metal, chlorine a halogen, argon a noble gas, and so on.

At this stage we might have expected to complete the third period with the [3,2] sub-shell, but, as I have already implied, this does not happen. Instead, because of the strong dependence of energy on l as well as n, the [3,2] sub-shell is built up *after* the [4,0] sub-shell. The fourth period of the Periodic Table consists of the building up of the [4,0], [3,2] and [4,1] sub-shells in that order. The [4,0] and [4,1] sub-shells resemble strongly the [3,0] and [3,1] sub-shells above them in the Periodic Table; the [3,2] sub-shell forms a 'transition' group between them.

The fifth period is analogous to the fourth – [5,0], [4,2], [5,1]. The sixth is more complicated – [6,0], [4,3], [5,2], [6,1]. The seventh begins as [7,0], [5,3], but is not completed, as there are no further stable atoms. This all looks a little heavy, and even here I have simplified somewhat, but I would invite you to notice the continuing trade-off between n and l in the order of filling-up of sub-shells.

There is an immense amount of chemistry and physics in the above. I shall just mention a few points. First let us consider, as a typical alkali metal, sodium, which has atomic number Z, the number of electrons, and also the number of protons in the nucleus, equal to 11. So sodium has a nucleus of charge 11e, two closed shells of charge −2e and −8e, and one valence electron. It is always a good simple picture, although an approximation, to think of electrons of a particular n being further from the nucleus than those of lower n. So the valence electron will orbit 'outside' a net charge of (11e − 10e) or just +e. It is thus very similar to the electron in the hydrogen atom, and its energy levels are close to those of the Bohr atom, except that, because of the exclusion principle, its value of n cannot be 1 or 2.

In the transition group, such as that where the [3,2] sub-shell is being built up, even though these electrons have higher energy than the [4,0] electrons already present, the approximation above tells us that we may think of the [3,2] electrons as lying closer to the nucleus. Since *chemical* properties are determined almost entirely by the outermost electrons, in this case by the [4,0] electrons, we should expect that elements in a transition group will differ very little from each other chemically. In this case, they are broadly similar, not particularly reactive metals – chromium, manganese, iron and so on. It is the *incomplete inner shells* of atoms of these transition elements that can give these elements interesting *magnetic* properties; those of iron and its compounds are particularly well known.

To conclude this discussion, I would point out that, though I have been using the notation of the modern quantum theory, this understanding of the Periodic Table was effectively built up in the days of the old quantum theory, and perhaps represents its greatest achievement. There is a wealth of knowledge involved, and though it was the work of many people, it is fair to say that virtually all this work was inspired by Bohr.

Heisenberg and matrix mechanics

By 1925, it was clear that, for all its considerable successes, the old quantum theory stood no chance of evolving to a complete rigorous form. It failed to handle helium and heavier atoms in a satisfactory way, and the correspondence principle was not even a starting-point for discussion of photons or the wave nature of electrons. There was a growing feeling that a totally new theory was essential, but a fear that it might be a long time coming. What was a great surprise was the rapid emergence of not one successful approach but two.

The central work of the first of these was carried out by Werner Heisenberg [44] at the age of 23. Heisenberg had worked with Sommerfeld, and was with Born at Göttingen at the time of his breakthrough, but always regarded Bohr as his intellectual parent, though the two had occasional sharp differences, as will emerge. Others took part in the preliminary work – Kramers in particular, and in the immediate follow-up – Born, Pascual Jordan, Pauli, Paul Dirac.

It seemed inevitable that the new theory would have to forsake the detailed physical *models* of the old quantum theory, which stemmed from classical physics and blended in quantum conditions with a wary eye for the correspondence principle. It seemed likely that the new theory would be more abstract, more specifically mathematical in its nature, giving correct answers, but no direct physical *picture* of the phenomenon.

Heisenberg was certainly ready for this – with Pauli he considered the concept of the electron orbit to be a major defect of the old quantum theory, since it was not directly observable. He retained the idea of correspondence, but, as Jammer [39] points out, at rather a higher conceptual level than previously. Rather than applying correspondence by a combination of skill and guesswork for each separate problem, Heisenberg enshrined it once and for all at the root of the theory.

The lead-up to matrix mechanics was through the study of optical dispersion; light is incident on an atom, and the intensity of the light emitted is analysed as a function of incident frequency. Classically the dispersion must be related to frequencies of electronic motions inside the atom, and studies using the old quantum theory almost inevitably led to these frequencies being replaced by orbital frequencies themselves, which is not correct. What *should* appear are essentially *differences* between frequencies, as Bohr's rule for transitions suggests.

This is why Slater, and following him BKS, replaced the motion of the electrons as actually given by the atomic model by a hypothetical (or 'virtual') set of oscillators, vibrating and absorbing energy at precisely the required frequencies. As early as 1921, Rudolph Ladenburg had applied the same trick to dispersion. In 1924, Kramers pointed out that, to obtain agreement with the correspondence principle, the virtual oscillators had to be regarded as emitting as well as absorbing energy. However, rather an awkward minus sign in his mathematics led him to regard his atom as behaving as a sum of 'positive virtual oscillators' of positive strength, and 'negative virtual oscillators' of *negative* strength.

One has already retreated a long way from the straightforward use of physical models to the independent and physically obscure use of mathematics, and its arbitrary modification to attain specific objectives. Kramers [41] himself pointed this out in his paper – and with apparent satisfaction rather than remorse!

Kramers had 'backed into' a correct treatment of the interaction between an atom and radiation. It was left to Heisenberg to create a general formal structure. Still in 1924 he assisted Born in the development of a scheme to extend Kramers' method to interactions between electrons. He then worked with Kramers himself in Copenhagen on an important paper considerably generalising the scope of Kramers' previous work.

Then in April 1925 he took up a post in Göttingen, but, suffering from hayfever, left for the island of Heligoland on 7 June. Here he produced *matrix mechanics*, the first form of rigorous quantum theory. The mathematics involved is fairly sophisticated, but I shall sketch the essential ideas.

Heisenberg determined to restrict himself to quantities amenable to experience – thus not orbital frequencies, but frequencies of emission of radiation. The most convenient way of representing these for a particular system is in a two-dimensional array. Using the simple harmonic oscillator (SHO) as an example, we may list along the top row of the array all the frequencies of transitions between the state with $n = 0$ and each of the other states (with values of n as on Fig. 4.3). We may write these as f_{01}, f_{02}, f_{03} and so on. Actually, we shall include at the beginning a space for the 'transition' *from* $n = 0$ *to* $n = 0$, f_{00}, for which the frequency is, of course, zero. Our second row will include f_{10}, f_{11}, f_{12} and so on, and further rows may be built up in a similar way. (In practice, and for more general systems than just the SHO, many of these transitions do not take place because of selection-rules; the frequencies may be regarded as observable 'in principle'.)

Heisenberg then proposed to write down other physical quantities as similar two-dimensional arrays. For a position coordinate, x, one will require an x_{mn}. At first sight it is not at all obvious that such a quantity may be regarded as 'observable', even in principle. However, it is, in fact, very closely related to the intensity of light emitted in a transition between states m and n. Note also that quantities such as x_{mn} turn out to be time-dependent.

To obtain an expression for this intensity of light, or, more immediately, to obtain the energies of the system, Heisenberg required the *square* of x, and had to come to an important decision on how to multiply together two of these arrays. His decision was satisfactory for his immediate purpose, but he was extremely uneasy about the fact that, with a and b general arrays of this nature, and the use of his rule, ab was not usually equal to ba. (In technical terms, his mathematics was *non-commutative*; when ab is equal to ba, as with normal numbers, it is said that a and b *commute*.)

Heisenberg was able to construct a dynamics on this rather strange basis, and he found the energies of a number of simple systems, including the SHO. The energies had two nice properties. First, although energy, like other quantities, was represented by a square array, only the *diagonal elements* (E_{00}, E_{11}, E_{12} and so on) were non-zero; it was natural to re-christen these as E_0, E_1 and E_2, and regard them as the allowed energies of the system. The second property was that these energies, unlike the other quantities involved, were independent of time.

For the SHO itself, his result was on the one hand re-assuring, but on the other a little strange. The energies were given by $(n + \frac{1}{2})hf$, with n equal to 0, 1, 2, . . ., compared with Planck's original expression of nhf. Each energy was shifted up relative to Planck by hf/2. This does not affect the radiation emitted by the oscillator, because that depends only on the *differences* between energy-levels. The hf/2 is called *zero-point energy*, and its presence indicates that, even at the lowest temperatures, an SHO always retains this energy; this is, of course, in stark contrast with classical ideas. A year previously, however, Robert Mulliken had demonstrated the existence of zero-point energy in molecular band spectra.

Another strange aspect of Heisenberg's results was that the numbers in his arrays were mostly (mathematically) *complex*. To explain what this means, I remind you that since −1 and 1 both give 1 when squared, elementary mathematics has no square root of −1. In more advanced work, such a quantity is introduced, given the symbol i, and used in a fairly straightforward way. A number such as 3i is called *imaginary*, whereas 3 itself is called *real*, though mathematicians would insist that these terms reflect the historical approach, but should not be taken literally. A number such as 2 + 4i is called *complex*.

Before Heisenberg, physicists often *used* complex numbers, but only for mathematical convenience. (The reason why it *is* convenient to use them is that often simpler equations could be used.) It was understood that, whenever a complex number was written down, the relevant quantity was actually represented by a real number, usually the real part of the complex number being used. (Sometimes, for a complex number a + ib, the relevant real number might be what is called the *modulus* of the complex number, given by $\sqrt{(a^2 + b^2)}$.) With Heisenberg it seemed that complex numbers played a more generic part in the theory, though any number that represented the result of a measurement still had to be real.

At the end of his first paper on these ideas, Heisenberg recognised that, successful as his method had been in practice, a more rigorous understanding was required. Such was not long coming. His own superior, Born, realised that Heisenberg's 'arrays' were quantities well known to mathematicians – *matrices*. In mathematics, matrices need not have the same number of rows and columns. However, the ones useful in quantum theory are usually either *square* or just single columns – the latter are often just called *vectors*. (I should mention that many of the matrices actually have infinite numbers of rows and columns – like the ones already discussed in fact; in a formal mathematical treatment this is an immense complication, but I shall gloss over it here.)

A simple example of use of a matrix outside quantum theory occurs if two quantities, f and g, say, are expressed in terms of two others, x and y. If f and g are given by $ax + by$ and $cx + dy$, respectively, in matrix algebra we may say that a column vector containing f and g is the product of a matrix containing the elements a, b, c and d with another column containing x and y. From such considerations, the way to multiply matrices may easily be obtained – it was the rule Heisenberg had already deduced for himself. From this rule it is clear that matrices do not generally commute, which had been Heisenberg's initial worry.

Born and his research student Jordan then wrote an account of Heisenberg's ideas in formal matrix language. While Heisenberg had written down only a matrix for the position coordinate, x, Born and Jordan also wrote one down for momentum, p, and they discovered a rule of great importance. The matrices for p and x do not commute; rather, in matrix language $xp - px$ is given by $i\hbar$ times I, where I represents the so-called *unit matrix*, the matrix with all diagonal elements equal to unity and all off-diagonal elements zero. This rather strange result stands at the heart of quantum theory; it indicates that non-commutation and imaginary numbers are here to stay.

The Born–Jordan paper was submitted for publication in September 1925, only two months after Heisenberg's initial work, and, rather strangely, at about the same time, an unknown British physicist, Paul Dirac, submitted a paper containing much of the same material. Dirac, even younger than Heisenberg, had read Heisenberg's original paper, and quickly produced a mathematical account that was actually in some ways more sophisticated than that of Born and Jordan. Much more would be heard of him [45].

Heisenberg meanwhile had been studying matrix mathematics feverishly; the follow-up paper to that of Born and Jordan, submitted in November, had all their three names on it. It was a comprehensive account of the methods and some of the applications of the theory. All that is required in this book will be discussed later in this chapter.

Of the founding papers of matrix mechanics, the last was by Pauli, who, in a computational *tour de force*, managed to solve the problem of the hydrogen atom. He was able to deduce the existence of the three quantum numbers n, l and m_l and the range of values each of them could take, and also to obtain an expression for the energy-levels that was identical, of course, to that of Bohr, which already agreed with experiment. (Spin did not emerge from this analysis.) Within six months of Heisenberg's initial ideas, the matrix form of quantum theory had been put into shape as a rigorous and powerful theory, ready to tackle whatever problems would be presented to it.

It gave, of course, virtually no physical picture of the system it analysed. There have been strong suggestions [46] that the physicists who produced the theory, and particularly the Germans among them, were motivated by contemporary philosophical trends (particularly towards non-determinism) in the Weimar republic. My belief is that physical motivation was enough to send Heisenberg down his particular path; nevertheless it must have appeared to him and his followers that one of the great achievements of their theory was to show that physics must restrict itself to mathematical analysis and eschew any attempt to provide a picture or model of the physical system.

This was an approach Bohr supported – at least up to a point. (See the following chapter.) Einstein did not; he admitted, of course, the successes of the theory, but was unwilling to accept it as in any way a final solution. Heisenberg protested that he had merely followed in Einstein's footsteps by restricting the theory to what was observable; Einstein replied that, on the contrary, it was the theory that *determined* what was observable. Heisenberg was very shortly to profit from this remark.

Schrödinger detested the whole theory.

Schrödinger and his equation

Schrödinger was discouraged by the difficulty of matrix methods, but, more significantly, he was repelled by the lack of visualisability provided – precisely the feature which Heisenberg and cohort regarded as the system's main conceptual advance. (Schrödinger was not alone; Enrico Fermi, at the start of a magnificent career in physics, was so ill at ease with the lack of physical interpretation of matrix mechanics that he contemplated giving up the subject altogether [39].)

Schrödinger was able to do much more than object. In an outstanding and sustained burst of achievement through the first half of 1926, he constructed what appeared, at any rate, to be a totally new version of quantum theory, completely independent of Heisenberg's, though just as successful. His formalism reverted to mathematical methods much more familiar to physicists, and, even more significantly, there seemed to be a return to physical understanding, rather than mere manipulation of mathematical symbols. This physical understanding, though, was of a very different nature from that provided by classical physics.

The mathematician Hermann Weyl, who was a close friend of Schrödinger, said that this great work was performed during a 'late erotic outburst' [27]. As Moore [6], Schrödinger's biographer, has detailed, he had many close relationships with women, and a particular, but unidentified, 'lady of Arosa' may have inspired much more than she could have imagined. (All the same, one might be reluctant to advise young physicists that the quickest way to a Nobel Prize is via the adoption of a high sexual profile.) As for 'late', Schrödinger was 38 – 14 or so years older than Heisenberg, Dirac and Pauli.

Schrödinger's work was based firmly on the ideas of de Broglie, though he was directed onto this route only by Einstein's strong endorsement of de Broglie's work. '[T]he whole thing would not have started . . . (as far as I am concerned) had not your second paper on the degenerate gas directed my attention to the importance of de Broglie's ideas', Schrödinger wrote to Einstein in April 1926 [39]. (Remember that this was well before any convincing experimental evidence for de Broglie's speculations.)

Schrödinger was also encouraged by Debye, who suggested that de Broglie's talk of waves was rather childish; instead, as for waves on a string or a sound wave, as for Maxwell's work, the fundamental thing should be a *wave equation* [6]. Schrödinger found one!

What he required was an equation that included V, the potential energy of the system, as a function of position. When V was zero everywhere, the particle would be free, and we should have the de Broglie case, the solution being a travelling wave. But when V was not zero, in particular when it was given the form for the simple harmonic oscillator (SHO) or hydrogen atom, somehow the energy levels had to emerge from the equation, together with information on angular momentum, selection-rules, transition-rates and so on.

De Broglie's work had been relativistic, and Schrödinger initially produced a relativistic equation – which did not, however, agree with experimental results for hydrogen. (It was actually a perfectly good equation, but for particles without spin, *not* for electrons; the equation was later published by other physicists, and became known as the (Oskar) Klein–(Walter) Gordon equation.) Schrödinger was naturally disheartened, but a little later took a non-relativistic approximation and obtained good agreement with experiment (as will be explained shortly).

I shall delay a little suggesting how Schrödinger may have found his equation, and instead make a few comments on its structure. An equation for a wave on a string, for example, is an equation of a displacement (the displacement of the string from its undisturbed position at a particular time and place). The equation relates rates of change of the displacement in time to rates of change in space. Similarly Schrödinger's equation was of a quantity known as the *wave-function* and always written as ψ (psi). Strangely it was not necessary for Schrödinger to know at the outset what ψ represented; he could make perfectly good progress without that knowledge. Like the equation for waves on a string, his equation related rates of change of ψ in time to rates of change in space.

The most immediate way of obtaining information from the equation for a particular form of potential is to search for the normal modes of oscillation that follow from the equation when the appropriate form of V is inserted. Mathematically, this is along the lines of the wave-on-a-string problem discussed in Chapter 2 and of the Rayleigh–Jeans work described earlier in this chapter.

For cases where the particle is bound – unable to escape totally from the potential, the kind of answer obtained is a discrete series of modes, each oscillating at a particular frequency. In a quantum context, it is natural to assume that the energy of this particular mode is just the product of h with this frequency – and this gives directly the energy-levels of the system. Schrödinger performed the necessary calculations for the SHO and the hydrogen atom. For the SHO, the energies of the various modes were exactly the same as those found by Heisenberg, namely $(n + \frac{1}{2})hf$, including, of course, the zero-point motion. For hydrogen, the energies were precisely the Bohr levels, though, in addition, Schrödinger's theory gave the *continuum* – the continuous energy-levels for *positive* values of energy, where the electron is *not* bound to the nucleus.

· Already Schrödinger's is seen to be an immense achievement. But there was a lot more information easily available from the modes themselves. These are mathematical functions of x for the SHO and of x, y and z for hydrogen, and they are important functions for a rather different way of looking at the Schrödinger equation, that of *eigenvalues* and *eigenfunctions*. (These German, or rather, hybrid words are those always used; the English expressions would be *characteristic values and functions*.)

In this approach, a quantity such as energy or angular momentum (often called an *observable*, though that term needs to be interpreted carefully rather than being taken too literally, as will be seen in later chapters) is represented by a (mathematical) *operator*. (In matrix mechanics it will be remembered that it is represented by a matrix.)

An operator is something other than a direct multiplier (a straightforward number) that *operates* on a function. An example from a different context is as follows – if a function of x represents the height of a road as one moves in that direction, a particular operator (the gradient operator), acting on that function, will give us the *gradient* or *steepness* of the road at x. It is worth pointing out that operators, like matrices, do not generally commute; applying to your foot the operator 'put your sock on', followed by the operator 'put your shoe on', is different from the same operators applied the other way round.

In quantum theory, the operator for momentum follows fairly directly from de Broglie's ideas – from a mathematical point of view it is related to the gradient operator just mentioned.

Operators for angular momentum, kinetic energy and the various contributions to potential energy all follow in a reasonably straightforward way. Since an operator is *not* just a multiplier, it follows that, acting on a general function, it will *not* usually result in a constant times that function. Eigenfunctions of a particular operator, then, are those special functions for which that operator acting on an eigenfunction *does* give a constant times the eigenfunction, and the constant is called the eigenvalue. This little bit of mathematics is of great importance in quantum theory.

Indeed, we may return to the genesis of Schrödinger's theory. While we may not claim to know the precise sequence of ideas that led to his conceiving his famous equation, it is interesting that it may be expressed in the form of the sum of operators for kinetic and potential energy applied to ψ being equal to a total energy operator also applied to ψ.

It follows that the 'modes' I talked of above are, from this point of view, just the eigenfunctions of the operator for total energy – the *Hamiltonian* operator, as it is called, after William Hamilton, an Irishman who had invented much of the necessary mathematics a hundred years before. The corresponding eigenvalues are just the energies already discussed above.

I may also try to explain the connection between these eigenfunctions and the wave-function, ψ. The wave-function gives us information on the state of the system. If it is equal to one of the energy eigenfunctions, we may say that the system is 'in the corresponding state', and its energy is equal to the corresponding energy eigenvalue. (This is rather analogous to the electron being in a particular orbit in the Bohr case.) More generally, though, the wave-function will not be equal to any one of the eigenfunctions, but to a sum of two or more. I shall explain what we may say about the energy of the system in this case later in this chapter.

For the hydrogen atom there are other important observables and corresponding operators. Each of the energy eigenfunctions obtained by Schrödinger is *also* an eigenfunction of the operator for the *total* angular momentum, and that for the *z-component* of angular momentum. In the old quantum theory there are

Fig. 4.11. The three possible orientations of the angular momentum vector for the case where l is equal to 1. The length of the vector is $\sqrt{2}\hbar$, and its z-component is equal to $\hbar$, 0 and $-\hbar$, for the three possible values of m_l. Thus the vector can never point along the z-axis. The vectors have been shown precessing about the z-axis to indicate that the x- and y-components may be said to average to zero.

different orbits with the same energy (for hydrogen, and the non-relativistic case), and similarly here there are eigenfunctions with the same energy. (For both, the exceptional case is $n = 1$.) The number of eigenfunctions for a particular value of n is just n^2, or $2n^2$ if one includes spin; or we may say that there is a total *degeneracy* of $2n^2$.

Hydrogen is a three-dimensional system, and each energy eigenfunction is characterised by *three* quantum numbers – n, l and m_l. We may write an eigenfunction as ϕ_{n,l,m_l} (where ϕ is the Greek letter 'phi'). The ranges of l and m_l have been given earlier. It is true, of course, that a reasonable understanding of these numbers had been put together in the old quantum theory, but in Schrödinger's theory, they came directly and unambiguously from the basic assumptions.

In terms of these quantum numbers, it follows directly that the magnitude of the angular momentum is given by $\sqrt{l(l+1)}\hbar$ and its z-component by $m_l\hbar$; these are the eigenvalues of the corresponding operators applied to ϕ_{n,l,m_l}. Again this was to an extent anticipated in the old quantum theory – but the $\sqrt{l(l+1)}$ rather than l was a bit of a shock. It meant that the angular momentum vector can never point along the z-axis. Consider, for example, the cases with $l = 1$, so that m_l may be equal to 1, 0 or -1. So the length of the angular momentum vector is $\sqrt{2}\hbar$, while the z-component is $\hbar$, 0 or $-\hbar$. These possibilities are indicated on Fig. 4.11. (On this figure I have shown the vectors precessing about the z-axis to indicate that the x- and y-components may be said to average to zero – in a way to be discussed shortly.)

Spin cannot be handled directly by Schrödinger's theory (or Heisenberg's), but it may be tacked on to the wave-function in the manner suggested by Goudsmit and Uhlenbeck.

Such is a taste of the content of Schrödinger's *theory*. It could scarcely be denied that it handled its problems extremely successfully. It may easily be used, for example, to predict the polarisations of the photons produced in transitions, and the relevant selection-rules and transition probabilities. Indeed, it, rather than matrix mechanics, is nearly always used to introduce students to quantum theory. This is because its mathematical methods – wave equations, eigenvalue equations, are familiar. The calculations for the hydrogen atom, for example, are not exactly simple, but they *are* straightforward – much more so than Pauli's labours with matrices on the same problem. Students are also taught that Schrödinger's methods are easier to understand, though, as Edmund Whittaker observed, and the rest of this book may confirm, they are really just easier to misunderstand.

Heisenberg and Schrödinger: the debate

This brings me to the fact that, along with his *theory*, Schrödinger produced an *interpretation*, which restored visualisability to the atom, albeit in a totally different form from that in classical physics or the old quantum theory. In Schrödinger's interpretation, the electron is *identified* with the wave, that is to say it is some sort of density distribution or 'wave-packet', rather similar to that shown in Fig. 4.7. In contrast, Heisenberg's theory, although not providing a detailed picture of the phenomenon, is to be regarded as a theory of particles.

There was another point of intense disagreement. For Heisenberg, following Bohr and his atomic transitions, *discontinuity* is unavoidable. Schrödinger believed that he was able to restore *continuity*. He had available eigenfunctions for Bohr's 'stationary states' – ϕ_1 and ϕ_2 say. But he also had mathematically rigorous non-stationary states, $a_1\phi_1 + a_2\phi_2$, say, and could imagine a *continuous* transition from ϕ_1 to ϕ_2 in which a_1 decreases continuously from one to zero, and a_2 increases continuously from zero to one. Thus Schrödinger claimed that he had eliminated discontinuity in the form of 'quantum jumps'.

Indeed, he actually went further by claiming to eliminate the need for energy-levels altogether; frequencies of emission were equal to differences of frequencies of oscillation of the wave-function in the two states, and the 'emission' could be interpreted as a resonance phenomenon involving waves only.

While virtually everybody admired Schrödinger's formalism, there were very differing views on the interpretation. Bohr could not accept it. Pauli was critical of it. Heisenberg detested it. Yet others felt that at least the formalism reached deeper into the meaning of quantum theory than matrix methods. That these included Einstein, who had stimulated Schrödinger in the first place, was no surprise. Yet initially they also included Born, later a strong supporter of the Bohr position. (His intellectual journey is discussed lucidly by Beller [47].) Born, in fact, quickly made very important contributions to the theory of atomic collisions following the Schrödinger formalism.

It soon, though, became clear that Schrödinger's *interpretation* (in contrast, of course, to his mathematical formalism) would not work. His 'wave-packet' model of a particle was a failure; the wave-packet could not hold together except in special cases. He could not explain such phenomena as the photoelectric and Compton effects using only waves. Bohr in particular argued ferociously that Schrödinger's formalism could *not* eliminate quantum jumps, which were required for direct detection of individual atoms on screens for example, and even to obtain the Planck radiation formula; Born's work on collisions mentioned above confirmed this point. (Schrödinger ruefully commented that, if he'd known we still had to put up with these 'damn quantum jumps', he was sorry he ever had anything to do with quantum theory.)

There were more basic arguments against at least the most facile interpretation of Schrödinger's waves. First, let us move from one particle to two. The wave-function for one particle is three-dimensional – one wave in three-dimensional space. We would hope that the wave-function for two particles could be interpreted as representing two waves in the same three-dimensional space, but, stubbornly, the mathematics tells us that, instead, it represents *one* wave in a mathematical *six-dimensional* space. This clearly suggests that wave-functions were 'mathematical waves', in the sense of being extremely useful for calculating experimental results, but not directly representing anything physically real. In terms of any ability to provide a picture of the quantum system, Schrödinger's methods were actually no more successful than Heisenberg's.

Another argument ran as follows. I have talked about frequencies of oscillation as eigenfunctions of energy, the frequency being related to the value of energy via Planck's constant. But, non-relativistically, energy can be defined only to within a constant, because there is an arbitrary zero of potential energy. Thus the frequency of oscillation is also arbitrary, which again makes it look as though the oscillation should be regarded as mathematical, rather than a real oscillation of a real quantity.

Yet another stubborn fact was that the wave-function was (mathematically) complex. At first Schrödinger had hoped that, as in much previous mathematical physics, one would find that this complex nature could be regarded as a mathematical convenience, but one that could always be disposed of in practice. Unfortunately this turned out not to be the case.

While there was intense conflict between Heisenberg and Schrödinger over interpretation of their formalisms, it soon, perhaps surprisingly, emerged that, totally different though they seemed, the formalisms were actually equivalent in terms of mathematical structure. This was shown by several people, including Pauli, Schrödinger himself and Carl Eckart.

I shall attempt to give at least some understanding of this seemingly rather strange statement. I have already said that observable quantities – position, momentum and so on – may be represented as a matrix or as an operator, by Heisenberg and Schrödinger, respectively. Neither matrices nor operators commute, and, in quantum theory, the same rule, that the difference between xp and px is $i\hbar I$, applies for matrices *and* operators. (For the matrix case, I must be a unit *matrix*, as already stated; for the operator case, it is just the number 1.)

The eigenvalue idea, which I have described as applied to the Schrödinger formalism, also holds for matrices. In matrix algebra, one may multiply an $n \times n$ matrix by a vector – an $(n \times 1)$ column; the answer will be another $(n \times 1)$ vector, but not usually a simple multiple of the initial vector. Those special cases for which it *is* a multiple of the initial vector are called *eigenvectors*, and the corresponding multiplying factors are *eigenvalues* (as in the Schrödinger scheme). This gives us the clue that an eigenvector plays the part in the matrix formulation played by the eigenfunction for Schrödinger. The Schrödinger wave-function, which represents the *state* of the system, also corresponds to a vector in matrix work, the vector being called the *state-vector*.

One may, in fact, start from a Schrödinger formulation – concentrating either on time-dependence and examining how a wave-function evolves in time, or on a time-dependent situation, calculating eigenfunctions and eigenvalues for a given observable, and translate to Heisenberg language, or vice versa.

Indeed, within 1926 itself, Dirac and Jordan independently produced a more abstract formalism, within which theoretical analysis is carried on by rather formal manipulation of symbols. The formalisms of Heisenberg and Schrödinger may be seen as representations of this abstract scheme in terms of more widely known and explicit mathematical structures.

Thus we may talk of just one theory – *quantum theory*, and the terms 'matrix mechanics', always disliked by Heisenberg because it suggested that his ideas were mathematical rather than physical, and 'wave mechanics' may usefully be dropped. (The term 'wave-function', and also 'matrix-elements' relating in particular to strengths of transition between quantum states, remain to remind us of our heritage.)

I would point out that establishment of this mathematical identity of the formulations did not end the disagreements, partly because Schrödinger initially hoped that his own scheme had scope for extension far beyond that of Heisenberg, but fundamentally because the disagreement was over interpretational aspects. Differences in the mathematical formalism were still relevant, but only because they encouraged adherence to totally different ways of thinking about physics.

Finally in this section I shall discuss the identification of what ψ, the wavefunction, actually represented. Remember that, whereas ψ is usually (mathematically) complex, and may be written as $a + ib$, the *modulus* is real, equal to $\sqrt{a^2 + b^2}$. I shall write it as $|\psi|$.

Born's analysis of collisions, mentioned above, led him to believe that the square of the modulus of ψ, $|\psi|^2$, represented the *probabilities* that the collision process took various routes. It could further be argued that, if the wave-function of a particle is ψ, $|\psi|^2$ may be regarded as the *probability density*, the probability that the particle is to be found in a given region. Again, consider a form of potential such as the SHO with energy eigenfunctions ϕ_0, ϕ_1, ϕ_2, . . ., with eigenvalues E_0, E_1, E_2, . . . If, at a particular time, the wave-function of a particle in the potential is $a_1\phi_1 + a_2\phi_2$, the *probability* of its energy being E_1 is $|a_1|^2$; the *probability* of it being E_2 is $|a_2|^2$. (Since probabilities must sum to unity, this tells us that the sum of $|a_1|^2$ and $|a_2|^2$ must be unity. Note also that the probabilities themselves are real numbers, as they must be.)

From the point of view of classical physics, Born's use of probabilities was a dramatic new departure. Of course we have had preliminary warning that they were coming, in radioactivity, and in the transition probabilities of Bohr and Einstein for example; and I should point out that there are two possibilities. The first is that the probabilistic nature of quantum theory is fundamental; it is genuinely an indeterministic theory. That is the line taken by Bohr, and it may be described as the 'orthodox' position. The second possibility is that the behaviour is only superficially probabilistic, that there is a deeper level of *hidden variables* behaving deterministically and causing the apparently random variation of outcomes at the observable level. Both positions will be explored at length later in this book.

(I should also mention that I was very casual above; I spoke of particles 'having' a particular position or energy. At least according to orthodox views, one should only talk of an experiment *measuring* a position or other observable; in the absence of any measurement, we should not even discuss the values the observable takes.)

Beller [47] describes the rather tortuous manner in which Born reached his idea, which was to earn him the Nobel Prize more than 20 years later, and also an important contribution by Pauli. Heisenberg, Beller points out, was so opposed to the notion of electrons having paths inside atoms that he would have been reluctant to consider even *probabilities* of positions. Only in retrospect, perhaps, did the ideas begin to appear fairly obvious.

Born's work could not help being another blow for Schrödinger's interpretation. It became clear that $|\psi|^2$ represented not a charge or particle density, but a density of probability.

Uncertainty and indeterminacy

The Schrödinger–Heisenberg debate continued into 1927. On the one hand, it may be regarded as being between those for and those against visualisation in physics; on the other, between wave and particle concepts. Born had made a definitive contribution to the argument, but much still remained obscure. Then in 1927 came Heisenberg's major contribution of what is most often called the *uncertainty principle*.

The principle may most simply be introduced directly from the de Broglie idea (though this route was taken only some time later, and then by Bohr). A straightforward de Broglie wave is of infinite extent in space and in time. It has a precise wavelength, and hence a precise momentum (since λ is given by h/p). But suppose that one were to ask the following question: where is the particle? Since the wave extends everywhere, one may reply that the particle has equal probability of being anywhere! (At first sight it might appear that the probability amplitude should just be the square of an oscillatory function, and so vary periodically between a maximum and zero. But actually the de Broglie wave is complex mathematically, and $|\psi|^2$ is constant.)

This does not look much like a particle! If we wish to localise it, we may add in waves with a spread of wavelengths centred on the initial wavelength. If we arrange for *all* the waves to have maxima at one particular position, we will have a region of high probability density in this vicinity, and interference between the various waves will create low probability density elsewhere. We may call the width of the region of high probability density roughly d_x, which we may call an *uncertainty* in x. However, our manoeuvre, by bringing in a spread of wavelengths, has introduced a spread of momenta, which we may call an *uncertainty* in p, d_p.

We can, in fact, make d_x as small as we wish, and thus localise the particle to whatever extent we choose – but only at the expense of having d_p correspondingly large. There is a basic mathematical relationship telling us that the product of d_p and d_x must always be greater than, or equal to, $\hbar/2$. This is the mathematical content of Heisenberg's principle, and mathematically it is an unambiguous and non-controversial component of quantum theory.

Its meaning is more controversial, even more than 70 years later. While much of this controversy will come up in later chapters, I note here that I used the word 'uncertainty' above; this is the most common terminology, but the description that goes along with this name often suggests that it is *measurements* that are the problem. One often reads that simultaneous measurement of position and momentum is impossible to a greater precision than the above, or that measurement 'disturbs the system'.

Orthodox interpretations of quantum theory take a different view, in which measurement is not involved. The quantitites d_x and d_p are looked on as intrinsic *indeterminacies* in the values of x and p, prior to, and independent of, any measurement. This point of view comes fairly directly from the Born postulate of the previous section, if it is supplemented by the (orthodox) assertion that there are no *hidden variables* (or *hidden parameters*) to give additional information on where the particle really is. Thus, for the orthodox, a better name is the *indeterminacy principle*.

I now, though, return to the mathematical formalism, and start with a slightly different approach to the principle. For Schrödinger, ψ, a mathematical function of x, was all-important. Born tells us that $|\psi|^2$ gives the probability distribution of a measurement of x. The work of Dirac and Jordan, though, tells us that we may equally well define a mathematical function of momentum p – I'll call it F – such that $|F|^2$ gives a probability distribution for a measurement of p. For a given ψ, a mathematical transformation will give F, and vice versa.

Three examples are given in Fig. 4.12. In (a), we imagine a particle completely localised; $|\psi|^2$ is totally sharp at a particular value of x, x_0. The uncertainty principle tells us that $|F|^2$ must be flat; we have *no idea* what value of momentum will be obtained in a measurement. In (b), however, we have a situation where p is known perfectly, so $|F|^2$ is sharp; in this case it is $|\psi|^2$ that must be flat. An intermediate case shown in (c) is for the ground-state of the SHO; both $|\psi|^2$ and $|F|^2$ have a certain width – non-zero and not infinite; the product of these widths d_x and d_p obeys Heisenberg's principle, of course.

I briefly note that the fact, demonstrated by Dirac and Jordan in this way, that ψ was not a unique, or even fundamentally significant, way of describing the system was another blow to Schrödinger's interpretation in which ψ represented a physically real wave, and also to his belief that his wave mechanics could be a more fundamental theory than matrix mechanics.

The uncertainty principle is really just one important example of a general quantum-mechanical fact. If we consider two observables A and B, and their corresponding operators (or matrices) do *not* commute, we do not have any (complete range of) functions that are simultaneously eigenfunctions of the two operators.

Fig. 4.12. The probability distributions of x and p for three cases. In (a), x is precisely equal to x_0, so, according to the uncertainty principle, nothing is known about momentum p. In (b), the reverse is the case; p is precisely equal to p_0, so nothing is known about x. In (c), both x and p are given by distributions centred around zero; thus there is a range of uncertainty for each of x and p, given roughly by d_x and d_p in the diagram, and these must obey the uncertainty principle.

I shall make two comments about the above paragraph. First, I could paraphrase the last few words by saying that we do not have any (complete range of) quantum-mechanical states for which both A and B have definite values. Secondly, the words in brackets are there because, in some cases, though not that of p and x, there are a *few* states with values of A and B in common.

I shall now consider the example where the two observables are the z- and x-components of angular momentum, l_z and l_x. This is actually easier to discuss than the case of p and x, because the values taken by l_z and l_x are discrete, whereas those taken by p and x are continuous.

No two of the operators for l_x, l_y and l_z commute with each other (though each of them commutes with the operator for the *total* angular momentum). Despite this fact there are a *few* simultaneous eigenfunctions of the three operators, though *not*, as stressed above, a complete range. These special (and rather boring) cases are those for which quantum number l is equal to zero, so the total angular momentum is zero, together, necessarily, with each of its components.

The case with l equal to 1 is much more interesting. There must be three functions that are simultaneously eigenfunctions of the operator for total angular momentum (each with eigenvalue $\sqrt{1(1+1)}\hbar$ or just $\sqrt{2}\hbar$) *and* of the operator for l_z with eigenvalues $m_l\hbar$, and m_l equal to 1, 0 and -1 for the three functions. These functions are *not* eigenfunctions of the operators for l_x or l_y. However, there *are* three functions that are simultaneously eigenfunctions of total angular momentum *and* l_x; these will *not* be eigenfunctions of l_z. But each of the eigenvectors of l_z may be written as a mathematical combination of the three eigenfunctions of l_x, and vice versa!

Such is the mathematically intricate nature of quantum theory. It is a compact and beautiful formalism, if probably somewhat daunting to the reader lacking in mathematical confidence.

What is more important in this book is that it raises considerable problems in *interpretation*, in particular of a *measurement*. Suppose that the wave-function of a particle is equal to one of the eigenfunctions of l_z; we may say that the system is in an *eigenstate* of l_z – and let us say specifically that it is the one with m_l equal to 1. A measurement of l_z would be *bound* to give the value $m_l\hbar$, or just $\hbar$. But let us suppose that the measurement instead is of l_x. In this case, theory tells us that the measurement may yield the value $\hbar$, 0 or $-\hbar$, with probabilities $\frac{1}{4}$, $\frac{1}{2}$ and $\frac{1}{4}$, respectively. But how does this happen? What occurs during the measurement? These are the fundamental problems discussed in the second part of this book.

Beller [47] stresses (and I should really say – according to orthodox interpretations of quantum theory) that, with his principle, Heisenberg had taken a step considerably beyond Born. Beller has written 'Quantum theory is essentially a probabilistic theory not because quantum laws, as opposed to classical ones, are statistical in Born's sense . . . , but because one needs probabilities to describe fully the state of the system. The future cannot be known in all its details not because quantum laws are statistical, but because the present cannot be known in all its details, and therefore has to be described probabilistically.'

Squires [16] makes a similar point when he says that 'the most significant revolution brought about by quantum theory is *not* the breakdown of causality [or
loss of determinism]'. Rather, as discussed in detail in the next three chapters, it
is loss of realism. Later in this book I shall consider how other interpretations of
quantum theory make a deliberate attempt to avoid this conclusion – ensemble
interpretations and many-worlds interpretations for example.

I shall now give one further example of the general mathematical formalism –
an important one in the rest of the book. It concerns spin angular momentum –
an electron with $s = \frac{1}{2}$, or just *spin* $\frac{1}{2}$. On the one hand this example is simpler
than the $l = 1$ orbital angular momentum case just discussed, because measure-
ments of s_x, s_y or s_z will give one of only *two* values, $+\frac{1}{2}\hbar$ or $-\frac{1}{2}\hbar$, rather than
three. From another point of view, it is more tricky, being more abstract, because
there are strictly no wave-functions for spin.

Let us start with s_z. There are two eigenstates for this observable. I shall write
them, using the so-called Dirac notation, as $|s_z = \frac{1}{2}\rangle$ and $|s_z = -\frac{1}{2}\rangle$. (*Please* don't
worry too much about these signs; just follow me as I use them.) Any particular
electron has a state-vector for its spin. If the state-vector happens to correspond
to one of the above states, $|s_z = \frac{1}{2}\rangle$ or $|s_z = -\frac{1}{2}\rangle$, then we *know* what the result
of a measurement of s_z will produce, $\hbar/2$ or $-\hbar/2$, respectively. (This is what we
meant by saying that they were eigenstates of s_z.)

More generally though, the state-vector for the electron spin may be written
as a mathematical combination of the two eigenstates, namely it is given as
$\{a_+|s_z = \frac{1}{2}\rangle + a_-|s_z = -\frac{1}{2}\rangle\}$, where a_+ and a_- are two numbers, in general mathematically complex. When one performs a measurement of s_z on a system with
this state-vector, one does not know what result will be obtained. There is a
probability $|a_+|^2$ of obtaining the result $\hbar/2$, and a probability $|a_-|^2$ of obtaining
$-\hbar/2$. This is an example of what is called the Born rule. (Since probabilities
must sum to unity, the sum of $|a_+|^2$ and $|a_-|^2$ must be equal to 1.)

What about s_x? Since the *x*- and *z*-axes have the same status, there must be
two eigenstates of s_x, $|s_z = \frac{1}{2}\rangle$ and $|s_z = -\frac{1}{2}\rangle$. If the spin is in one of *these* states,
we know what result we shall find in a measurement of s_x.

The eigenstates of s_z and s_x are intimately related mathematically. In fact
$|s_x = \frac{1}{2}\rangle$ is equal to a constant, $1/\sqrt{2}$, times the *sum* of $|s_z = \frac{1}{2}\rangle$ and $|s_z = -\frac{1}{2}\rangle$,
while $|s_x = -\frac{1}{2}\rangle$ is equal to $1/\sqrt{2}$ times their difference. Let us now consider a
spin with state-vector $|s_x = \frac{1}{2}\rangle$. If we measure s_x, we know we will obtain $\hbar/2$.
If, though, we measure s_z, we do *not* know what value we will obtain; there is
a probability of $(1/\sqrt{2})^2$, or just $\frac{1}{2}$, of obtaining $\hbar/2$, and an equal probability
of obtaining $-\hbar/2$. (If we were to bring in s_y, there would be a very similar
relationship between it and either s_x or s_z.)

The above scheme makes the results of a wide range of experiments easy to calculate. What it all *means*, how we may *picture* the spin (if at all) will be left for later chapters.

To conclude this section I shall point out that Heisenberg's own route to the principle was not primarily the formalism-based one we have followed, but a more physical one. The famous example of this is the so-called *Heisenberg microscope*, with which one tries to locate an electron using a photon. It is not physically realisable but a thought-experiment (*Gedankenexperiment*).

The version of this in today's textbooks works in terms of a collision between the electron and the photon (*à la* Compton). The photon then bounces off into a microscope. However, the microscope has only a finite *resolving power* (a technical term from optics meaning how accurately the microscope may locate a particle). Thus there is an uncertainty in position, d_x, which increases with the wavelength of the light, but decreases with the aperture of the microscope lens. Also, since the lens does have this finite aperture, and we do not know at what point the photon enters it, the photon must have an uncertainty in its momentum following the collision; it has thus given the electron a momentum with the *same* uncertainty, d_p. And the product of d_x and d_p is rather greater than $\hbar/2$.

Many such thought-experiments have been designed to try to beat the uncertainty principle. Einstein thought up a good number of them in his arguments with Bohr. (See Chapter 6.) Students of later generations have followed suit, but none has succeeded, and the argument from the mathematical formalism suggests that they are most unlikely to do so! Actually the physical mode of analysis may be thought to be fundamentally flawed in principle; it may be said to *use* quantities as if they have definite values, in order to prove that they don't! I would suggest that the physical arguments should be regarded as interesting and thought-provoking, but not particularly rigorous.

My main purpose in describing the Heisenberg microscope is to point out that it led to a conflict between Heisenberg and Bohr of great significance. By 1927, it was fairly clear that Schrödinger's interpretation of quantum theory in terms of real physical waves was untenable, for the many compelling reasons already given. (Though Schrödinger continued to reject the orthodox views of Bohr, it is fair to say that he never developed a serious competing position of his own.) Heisenberg's *original* discussion of the microscope was different from the one just given; he used *only* particle-like ideas (the d_x being discussed in terms of recoil of the electron) so that the wavelength of the photon did not enter the analysis. Thus it seemed to be the ultimate triumph of particle-like ideas.

Bohr disagreed. He produced the account given above in which particle-like notions caused the d_p, but wavelike ones caused the d_x. Bohr and Heisenberg had intense arguments on the matter, Heisenberg being persuaded only gradually and reluctantly to Bohr's point of view.

In terms of the struggle between Heisenberg and Schrödinger, between particle and wave, between theories that spurned visualisability and those that sought it and claimed to achieve it, Bohr struggled for the middle ground [47]. His framework of *complementarity*, discussed in detail in the following chapter, used particles *and* waves; it retained classical ideas, though recognising that, for them to be applied consistently, severe restrictions must be placed on their use.

Bohr felt he had recognised the achievements and arguments of Schrödinger, and behind him, Einstein, sufficiently to hope for and indeed to expect them to respond favourably to his own position. This is why he was so hurt by Einstein's long-standing opposition.

To conclude this section, I refer to the so-called *time–energy uncertainty principle* (as opposed to the *position–momentum uncertainty principle* we discussed above). To obtain one from the other, position x is replaced by time t, and momentum p by energy E. The two principles may often be *used* in much the same way – and we shall do this in the next chapter. But I should warn you that, at a fundamental level, the meanings of the two principles are very different. There is no operator for time in non-relativistic quantum theory, and we must not talk of an 'uncertainty in time'; the quantity d_t relates instead to the time available for a given measurement of energy. Then the greater d_t is, the lower d_E, the uncertainty in the result of the measurement, may be.

Symmetry and statistics

It will be remembered that one of the great difficulties of the old quantum theory before 1925 had been the helium atom. Indeed, the problem was really that the old quantum theory could not cope with *any* problem where the properties of two or more electrons had to be considered.

Heisenberg showed that the difficulty was related to the handling of *indistinguishable particles*. The two electrons in the helium atom are said to be indistinguishable; there is *no* way to tell them apart. In classical physics this would not create problems. We could call one of them 'number 1', the other one 'number 2', and, if desired, keep a close enough eye on them so that we always know which is which.

In quantum theory this is not possible because the uncertainty principle limits how closely we can monitor the electrons. It turns out – I don't think it actually *follows* from the above, but it is clearly related to it – that we must never write down a wave-function for a number of indistinguishable particles that implies we know which particle is which.

Consider a helium atom and suppose we know that one electron has wave-function ϕ_a and one ϕ_b. We might be inclined to write down the combined wave-function as $\phi_a(1)\phi_b(2)$, where the numbers in brackets refer to the first and second electrons. But clearly this would be wrong – it implies that we known it is electron number 1 in state a, and electron number 2 in state b. Another way of seeing this is that, if we interchange (or *exchange*) 1 and 2, the form of the wave-function is changed.

There are two combined wave-functions we can write down that do obey our condition (at least as the condition is clarified below). The first is $c\{\phi_a(1)\phi_b(2) + \phi_a(2)\phi_b(1)\}$. Here c is a constant needed for mathematical consistency – it is actually $1/\sqrt{2}$. The algebraical part clearly does not attempt to say whether it is electron 1 or electron 2 in state a (or state b). Again, if 1 and 2 are interchanged, the wave-function is unchanged. We may say that the wave-function is *symmetric under exchange* of the two particles.

The second form of combined wave-function is $c\{\phi_a(1)\phi_b(2) - \phi_a(2)\phi_b(1)\}$, where c is again $1/\sqrt{2}$. In this case, the wave-function itself clearly *does* change if 1 and 2 are interchanged – it changes sign. We may say that the wave-function is *anti-symmetric* under exchange, but the quantity which has physical meaning, the probability density, is given by $|\psi|^2$, and this remains the same under interchange of particles.

So we have two rules for combining individual wave-functions, and different types of particle follow different rules; the determining factor is the spin of the particle. Particles with half-integral spin ($\frac{1}{2}$, $\frac{3}{2}$, ... in units of $\hbar$) obey the second rule, and must have the minus sign. We already know that electrons have spin $\frac{1}{2}$; so do protons and neutrons among well-known particles. It is interesting that this leads directly to Pauli's exclusion principle, for if you try to put particles 1 and 2 *both* into the same state, state a or state b, the total wave-function becomes identically zero.

Heisenberg used this form of wave-function for the two electrons in the helium atom, and obtained reasonable agreement with experiment. To be a little more precise, it is the *total* wave-function that must be anti-symmetric under exchange, and this is the product of the spatial part (the wave-function without spin) and the spin part itself. So *one* of these parts must be symmetric, *the other* anti-symmetric, under exchange. The two types of helium – with spin function symmetric or anti-symmetric under exchange – have very distinct atomic spectra. It is obvious how far these mathematical ideas have moved on from the physical models of the old quantum theory, which clearly could not have succeeded with helium in a million years.

Particles with integral spin, on the other hand, must have the plus sign in the combined wave-function. The important example for us is the photon, which has a spin of 1. Note that photons therefore do *not* have to obey the exclusion principle.

The differences between the two types of particles and their wave-functions come out very clearly in statistical mechanics, where the properties of large numbers of particles are concerned, and symmetrisation or anti-symmetrisation must be enforced. Thus two new forms of statistics are produced, different from those of Boltzmann, though, of course, becoming identical to them in any suitable classical limit.

For the case of particles of half-integral spin like electrons, the relevant statistics are known as *Fermi–Dirac*, and are very important, for example, for the behaviour of electrons in metals and semi-conductors. For particles of integral spin, such as photons, the relevant ideas had already been produced by Bose and Einstein, as already outlined, though much of the rigorous justification of their methods became clear only with modern quantum theory and the idea of indistinguishability. These particles are said to obey *Bose–Einstein statistics*.

I shall now explain a related point that we shall often meet later in this book. Let us consider a decay-process in which a particle of spin zero decays into two particles each of spin $\frac{1}{2}$. I shall write down the *spin-part* of the particle's wave-function, and I shall write α_+ and α_- for individual spin eigenstates of s_z with eigenvalues $\hbar/2$ and $-\hbar/2$, respectively. For conservation of spin angular momentum, it is clear that one particle must be in state α_+, one in α_-, but we must not pretend to know which is which. It turns out that the total spin-function must be written as $(1/\sqrt{2})\{\alpha_+(1)\alpha_-(2) - \alpha_-(1)\alpha_+(2)\}$, where (1) and (2) denote the individual spins. This is easy to write down, but difficult to understand in a physical sense; in particular the problem that will concern us later on is the question of what happens to the particles and the spin-function itself when a measurement of s_z of *one* of the particles is performed? This is the heart of the EPR problem as discussed in Chapter 6, where we will learn to call the total state an entangled state.

The scope of the quantum

In this section I shall give an extremely condensed account of the applications of quantum theory – most of the rest of the book from now on is to do with interpretation.

One limitation, of course, is that the theory presented by Heisenberg and Schrödinger was non-relativistic. In 1928, Dirac [45] published his most important work – a full and rigorous relativistic quantum theory of the hydrogen atom. I shall not attempt to describe the mathematics – it is not really more difficult than what we have already had, but some new ideas are introduced. Dramatically, spin does not have to be added to the theory in the rather artificial way we have done so far; rather, it emerges from Dirac's theory, correct in every detail. (So logically we should drop the word 'spin' from our vocabulary – the electron is not really spinning. In practice, it is convenient to retain the word, without, however, taking its meaning literally.)

Another interesting prediction of Dirac's theory is the *positron*, though it took five years of struggle for the theoretical ideas to be ironed out, and for the positron itself to be detected experimentally (by Carl Anderson). The positron is a particle of the same mass as the electron, but of opposite charge. If they meet, a positron and an electron will annihilate each other, their masses being transformed to energy according to Einstein's famous equation and the energy coming off as a pair of photons; it is for this reason that the positron is known as the *anti-particle* of the electron. All particles have anti-particles, the *anti-proton*, for example, having a negative charge. The neutron and *anti-neutron* both have zero charge, but have contrasting magnetic properties.

Dirac's work brought relativity into the quantum domain, providing, for example, relativistic corrections for atomic energy-levels, but the co-existence of relativity and quantum theory is limited. It has not been possible to quantise general relativity – there is no quantum theory of gravity, and though both quantum theory and relativity play important parts in cosmology and studies of the origin of the Universe [48, 49], there remains a fundamental incompatibility, a blemish on modern physics, which physicists would love to eliminate.

So one must regard quantum theory as a theory with limitations; what is striking is how consistently successful it has been within these limitations. Let us start with atoms. An important point not stressed so far is that very few problems may be solved *exactly* by quantum theory. (Hydrogen and the SHO are almost the only non-trivial examples.) However, straightforward approximation techniques that enable reasonably precise answers to be obtained for other problems are available; their precision may always be increased, but only at the expense of rapidly increasing amounts of work. Usually one is satisfied with reasonable agreement with experiment obtainable from a first or second approximation.

An exception was helium, the smallest atom apart from the exactly soluble hydrogen. Heisenberg's solution above showed clearly the general nature of the eigenfunctions, and gave approximate, not exact, energy-levels. Remember, though, that one important limitation of Bohr's approach was that it could deal only with one-electron problems. It seemed desirable to demonstrate that, for at least one problem, modern quantum theory could clearly and quantitatively do better. Egil Hylleraas took on this task in 1930. By prodigious labour, he obtained a value for the ground-state energy of helium agreeing with experiment to better than one part in 10 000.

Indeed, slowly before the computer, much more quickly afterwards, quantum theory has tackled with great success the problem of atomic energy-levels for atoms of all sizes. It has moved on to molecules, and then further to the much more complicated problems of solids. It has been able to explain, at first qualitatively, and now with modern computers much more quantitatively, a huge range of properties of various types of solid. From the transistor of the 1950s to the modern 'chip', it is fair to say that technological progress has depended on the understanding of solids provided by quantum theory [50, 51].

Another important early achievement of quantum theory was the 1928 analysis of α-decay performed by George Gamow, and independently by Ronald Gurney and Edward Condon. This showed that quantum theory was not restricted in application to atomic physics, but could also tackle nuclear problems. The fact that, over the last 75 years, further progress in nuclear physics has been steady, rather than dramatic, is to be attributed not to any failing of quantum theory, but to the incomplete knowledge of nuclear forces. While forces between electrons in atoms are straightforward – there is just the Coulomb interaction – forces between neutrons and protons in nuclei are complicated, and still not entirely understood [27].

Though we have discussed radiation and fields a good deal in this chapter, the quantum theory described so far has been for particles only. It was natural to wish to develop a quantum theory of the electromagnetic field – *quantum electrodynamics* (QED), and of other fields (nuclear fields in particular) – *quantum field theory*. During the 1930s, physicists like Dirac, Pauli, Heisenberg, Jordan, Robert Oppenheimer and many others expended a good deal of effort on the study of QED, with some success, but the theories failed at high energies. It seemed that infinite quantities inevitably emerged from the mathematics. Around 1947 a number of theoretical physicists – Julian Schwinger, Richard Feynman and Sin-Itiro Tomonaga – invented a novel mathematical technique called *renormalisation* that disposed of these infinites [27], though Dirac, for one, thought that this approach was arbitrary and inconsistent [45]. We will meet Feynman a great deal in Chapter 9, for he was one of the pioneers of quantum computation.

Lastly in this section I turn to the area of *particle physics*. From the mid 1930s on, it became clear that there were many seemingly *elementary* or *fundamental* particles (in the sense of their having no constituents) besides the proton, electron, neutron and photon – there were neutrinos, mesons, muons and many more. Since the 1960s, and the work of Murray Gell-Mann in particular, the idea that neutrons, protons and many other apparently elementary particles actually consist of a number of *quarks* has developed. Initially there were supposed to be only three different types of quark; now there are quite a lot more . . . Three quarks of various types make a proton or a neutron; for this to be possible, quarks must have charges of $\frac{2}{3}e$ and $-\frac{1}{3}e$, quite different from any particles we actually encounter, which have charges of integral numbers of e. Free quarks are never found [27].

Quantum theory has, of course, no power to predict these novel features of the subject. Neither is it contradicted by them, though, and indeed it still provides the basic set of laws and concepts which any theory, however bizarre, must keep to.

Back to Bohr and Einstein

Let us conclude this chapter by returning to our two main characters, and examining and comparing their contributions to the rise of the quantum. It may be said that they complemented one another – Einstein being particularly strong on matters to do with radiation, while Bohr was the expert on the atom.

Some of Einstein's ideas – in particular his advocacy of the photon, but also his championing of de Broglie, and the resulting encouragement of Schrödinger – appear, even today, to have been almost literally inspired. Here, as with general relativity, his ability to sweep away preconception, and home in on a crucial insight, appears greater than that of any other scientist of any time – Newton, Bohr, anybody!

It is the mark of Bohr that he suffers little by comparison. His quantum model of the atom moved the focus of quantum thought to an area where it could develop quickly and widely. His correspondence principle established the important connection between quantum and classical; and the old quantum theory, by its eventual definite failure, cleared the decks for the novel ideas of Heisenberg and Schrödinger.

Orthodoxy would tell us that from now on Bohr was triumphant, and it was all downhill for Einstein, who, it would say, had reached his mid forties, and was no longer able to cope with the new and challenging ideas generated by Bohr to explain the quantum. Such is not exactly the position of this book, as will emerge in the remaining chapters.

5

Bohr: what does it all mean?

Wonderful news from Copenhagen

Heisenberg and Schrödinger invented mathematical formalisms that provided correct answers to all the various problems of (non-relativistic) quantum theory. It was Bohr who uncovered the underlying significance of the theory; in particular he showed how the profound conceptual difficulties encountered as the theory developed – the so-called wave–particle duality, the totally unclassical nature of the uncertainty principle – could be neutralised by a revision, or more accurately a generalisation, of our use of physical concepts.

Such was the generally accepted view of things from soon after 1927, when Bohr first put forward his ideas on *complementarity* at an international meeting of physicists at Como, through the 1930s, when it was felt that Bohr destroyed Einstein's contrary opinions, and certainly up to the time of Bohr's death in 1962. Bohr's analysis is usually spoken of as the *Copenhagen interpretation* of quantum theory, though it is instructive to realise that such a term was frowned upon by those closest to Bohr, since it appears to suggest that the interpretation is just one among (conceivably) many.

Let us examine, for example, the words of Léon Rosenfeld, Bohr's disciple and long-term collaborator, as expounded at a conference in 1957 [52]. Rosenfeld maintained that any idea of 'interpreting a formalism' was a 'false problem', that in a good theory, the 'ordinary language (spiced with technical jargon for the sake of conciseness)' in which it is described, is '*inseparably* united . . . with whatever mathematical apparatus is necessary', that 'we are here not faced with a matter of choice between two possible languages or two possible interpretations, but with a rational language intimately connected with the formalism and adapted to it, on the one hand, and with rather wild, metaphysical speculations . . . on the other'. For Rosenfeld, Bohr's ideas had become a genuine and uncontroversial part of the actual structure of quantum theory, not an interpretation of it.

More than 30 years later, Rudolf Peierls, a leading quantum physicist for practically 70 years, made similar remarks. In his contribution to a book of short and readable essays by physicists with different approaches to the meaning of quantum theory [53], he wrote that use of the term Copenhagen interpretation 'sounds as if there were several interpretations of quantum mechanics. There is only one. There is only one way in which you can understand quantum mechanics . . . so when you refer to the Copenhagen interpretation of the mechanics what you really mean is quantum mechanics.'

Even during the heyday of the Copenhagen interpretation, textbooks usually restricted themselves to giving a brief mention of Bohr's approach, and then moving swiftly on to their real concern – the solution of physical problems. The book which was perhaps the best-known and most influential in these years, that by Leonard Schiff [54], actually does somewhat better. It provides about 200 words on the 'Complementarity Principle' in a section on 'Uncertainty and Complementarity', which is followed by another titled 'Discussion of Measurement'. I shall briefly return to Schiff's account later in this chapter – it could not be said to give a uniformly helpful picture of Bohr's views. Here I just note that his discussion of complementarity is somewhat overwhelmed by that of the uncertainty principle itself; indeed, in many accounts, complementarity is seen as little more than a sophisticated way of describing the uncertainty principle, and this is quite wrong.

And though texts usually paid lip-service to Bohr, any details they give about what they call the Copenhagen interpretation are often closer to the views of John von Neumann; I shall discuss von Neumann's work later in this chapter, but I shall regard it as very much distinct from that of Bohr.

Despite the wishes of Rosenfeld and Peierls, Bohr's position *has* lost its total pre-eminence over the last four decades or so. Other interpretations have at least been allowed into the ring; whether they have been able to land knockdown blows on the champion is much debated. The remaining chapters in this book cover this question in some detail, and Chapter 7 deals in particular with the issue of hidden variables, in some ways the most thorny area in the whole subject. Bohr's view on hidden variables was totally negative, for reasons that will be discussed in the following section.

Bohr and hidden variables

The hidden-variable question may be posed using the following example. The formalism of quantum theory allocates a precise value to at most one of the three components of spin. If s_z has a precise value, the formalism will provide *no* value for either s_x or s_y. But that does not necessarily mean that s_x and s_y actually *have* no values. In some ways a more natural assumption might be that s_x and s_y *do* have values, but they are what are known as *hidden variables* or *hidden parameters*, which just means that they are not present in the mathematical formalism.

A particularly neat assumption might be that a quantum-mechanical wave-function, such as $|s_z = \frac{1}{2}\rangle$ for the spin-$\frac{1}{2}$ case, which we met in the previous chapter, may relate to a group or *ensemble* of spins, each with the same value of s_z, in this case $+\hbar/2$, but with different values of s_x and s_y. (In later chapters I shall refer to this idea as a *Gibbs ensemble interpretation* – an ensemble interpretation with precise values for all observables at all times.) A slightly less elegant but probably more robust idea might be that, though values of s_x and s_y may not exist, there is a hidden variable that dictates what values will be obtained *if* they are measured.

I think I have said enough to show that consideration of the possible existence of hidden variables would have seemed natural. As a matter of fact, hidden-variable theories *are* possible (though it should be added that they have rather unwelcome properties of their own, so they are not necessarily preferable to those without hidden variables; all this is in Chapter 7).

The striking point, though, is that Bohr does not even seem to have considered the option. The reason is almost certainly to be found in his own great quantum theory of 1913. Here there were the same two possibilities for transitions between energy-levels – either they were spontaneous and random, or they were controlled deterministically by a sub-stratum of hidden variables. Indeed, this was not the first such case as physics developed; the same two possibilities had applied also for radioactive decay. For these cases, though, the hidden-variables option does seem more contrived and awkward.

Bohr himself was reluctant to commit himself explicitly to randomness in the years following 1913. When asked about the mechanism of the transitions, he would reply that this was a part of the problem that not yet fully been resolved. This was almost certainly a means of keeping attention on the undeniably successful aspects of his theory, in particular the agreement with experiment. Bohr did not want this success to be overshadowed by contentious debates on determinism.

Even if he ever genuinely believed that a deterministic explanation was possible, it does not seem that he sought one, and it soon became the accepted wisdom that the transitions should be thought of as occurring at random, wisdom not accepted by Einstein perhaps one should add. In the important paper discussed in the previous chapter in which Einstein added the process of stimulated emission to those of spontaneous emission and absorption, the formalism suggested that the transitions occurred in a random way, but Einstein insisted that this idea was provisional. At a deeper level he believed that determinism would be retained [39].

In the years after 1925, it was extremely natural for Bohr to take over the same lack of determinism to the new quantum theory – though here there was the added implication not just that processes were random, but that physical variables did not usually possess specific values in the absence of experiment. Such was Bohr's prestige that this became the orthodox belief, a belief that was strengthened by work of von Neumann discussed towards the end of this chapter. Early opposition to Bohr's views by de Broglie was crushed by Bohr's supporters, particularly Pauli.

By the time hidden-variable theories *did* begin to receive attention again in the 1950s, it was natural for supporters of Bohr to criticise them as unwanted, unneeded, unnatural, . . .; it is such theories that Rosenfeld characterised as 'metaphysical' in the quotation above. ('Metaphysical' is a good slur-word for theories you don't like but can't prove wrong; it's rather like using the word 'alcoholic' for 'someone I don't like who drinks as much as I do'.) Slowly, however, interest in such theories has risen to a level where at least they are widely discussed, although there is no general agreement on their fundamental significance.

To conclude this section, I should explain two points that may have occurred to the reader. I have used the term 'hidden-variable *theories*' rather than 'hidden-variable *interpretations*', because, although hidden variables may start as an attempt merely to interpret the formalism, their properties must be developed as a genuine *addition* to it. Thus the word 'theory' seems more appropriate.

Secondly, for the Bohr atom case in particular, I have assumed that the hidden variables should be deterministic. In this case, it is certainly the reasonable view to take; the perceived problem is lack of determinism, and the whole point of the hidden variables is to restore it. In other areas of quantum theory, it may be helpful to consider hidden variables which are themselves probabilistic – *stochastic* hidden variables.

Bohr's ideas – static or evolving?

Hidden-variable theories were only one aspect of the growing questioning of the Copenhagen position from the 1950s onwards. Interestingly, it was philosophers who were to move, if not to Bohr's defence, at least to stress that his ideas, although often expressed in philosophically naive language, deserved serious and sustained scrutiny. These included Henry Folse [55], Dugald Murdoch [56], John Honner [57] and Jan Faye [58]; in the 1980s and early 1990s they all produced extensive studies of Bohr's work. Further interesting books have been produced in more recent years [59–62], while Bohr's own philosophical writings have been updated and reprinted by Ox Bow Press [63].

Their main aim was to present Bohr's approach in a systematic and holistic way – making the best possible case for complementarity, rather than dwelling on inconsistencies and infelicities in Bohr's presentation. Only thus could the real strength of the argument be put to the test. The approach of this chapter must be along the same lines.

Yet such an approach has one major potential drawback. It tends to rule out the study of whether and how Bohr's doctrines varied over time – an important matter because, if Bohr was forced to change his views from time to time in response to the arguments of others, it becomes much more difficult to present him as the once-and-for-all illuminator of the quantum theory.

That such changes did take place, and were very much required, was particularly insisted upon by Paul Feyerabend [64]. He accused the advocates of complementarity of vagueness, which 'allows [them] to take care of objections by *development* rather than by *reformulation*, a procedure which will of course create the impression that the correct answer has been there all the time and that it was overlooked by the critic. Bohr's followers, and also Bohr himself', he said, 'have made full use of this possibility even in cases where the necessity of a reformulation was clearly indicated. Their attitude has very often been one of people who have the task to clear up the misunderstandings of their opponents rather than to admit their own mistakes.'

Feyerabend was concerned with the hidden-variable issue, and also with Bohr's response to the famous Einstein–Podolsky–Rosen (EPR) argument of 1935. Did Bohr actually change his position in response to EPR or just tighten his terminology? I shall examine the question later in this chapter and in the following one.

The problem – as perceived by Bohr

As we move towards Bohr's actual ideas, it will be helpful to remind ourselves of the problem he was trying to solve, because for others, in particular Heisenberg, there really was no problem. Discussing his argument with Bohr 35 years later, Heisenberg reported [55] that 'I . . . would say, "Well we have a consistent mathematical scheme and this consistent mathematical scheme tells us everything that can be observed. Nothing is in nature which cannot be described by this scheme".' So it would be pointless even to think about the actual nature of the system under investigation; all that was required was correlation of experimental results with mathematical predictions. One did not need to analyse the reason for the failure of the classical concepts, just to note the extent of this failure, given by the uncertainty principle, and settle for (or retreat into) the mathematical scheme itself.

Around 1927, this position would also have been held strongly by Kramers, and, following his intellectual journey described in the previous chapter, by Born, probably a little less emphatically by Pauli. The other main component of Heisenberg's ideas at this time was that Schrödinger's formalism, and also his waves, were useful merely as a calculating device; it was never necessary to go beyond particle-like ideas in considering actual experiments.

Bohr strongly disagreed with both aspects of this position. He maintained that, even to describe ourselves as confirming mathematical predictions, it is essential to include in the experimental account, not only our evidence for the assertion – the flash on a screen, or black mark on a strip of paper – but also the interaction between the atomic system and the macroscopic observing system. This interaction, he said, could be described only in classical terms, which require picturing systems as waves *or* particles in different experimental contexts. Until we understand how these classical terms may be related to the atomic system itself, we should not claim to have produced answers to the riddles of quantum theory. Heisenberg could show no such thing. And while Bohr had no more respect than Heisenberg for Schrödinger's interpretation of quantum theory, which *only* referred to waves and excluded quantum jumps, it was clear to him that the wavelike nature of atomic systems *did* have to be included, together with the particle-like one, in any complete account of quantum theory.

It is easy to see that Bohr's approach to these matters was bound to contain as much opposed to Heisenberg's views as to those of Schrödinger (and, to emphasise the point, as much supporting Schrödinger as Heisenberg). Indeed Beller [65] goes further, claiming to show that, in Bohr's first report on complementarity in Como [63, 66], 'each sentence is an implicit argument with leading physicists of the time (Einstein, Heisenberg, Schrödinger, Compton, Born, Dirac, Pauli and the lesser-known Campbell)'. (Norman Campbell had suggested that the difficulties of quantum theory could be removed by making the concept of time statistical; this was rejected by Bohr, but Campbell's ideas played a part in stimulating Bohr's own thinking.)

Since our interest is in Bohr's overall work, not just the Como paper, I do not discuss this aspect of Beller's paper in detail. But I do take her point that the 'history' that has come down to posterity has been very different. Subsequent to Bohr promoting his ideas on complementarity, those close to him were, painfully or otherwise, drawn back into the fold. In particular, of course, I am referring to Heisenberg, who, after intense disagreements with Bohr during this period, returned to remain a faithful supporter for the remainder of his life, though admittedly his pronouncements were a little erratic; Born and Pauli would also come into this category. In contrast, of course, Schrödinger and Einstein remained in almost total disagreement with Bohr's position.

It is not surprising, therefore, that there was a closing of ranks on Bohr's side, past disagreements being censored. The fact that Born had initially been very attracted by some of Schrödinger's philosophy, as well as his physics, was conveniently forgotten. The fact that Heisenberg and Bohr had argued persistently and emotionally over the uncertainty principle and over complementarity was distorted; while Heisenberg's tears became part of the folklore, the reason for them ceased to be thought of as a matter of important principle, but came to be regarded as little more than a misunderstanding, or perhaps just youthful over-excitement on Heisenberg's part.

This kind of distortion led to a common but highly unfortunate misunderstanding of Bohr's position on quantum theory. For if there was never a genuine clash of opinions between Bohr and Heisenberg, then it must surely be legitimate to assume that Bohr's rather difficult ideas are equivalent to the rather straightforward, not to say naive, ones of Heisenberg outlined above. The consequences follow into the next section.

I close this section by mentioning that I suspect many people have been put off Bohr's views because their motivation is so different from his. Today most physicists come to accept readily the mathematical approach to quantum theory, and are unaware or even dismissive of any conceptual worries about working with it. What they require from an interpretation is some sort of *picture* of what is going on, to give at least a partial explanation of the very perplexing aspects of the theory, or, it might be said, to change mathematics into physics. To these people, the Copenhagen interpretation probably seems little more than unenlightening, pedantic and frankly uninteresting logic-chopping. In contrast, some of the later interpretations discussed in the next few chapters meet such people's demands more closely.

It is important to remember that Bohr's own aims were very different. He saw quantum theory as demanding fairly drastic reformulation of our usual physical concepts; this was required, not in order to provide a physical picture or any sort of 'explanation' of quantum theory, but just to avoid logical confusion and to give us a rigorous basis for discussion of experimental results. It is with respect to *these* aims that his achievement must be measured, and a fair assessment may be that it is considerable, though not total, and certainly not final.

Popular views of Bohr's approach

How should one describe Bohr's approach? In books and articles, three very different descriptions may be found. I shall refer to these as

(1) philosophical, (2) conceptual and linguistic, and (3) physical. By (1), I specifically mean *either* that Bohr deliberately adopted a particular standard philosophical viewpoint, *or* that, less deliberately but just as clearly, his adopted position was identical to that of such a stance.

Such an idea about Bohr is extremely widespread. Folse [55] writes that 'each of the various schools of philosophy of science in this century has . . . claimed Bohr as both ally and enemy. He has been championed as a positivist, a realist, a materialist, an idealist, and a pragmatist, *and* at the same time criticised under at least as many labels'.

In the USSR, for instance, Bohr's views were taken as *idealistic* [67], since such basic aspects of his approach to quantum theory as interference of probability waves in a double-slit experiment appeared contrary to dialectical materialism. Thus, particularly through the decade following the Second World War, the Copenhagen interpretation and Bohr himself were heavily and officially criticised in the USSR; hidden-variable theories and ensemble interpretations (for which see more later) were in favour. However, the extremely influential physicist Vladimir Fock had always been a strong supporter of Bohr, and, as a result of his influence, Bohr's views became re-established in the USSR from about 1957 on.

Bohr had been charged, among other things, with encouraging a *subjective* approach, according to which the experiencing subject played a role in physical measurement. While Bohr stressed the need for the full experimental arrangement to be considered, the measuring equipment would be considered wholly *objectively*. Though unfortunately Bohr did occasionally use misleading terminology in this area, when alerted to the dangers he was very quick to point out emphatically [68] that there was no subjective component in his analysis. Since Bohr and Fock between them convinced the Russians on this point, it is unfortunate that philosophers such as Karl Popper [69] and Mario Bunge [70] have repeated the charge of subjectivity.

A much more serious charge is that of positivism and instrumentalism. Indeed, this is more than a charge – it is commonly taken as an established fact [45, 71]. *Positivism* may be said to state that observed phenomena are all that require discussion or scientific analysis; consideration of other questions, such as what the underlying mechanism may be, or what 'real entities' produce the phenomena, is dismissed as meaningless. *Instrumentalism* regards scientific theories purely as tools to calculate experimental results; non-observable theoretical concepts may be potentially useful, but are always disposable. As a variant, *operationalism* defines concepts entirely through their measurement.

This group of 'isms' may be said to be in direct opposition to *realism* – the view that the observations we make in science are related to a *real world* existing independently of our observations. When we analyse our results, and produce theories, which are often mathematically abstract and elaborate, we are not merely trying to discover *correlations* that may enable us to predict further experimental results; we are attempting to gain information about this ultimate reality.

Aristotelian and Newtonian physics were very different, but both were extremely compatible with realism. In both cases it was natural to conclude, indeed practically perverse to dispute, that what was observed was in total and direct correspondence with what actually existed, that theoretical terms in one's equation related directly to physical quantities. This position may be called *naive realism*.

As explained in the previous chapter, the gradual unfolding of the quantum theory inflicted grave damage on naive realism. It might be said that three responses could be made. First, one could struggle to limit the damage, refusing to let go of certain key elements of the position; this is at least an approximation to Einstein's approach. Secondly, one might renounce realism altogether. Thirdly, one might seek to retain realism, but to revise or restrict its nature (to make it less naive, we might say).

As I have said, many commentators have assumed that Bohr took the second position, giving up realism and embracing positivism. (Einstein probably assumed this himself.) Yet many of those who have analysed Bohr's work most carefully say the opposite. Folse [55] says that '[w]hat [Bohr] rejects is not realism, but the classical [naive] version of it', and that 'complementarity offers a realist interpretation of scientific theory in the sense that it provides knowledge about what it takes to be independently existing real atomic systems and their constituents'. Honner [57] suggests that 'implicit in Bohr's vision, there is the belief that somewhere along the line the theory is anchored in reality'. Carl von Weizsäcker (quoted in [57]) is insistent that 'Bohr was no positivist'. Murdoch [56] describes Bohr as an 'empirical realist' and a 'weak realist'.

It may be pointed out that the positivist approach to quantum theory would appear to provide fairly straightforward solutions to the various difficulties. One could concentrate on the experimental results, regarding the only goal as their being expressed in a simple mathematical form, and dismissing all deeper speculation as meaningless. Such a position would be a-priori philosophically and independent of any experimental result.

It is clear that Heisenberg was at least very close to this position from the time of his invention of matrix mechanics up to his work on the uncertainty principle. But we must remember his bitter confrontation with Bohr at the end of this period, while Bohr was working on complementarity. This makes it practically certain that Bohr's position must have been entirely different. It is strange that the 'closing of ranks' behind Bohr in the following years may have helped foster the idea that his position had been merely a variant of Heisenberg's original positivism.

Bohr's position *was* different, yet it is not necessarily easy to demonstrate this in a simple and clear-cut way. The volume of his collected works that contains papers published in this area between 1926 and 1932, for example, is a little like the Bible, in that individual sentences may provide support for practically any position one espouses. A *fair* reading, though, would undoubtedly suggest that Bohr invariably thought in terms of a *real* atomic system, the properties of which were admittedly difficult to elucidate; he would have ridiculed the idea that all he was seeking was correlations.

So of the 'isms' of (1), we must conclude that Bohr at least started from the realist position. One more 'ism' that Bohr may have found quite congenial was the *pragmatism* of William James, and, in particular, Harold Høffding [58]. Murdoch [56] sums up the pragmatic position on quantum theory as follows – 'The ascription of an exact position and an exact momentum to an object at the same time is meaningless, not principally because such a property is unobservable, but because the assumption has no practical consequences whatever . . . The meaning conditions, then, are not primarily epistemic, but pragmatic, having more to do with what we as agents can *do* than with what we can *know*.' Henry Stapp [72] wrote a short but influential article on the Copenhagen interpretation in which he suggested that it represented 'a shift to a philosophical perspective resembling that of William James'.

I suggest, though, that, while Bohr undoubtedly found pragmatism congenial, his approach was very different from what I have called above (1) philosophical, in the sense of applying a particular brand of philosophy, even pragmatism, to the quantum problem. Rather it was what I called (2) conceptual and linguistic; Bohr started from what he perceived to be the fundamental facts and difficulties of quantum theory, and considered how the usual forms of conceptual analysis, and the usual modes of language, must be modified in such a way as to render the discussion unambiguous. This is what I shall describe in the following sections.

Bohr was, however, a physicist as well as a philosopher, and he was tempted into regarding a (3) physical argument as an alternative to his conceptual one. This physical argument was to do with the 'disturbance caused by a measurement' which was supposed to lead to many of the quantum difficulties. This argument, which will be described later in this chapter, provided a pleasant physical picture of what was supposed actually to be happening, and, as such, it is very popular still with writers of textbooks.

Unfortunately for Bohr, in 1935 Einstein, in collaboration with Boris Podolsky and Nathan Rosen, produced the famous EPR paper, which effectively demolished the disturbance interpretation. Bohr was forced to retreat towards his conceptual argument. Indeed, his supporters could obtain sufficient justification to claim that the disturbance idea was never a central component of Bohr's ideas. (It was this policy of meeting criticism by insinuating that the critic had misunderstood the argument which so infuriated Feyerabend, as I mentioned above.)

For the physicist, Bohr's loss of this physical argument must be a strong negative point. It leaves him seemingly stranded in conceptual and linguistic argumentation of a rather abstract and technical nature. Even Folse [55], whose self-imposed task is to assemble Bohr's ideas into a coherent and philosophically respectable position, asks whether Bohr has backed himself into an anti-realist position at this point. Others, even those who recognise that Bohr's views were not intrinsically positivistic, wonder whether he has not reached a 'higher-order positivism'.

The fact that Bohr was forced to abandon his more physical argument in favour of conceptual ones, may explain why, though most physicists who merely *use* quantum theory certainly regard themselves as his followers, among those who study the foundations of the subject I think Bohr is treated more respectfully by philosophers than by physicists (more respectfully, for example, by Henry Folse than by John Bell).

Complementarity – the quantum postulate

For Bohr, all the problems of the quantum world began with discontinuity. From the point of view of language, he wrote [55] that 'every notion, or rather every word, we use is based on the idea of continuity, and becomes ambiguous as soon as this idea fails'. Discontinuity must inevitably bring about changes in the ways we can use words consistently.

From the more physical point of view, Bohr always commenced his discussions of discontinuity by using his own 1913 model of the atom, discussed here in Chapter 4, where there were what he called *stationary states*, in which the atomic system remained unchanged (the Bohr orbits), and between which transitions were caused by the discontinuous emission or absorption of a photon. In the absence of hidden variables, which were excluded by Bohr as I have explained already, discontinuity must inevitably imply a loss of determinism.

Bohr generalised this idea to his *quantum postulate*, which describes an interaction between any two objects as follows. For the most part, the two objects exist in a stationary state, in which there is no exchange of any type between the two objects. However, at certain random moments, there is a substantial instantaneous exchange of energy, momentum and perhaps angular momentum. Thus again Bohr emphasised *discontinuity*. An extremely important example of such an interaction is that between an atomic particle and a measuring device.

To understand the effects of discontinuity, I shall first analyse the case of a classical continuous interaction. This interaction may just be gravitational, but there may also be an electrical or magnetic interaction, and, as we saw in Chapter 2, it is convenient to describe this as taking place via an electromagnetic field. In the classical case the interaction is continuous in time; we may say that the electromagnetic field itself is continuous in time.

We would hope to provide a mathematical description of each of our objects at all times by stating its position and velocity. (Actually we usually use the momentum rather than the velocity.) Using more general terminology, we may say that we would like to define mathematically, and at all times, a *state* for each of the participating objects, where the state just tells us all we want to (and can) know about the object at that time.

For example, let us consider the measurement case, in particular a Stern–Gerlach measurement of a component of angular momentum of a given atomic object. We would like to trace the path of the object through the apparatus, a path that must terminate where the object reaches a screen, at a point where it leaves a permanent black mark. The final *state* of the measuring device just refers to the position on the screen at which the black mark occurs, and from this information it should be possible to deduce the initial value of the appropriate component of angular momentum of the atomic object.

Are we prevented from giving the detailed description of the measurement that we would like, by the fact that the objects *are* interacting? In the case of two interacting particles, the behaviour of each must be affected by the presence of the other; even if we wanted to, we could not switch off *all* interactions, including the gravitational. And, of course, for the measurement case, the interaction is central; we may scarcely hope to register a measuremental result without an interaction between measured object and measuring device.

Yet classically we may cope. Because the interaction is continuous, the momentum and energy of each particle will vary only a very little over a short period of time. Thus we might measure the momentum and energy of each particle at a particular time t, and be quite happy that, at times shortly after t, momentum and energy will have changed little, and this change may in any case be calculated.

Indeed, we may go further, taking advantage of continuity and determinism. For convenience, let us consider two electrically charged particles approaching one another. (I shall make them both positively charged, so that they repel each other.) We may wish to start our consideration of the problem at a time when the particles are so far apart that the force between them, though never of course actually absent, may be negligible. At this time we may note the *initial conditions* – values of position and momentum for each particle. We may then use our knowledge of the laws of electromagnetism to calculate the values of these quantities at all times through the interaction process, at the end of which the particles will again be very far apart, and almost non-interacting. (An analogous physical case would be a comet travelling towards, and being deflected by, a star.)

Since we may calculate these quantities, it is obvious that they exist! (It may seem superfluous even to raise this point, but see our discussion of the quantum case over the next few sections.) And again we may be sure that, in a classical measurement, the final state of the measuring device must be well defined, and it will give direct information on the initial state of the system being investigated.

Thus classically there is no problem in defining, and indeed calculating, the variations in time of the positions and momenta of two (or, in fact, more) particles interacting with each other, and so constituting a (composite) *system*. Bohr would say that this is the continuous case for which all our language has been designed, so it is scarcely surprising that the language handles it very well indeed.

However, sometimes we may *wish* to do things differently. When thinking about the Moon, we *may* consider it as a composite system of interacting particles, and think about the position and momentum of each one. But it is usually much easier to think of it as a single object with mass equal to the sum of the masses of all the interacting particles, and with a single position and momentum, that of the centre of the composite system. In that way we may forget the details of all the interactions between the various particles. (Of course, this is when we are well away from the Moon; an astronaut actually on the Moon will be very concerned with the actual bit of rock she is standing on, and with any possible motion with respect to other nearby bits.)

Now let us turn to the quantum case, where, as stressed by Bohr, we must face up to discontinuity. If, for example, the two objects interact electromagnetically, we must realise that they exchange energy and momentum, not continuously via a classical electromagnetic field, but in an intermittent and irregular fashion by passage of photons between them. Then the position and momentum of each particle will vary with time in an erratic and totally unpredictable way; they will certainly be impossible to calculate.

Discontinuity also has a drastic effect on the very idea of measurement. In our classical example, the particle was certainly affected by the electromagnetic field in the Stern–Gerlach measuring device, but because of continuity the effect may be precisely calculated. The final state of the measuring device thus gives direct information on the initial value of the component of angular momentum being measured, even though, in this case, the *final* state of the particle is very different from its initial state. Bohr would say that this is just the *meaning* of the word 'measurement', and again the meaning applies very well in this classical continuous case.

In the quantum case, however, the interaction between measuring and measured systems is discontinuous and certainly impossible to calculate. As a result it must be quite impossible to argue unambiguously from the final state of the measuring device to an initial value of the quantity being measured. Clearly what we mean by the word 'measurement' must be totally re-assessed.

The 'disturbance' trap

What I said in the previous section was correct as far as it went, but it left out a crucial piece of the argument. The analysis so far would lead one to say the following. Certainly discontinuity means that it must be impossible to predict the values of the position and momentum of each component of an interacting system throughout the interaction process. Certainly discontinuity means that, in a measurement, both measuring and measured systems will be affected in an uncontrollable way, and the relationship between the final state of the measuring device and the initial state of the measured system must become unclear. But so far I have not demonstrated – as I shortly shall – the far stronger result that such states simply cannot be defined, and just do not exist.

The argument at the moment is at the stage which may be referred to as 'the measurement disturbing the system'. As I said earlier in the chapter, when I called it Bohr's physical argument, it was very often used by him in his earlier accounts of complementarity to supplement his more conceptual analysis. However, it was a trap, and he had to relinquish it following publication of the EPR paper in 1935.

What I shall call the *disturbance interpretation* implies that an atomic particle which is having one of its properties measured exists in a perfectly well-defined state; this state cannot, however, be determined exactly because our apparatus is so much larger than the atomic system that it *disturbs* it, changing its state unpredictably to another well-defined one.

Something rather close to the disturbance interpretation was implicit in the Heisenberg microscope scheme of Chapter 4, as Heisenberg assumed his particles to be in precise states, which were then disturbed by the measuremental interaction. Bohr was critical of Heisenberg's initial formulation, and improved some aspects of it, but the disturbance idea was not affected. It was not surprising, then, that Bohr fell into the same trap himself, and I don't think he should be criticised too much for it, because these are certainly difficult conceptual areas to venture into, and Bohr was travelling essentially alone.

What is a little more disappointing is that the disturbance idea is still very prevalent in texts. Indeed, even outside physics altogether, one often hears or reads that the fact that 'the observer disturbs the system' is the central idea of quantum measurement.

That the disturbance interpretation is quite wrong is shown particularly clearly by Folse [55]. As he says, it would seem to imply that our best chance of performing good measurements in the quantum domain would be to use small-scale measuring equipment, if possible of atomic dimensions itself. In that way we could hope to disturb the system as little as possible.

Yet it is well known, and indeed an essential part of Bohr's analysis as we shall see, that measuring apparatus must be of macroscopic dimensions – one might say, as *large* as possible. It is only because there is an immense *difference* between atomic dimensions and the dimensions of ordinary human experience that, for a considerable period, it seemed that there was strict determinism. Thus the disturbance interpretation must be wrong, and I shall now proceed to discuss why this is the case.

Complementarity – the cut

What is crucially missing from the disturbance interpretation is acknowledgement that, in an interaction, the boundary between the interacting systems cannot be defined uniquely. In the measurement case, to which I shall largely specialise from now on, the division into measuring device and measured system is arbitrary. There *must* be a division between the two, usually called a *cut* or a *Heisenberg cut*, but there is no possibility of specifying uniquely where it should occur.

Bohr recognised this point in his very first account of complementarity given at Como [68]. He noted that 'the concept of observation is in so far arbitrary as it depends upon which objects are included in the system to be observed. Ultimately, every observation can, of course, be reduced to our sense perceptions. The circumstance, however, that in interpreting observations use has to be made of theoretical notions entails that for every particular case it is a question of convenience at which point the concept of observation involving the quantum postulate with its inherent "irrationality" is brought in.'

Note first in this quote an example of how Bohr's language sometimes did him a disservice, encouraging his opponents to ascribe to him views quite contrary to those he actually held. The word 'irrationality' was intended to indicate lack of classical predictability, certainly not an abandonment of logic or reason.

However, to move to the actual content of his statement, Bohr says that there can be no a-priori boundary between measuring and measured systems. Let us take as an example a typical experiment in which certain properties of an electron are to be studied. In such an experiment, the electron initially interacts with an atom, and this atom then creates a black image on a photographic plate. (It may be supposed that the precise location of this image depends on those properties of the electron which we wish to study.) A human being then views the photographic plate by use of a microscope.

In this experiment, what is the measured system, and what the measuring system? (Or what is the *observer*, and what the *observed*?) We might say that the electron, the real object of our interest, is the observed system, and everything else, from atom to brain of the human being, the observing system. There is nothing wrong in this point of view. However, since the atom that interacts with the electron is itself microscopic, it might seem more natural to regard electron and atom together as the observed system, and everything else as the observing system.

To go further, and particularly if there is some question about the way in which the atom forms an image on the photographic plate, we might wish to include those atoms on the plate which are in the blackened region as part of the observed system. Though it is much less natural, we could go further and include some components of the microscope, or the whole instrument, as part of the observed system. (This would entail detailed analysis of the passage of light through the microscope, which, if the microscope were considered part of the observing system, could be taken much more for granted.)

We could even, if we wished, consider parts of the human being as belonging to the observed system rather than the observing system, leaving maybe just the brain (mind? consciousness?) of the human as the observer, and everything else, including the eye and optic-nerve tract of the human, as the observed system. Such ideas may appear highly artificial, but need to be taken much more seriously when we turn to the work of von Neumann towards the end of this chapter.

Bohr himself would not have allowed all the possibilities I have included. He would certainly have refused to consider any macroscopic object as part of the observed system, because, if he did so, he would have to assign it a wave-function. Nevertheless there is still sufficient ambiguity to justify his remark [73] that 'this crucial point . . . implies the impossibility of any sharp separation between the behaviour of atomic objects and the interaction with the measuring instruments which serve to define the conditions under which the phenomena appear'.

This ambiguity implies the collapse of the disturbance interpretation. One *cannot* handle the quantum postulate by first defining states for measuring and measured systems, and then acknowledging discontinuity by recognising that these states change in an irregular and unpredictable fashion. We *cannot* say that the state of the measuring system disturbs that of the measured system. The division across which these uncontrollable exchanges take place is itself arbitrary! The only way we can make sense of this is to acknowledge that these states just do not exist. In this book I have pointed out a fair number of differences between quantum and classical physics, and this one is as striking as most.

We do have an alternative – the alternative already described for the classical case – of defining and working with a *combined* state for the total system including all interacting sub-systems. *For the measurement case, this implies that observing and observed systems may be handled only as a whole, not individually.*

Complementarity – the use of classical concepts

Our argument so far might suggest that classical ideas have failed, and so we must work towards the creation of a completely new set of physical concepts. Yet, almost paradoxically, it might seem, Bohr's second great starting-point for complementarity was that use of the classical concepts is the only possibility. In his Como lecture [66], he wrote that 'our interpretation of the experimental material rests essentially on the classical concepts'. Much later [73], he wrote that 'however far the phenomena transcend the scope of classical physical explanation, the account of all evidence must be expressed in classical terms'.

Bohr's argument rested on his view that, in order to justify itself, a novel theory such as quantum mechanics has to establish a contact between its mathematical content and the experimental results, and the only language we have to express the latter is that of the classical concepts.

The use of the term 'classical concepts' may be confusing and even off-putting. It has certainly been attacked vigorously by Feyerabend [64], who suggests that concepts cannot be independent of theory. If a new theoretical framework is produced, he argues, it will almost certainly entail a new set of physical concepts, and the older ones may be put to rest.

After all, he suggests, similar comments to those of Bohr could easily have been made about the Aristotelian theoretical framework, which had a fine array of concepts, and used ideas that were both close to everyday experience and exceptionally well entrenched. It could have been argued that Galileo might invent new descriptions of phenomena, but could not move outside the Aristotelian conceptual framework. Yet, as Feyerabend points out, Galileo's law of inertia does not even make sense in such a framework. Fortunately, Galileo had the ability to invent a totally new theory, which came complete with a fresh conceptual scheme. Why, Feyerabend asks, should Bohr attempt to inhibit physicists from making similar conceptual breakthroughs today?

The answer may be that Bohr's idea of classical concepts is a little more prosaic than is sometimes assumed. When Schrödinger wrote to him in 1935 to ask about his views. Bohr replied [6] that 'my emphasis on the unavoidability of the classical description of the experiment refers in the end to nothing more than the apparently obvious fact that the description of every measuring apparatus basically must contain the arrangement of the apparatus in space and its function in time, if we are to be able to say anything at all about the phenomena ... The argument is thus above all that the measuring instruments, if they are to serve as such, cannot be included in the actual range of applicability of quantum mechanics.'

The second sentence will be important later. The first would appear to mean that, by 'classical concepts', Bohr really means primitive notions such as positions of marks on photographic plates, or pointers on dials, or the path of the Sun through the heavens. These are facts about which, Bohr hoped, everybody, whether Aristotelian, classical physicist, quantum physicist, or believer in some physics yet to be discovered, would be bound to agree, and which actually exist prior to any genuine physical concept like velocity or force. A physicist of today will certainly – for convenience, for shorthand, for direct connection to theory – use notions such as energy or momentum when analysing experiments, but may, if challenged, retreat to the more primitive notions that Bohr really considered to be classical concepts.

Complementarity – the heart of the argument

Bohr had to reconcile the considerable difficulties laid in the path of classical ideas by the quantum postulate and the arbitrary position of the cut with his intense conviction that one must retain the classical terms. He did this by restricting the application of these classical terms so that they could be used unambiguously. He believed that such a re-assessment of the status and scope of concepts was essential at the time of any revolutionary development in science, such as that of the theories of relativity, and considered himself to be delivering an *epistemological lesson*, applicable, in fact, well beyond the confines of atomic physics.

(I feel it important to re-state that Bohr's accounts of complementarity always based it squarely on his deduction from experimental fact, the quantum postulate. There is a well-known story [58, 74] of a friend responding to an account of complementarity given by Bohr in 1929 with the comment that it was exactly what Bohr had been saying ten (or 20, accounts differ) years before. There is, inevitably and justifiably, great interest in tracing Bohr's early philosophical development and influences [55, 58], and seeking connections between early general ideas and later more specific achievements. Of course new ideas favour prepared minds, and the structure and presentation of novel arguments are bound to reflect previous knowledge and understanding. It must be said, though, that, according to anything Bohr ever wrote, complementarity rests not on any innate or inculcated beliefs he may have had, but on his analysis of experimental results.)

In complementarity, one removes ambiguity by restricting use of a particular physical quantity to experimental situations where evidence related to that quantity may be obtained. One may, for instance, wish to consider or discuss the value of the *position* of a particular atomic object. To do so unambiguously within the confines of complementarity, we must discuss the value in terms of an experiment set up specifically to measure the position.

Now a measurement of position requires a completely different experimental arrangement from a measurement of momentum. Furthermore (and Bohr established this in detail in discussions with Einstein to be described in the following chapter), the circumstances in which a precise measurement of position may be made are precisely those in which an indefinably large uncontrollable transfer of momentum takes place between observed and observing systems. So, in the context of a measurement of position, nothing at all may be said about the momentum of the atomic object. Similarly, if one wishes to consider or to discuss the momentum of the atomic object, one may do so only in the context of an experiment to determine its value. But in an experiment to measure momentum exactly, one totally loses any ability to locate the object in space, and so must say nothing about the position of the object.

Bohr himself [73] said that 'evidence obtained under different conditions cannot be comprehended within a single picture, but must be regarded as *complementary* in the sense that only the totality of the phenomena exhausts the possible information about the objects', and on another occasion [66], that the quantum postulate 'forces us to adopt a new mode of description designed as *complementary* in the sense that any given application of classical concepts precludes the simultaneous use of other classical concepts which in a different connection are equally necessary for the elucidation of the phenomena'.

Murdoch [56] says that the latter statement is as near as Bohr ever came to giving a definition of complementarity. As Murdoch says, there are two factors necessary for a complementary description – *mutual exclusiveness* and *joint completion*. Mutual exclusiveness tells us that, in the position/momentum case, if one of the quantities is examined experimentally or discussed theoretically, we are precluded from measuring or discussing the other. Joint completion means that, in the classical case, one may of course consider position and momentum simultaneously, and together they then provide a *complete description* of the motion of the particle. (The combination 'exhausts the possible information' in Bohr's words above.) For Bohr, the case of position and momentum, *kinematic – dynamic complementarity*, is the fundamental example of complementarity.

Another example of complementary phenomena would refer to angular momentum. Classically one requires knowledge of *each* of the three components of angular momentum in order to give a complete description. In the quantum case, experimental investigation of one component, say the x-component, precludes possession of any knowledge about the y- and z-components (although the total angular momentum may be known). So both conditions for the three components to constitute a complementary description of the system are satisfied.

From his Como article on, Bohr also referred to another type of complementarity – that between wave and particle models of light and matter. (It will be remembered that one of the main purposes of this article was to stress that particle-like and wavelike ideas were *both* essential in a complete quantum description.) In the article, Bohr says that wave and particle models provide 'different attempts at an interpretation of experimental evidence in which the limitation of the classical concepts is expressed in complementary ways' and also that we are dealing with 'not . . . contradictory, but with complementary pictures of the phenomena, which together offer a rational generalisation of the classical mode of description'.

So Bohr definitely sanctions the idea of *wave–particle complementarity*, but it should certainly be considered as secondary to that between position and momentum [55, 56]. While Bohr would suggest that position and momentum are concepts *necessary* for classical description, it is not absolutely clear that either wave or particle model is actually *necessary* for a description of light or matter. Neither is it clear that together they provide a complete description in the classical case. Indeed, it is not entirely obvious that one could not search for an altogether different model, which might, on its own, represent light and matter in a satisfactory way. So analysis of complementarity should always concentrate on the kinematic–dynamic, not the wave–particle, case.

After this account of complementarity, it should be reasonably clear that it enables one to avoid the problems that had beset quantum theory from the earliest ideas of 'wave/particle duality' through to those surrounding the uncertainty principle. These difficulties came from the possession of two alternative but mutually exclusive descriptions, or from the realisation that the existence of a precise value for one physical quantity meant that another could not have a precise value.

At a stroke, it might be said, complementarity removed the difficulties. Consideration of one description or the value of one quantity forbade one to consider the other. Of course the reader may feel that it is an evasive procedure – even perhaps obscurantist. One could not say that it *solves* the difficulties, or even *explains* them – it just *evades* them, but I would remind you that Bohr's aim was merely to obtain a procedure that worked.

Wholeness

Complementarity necessarily entails what we earlier obtained as the result of our argument from the quantum postulate – that a measuremental result cannot be regarded as relating only to the object being measured. Rather, we may say that the observed object and the measuring apparatus are inextricably linked. Bohr calls this feature of quantum theory *wholeness*. He says [68] that 'the essential wholeness of a proper quantum phenomenon finds itself logical expression in the circumstance that any attempt at any well-defined subdivision would require a change in the experimental arrangement incompatible with the appearance of the phenomenon itself'. The idea of wholeness was important in Bohr's response to the EPR argument, as will be described in the following chapter.

It may be suggested that this holistic aspect of Bohr's work has echoes of pre-scientific ideas, particularly those of Aristotle, who stressed the primacy of the whole over the parts. Such suggestions may be of interest. On the whole, though, I feel that the attempt to view modern physics as a triumph of ancient wisdom [75] shows little more than wishful thinking and the ability to conjure analogies out of rather thin air. Bell [76] comments that 'It seems to me irresponsible to suggest that technical features of contemporary theory were anticipated by the saints of ancient religions . . . by introspection.'

Another word usage introduced by Bohr to cover much the same ground as wholeness is that of the *phenomenon*. In the first few years of complementarity, Bohr [66] used this word in a very conventional way; in a measurement of the momentum of an electron, for example, the phenomenon under investigation is just the electron. In the disturbance interpretation, the phenomenon could be disturbed by the observing apparatus – 'the finite magnitude of the quantum of action [i.e. discontinuity] prevents altogether a sharp distinction being made between a phenomenon and the agency by which it is observed'.

However, particularly once the EPR paper had disposed of the disturbance interpretation, Bohr radically changed the way in which he used the word. The phenomenon became explicitly the behaviour of the whole system of measured object and measuring device. Thus an account of a phenomenon must involve a complete description of the experimental arrangement as well as the results that the observer obtains. For Bohr, the *phenomenal object* refers to the measured object *as it appears under particular experimental conditions*, and it is natural to talk of *complementary phenomena* – the behaviour of the same object under different experimental conditions, experimental conditions that exhibit complementary descriptions of the physics.

John Bell was especially appreciative of this contribution of Bohr. I would mention that Bell regarded himself as a follower of Einstein, and was not in general much in favour of complementarity, or of much of Bohr's efforts to understand quantum theory. He [76] accused Bohr of taking satisfaction in ambiguity, rather than being disturbed by it, and of revelling in contradictions. Bell liked to replace the word complementarity by *contradictoriness*.

He called complementarity one of the three *romantic world views* inspired by quantum theory. He meant the word 'romantic' in a derogatory sense – bizarre, extravagant, vague and, perhaps worst of all, unprofessional. He contrasted these romantic world views with three *unromantic* alternatives, which require mathematical work by theoretical physicists for their elucidation, rather than interpretation by philosophers. We shall meet all Bell's other world views later in this book.

As I have said, however, Bell [76] approved of wholeness, and Bohr's later use of the word 'phenomenon', very much – 'The word [measurement] very strongly suggests the ascertaining of some pre-existing property of some thing,' he wrote, 'any instrument involved playing a purely passive role. Quantum experiments are just not like that, *as we learnt especially from Bohr* [my italics]. The results have to be regarded as the joint product of "system" and "apparatus", the complete experimental set-up. But the misuse of the word "measurement" makes it easy to forget this and then to expect that the "results of measurement" should obey some logic in which the apparatus is not mentioned.'

To return to Bohr, it is clear how much of his analysis is *linguistic* or *semantic* (particularly once the alternative argument based on physical disturbance had to be abandoned). One is learning how one may use language unambiguously. The argument is not so much that if an object has a precise value for position its momentum has no value; rather, in the circumstance in which one may discuss the value of position, the concept of momentum is inapplicable. This kind of argument will go forward to Bohr's discussion of EPR. (Actually it might be mentioned that what I have here called linguistic or semantic is quite close to my earlier discussion of the *pragmatic* position, which helps to explain why several commentators have put Bohr into that camp.)

Bohr follows the linguistic path further by his introduction of the word 'wholeness', and his re-definition of 'phenomenon' to facilitate his analysis of quantum ideas. Again Bell has followed him, claiming that the word 'measurement' has been so abused that it should be banned altogether, and be replaced by the more neutral term 'experiment'.

Heisenberg described his own views on language in quantum theory in a letter to Stapp [72]. He wrote that 'it may be a point in the Copenhagen interpretation that its language has a certain degree of vagueness, and I doubt whether it can become clearer by trying to avoid the vagueness'. I am bound to say that I don't think this would represent Bohr's views at all. For Bohr, one had to learn afresh how to use language so that it might convey precise ideas. I don't think Bohr would ever have favoured vagueness, although his own quest for ultimate precision may have rendered parts of his writings somewhat obscure.

Complementarity – the problems

In this section I shall describe one major and one related minor problem, *internal* to the argument of complementarity (thus leaving aside the more general *external* issue – whether complementarity provides a satisfactory account of the quantum world – which is a question for the whole book).

Earlier in this chapter I stressed the importance to Bohr of the measurement system having at one end a microscopic, atomic region and at the other a macroscopic, classical region. Somewhere between these two regions is the *cut*, though, as I have said, its position is arbitrary. This arrangement may be regarded as a strength of complementarity, for without it complementarity could not exist!

Yet in other ways it is a weaknesss. First it restricts the kinds of situation that may be handled. Stapp [72] has expressed this graphically – 'In the Copenhagen interpretation the notion of an absolute wave-function representing the world itself is unequivocably rejected. Wave-functions . . . are associated with the studies by scientists of finite systems. The devices that prepare and later examine such systems are regarded as parts of the ordinary classical world. Their space–time dispositions are interpreted by the scientist as information about the prepared and examined systems.'

Of course, the experiments that led to the quantum theory were precisely of this type – the quantum systems under investigation were microscopic in nature – individual atoms or molecules. Even when solids were studied, the individual atoms behaved independently to a considerable extent, and hence could be treated as separate microscopic objects. As far as experiments go, this is still largely the case today.

In two important ways, however, physicists seek to go beyond this restriction. First, cosmology has developed enormously since Einstein's early ideas in connection with his general theory of relativity. Over the last 30 years or so in particular, it has become a scientifically respectable, even fashionable [48, 49], field. Scientists now seek to describe the whole Universe throughout its entire history quantum mechanically – to provide a wave-function for the Universe.

However, it is clear that the Copenhagen interpretation, as Stapp describes it, cannot hope to handle the Universe, which is not a restricted system set up by an experimenter. Though it may possibly be described as a vast experiment, it is clearly not one that may be terminated by a scientist on Earth in order to obtain a specific outcome. Most astrophysicists, like Stephen Hawking, prefer a different interpretation – the many-worlds interpretation, to be discussed in Chapter 8.

Down on Earth, it was always possible in principle that the chasm between microscopic and macroscopic might be bridged, or, more subtly, that the connection between microscopic and quantum on the one hand, macroscopic and classical on the other, could be broken. It must be conceivable that a macroscopic object with quantum properties – a large object that behaves as a single quantum system – could be created. Stapp has thought in terms of extra-large molecules, and Tony Leggett [77] has spent a great deal of effort studying superconducting devices that may behave in this fashion, as I shall discuss in Chapter 8.

Yet I feel that even these considerations might be felt by Bell to be interesting enough, but rather missing the central point, which is that quantum theory, as it stands today, is fatally flawed; one does not need to wait for future developments.

I should be fair and admit that Bell is prepared to back Bohr up on one central part of the argument – the need for the apparatus to be treated classically. He told Bernstein [78] that 'I disagree with a lot of what Bohr said. But I think he said some very important things that are absolutely right and essential. One of the vital things that he always insisted on is that the apparatus is classical . . . For him it was inconceivable that you could extend the quantum formalism to include the apparatus.'

Bell [76] had also a professional respect for what he called the *pragmatic approach* to quantum theory, which I think might justifiably be described as Bohr's general ethos stripped of what Bell would call the contradictoriness of complementarity. In the pragmatic approach, the ambiguity in position of the Heisenberg cut is dealt with by Bell's prescription that one should 'put sufficiently much into the quantum system [on the quantum side of the cut] that the inclusion of more would not significantly alter practical predictions'. Once one reaches obviously macroscopic objects it would not be *wrong* to move them to the quantum side of the cut, but it would cause no significant change, and would probably be inconvenient when the analysis is carried out. 'The pragmatic approach', Bell says, '. . . has undoubtedly played an indispensable role in the evolution of contemporary theory.'

He sought, if not to justify, then at least to rationalise, the use of the pragmatic approach as follows – 'As we probe the world in regions remote from everyday experience, for example the very big or the very small, we have no right to expect that familiar notions will work. We have no right to insist on concepts like space, time, causality, or even perhaps unambiguity. We have no right whatever to a clear picture of what goes on at the atomic level. We are very lucky that we can find rules of calculation, those of wave mechanics, which work.'

He continued that 'It is true that in principle there is some ambiguity in the application of these rules, in deciding just how the world is to be divided into "quantum system" and the "classical" remainder. But this matters not at all in practice. When in doubt, enlarge the quantum system. Then it is found that the division can be so made that moving it further makes very little difference to practical predictions. Indeed good taste and discretion, born of experience, allow us largely to forget, in most calculations, the instruments of observation. We can usefully concentrate on a quite minute "quantum system", and yet come up with predictions meaningful to experimenters who must use macroscopic instruments.'

Bell would say that the pragmatic approach works FAPP (as he called it) – *for all practical purposes*. The pragmatic philosophy, he suggested, is 'consciously or unconsciously the working philosophy of all who work with quantum theory in a practical way . . . when so working'. Bell would certainly have included himself in this category. The pragmatic approach, in fact, was the first of his three unromantic professional world views that he held in esteem – for the good professional reason that it had achieved 'great success and immense continuing fruitfulness'. He even considered it possible that 'in the course of time one may find that because of technical progress the "problem of the Interpretation of Quantum Mechanics" has been encircled. And the solution, invisible from the front, may be seen from the back.'

However, he went on to say that '[f]or the present, the problem is there and some of us will not be able to resist paying attention to it.' And again – '[those who work in the quantum theory] differ only in the degree of concern or complacency with which we view . . . out of working hours, so to speak . . . the intrinsic ambiguity in principle of the theory'. Or, more trenchantly – 'To ask whether such a recipe [the pragmatic approach] is also a satisfactory formulation of *fundamental* physical theory, is to leave the common ground.'

For Bell, a *fundamental* physical theory, as distinct from a good working pragmatic one, must leave no room for 'taste and discretion, born of experience'. It should be precise and specific. The fact that, once the Heisenberg cut is positioned sensibly, further movement makes no *significant* change should be an admission of failure, not a claim of success. If there is to be a cut, its position should be given exactly from first principles.

'But', he says, 'suppose we are sceptical about the possibility of such a division being sharp, and above all about the possibility of such a division being shifty. Surely the big and small should merge smoothly with one another? And surely in fundamental physical theory this merging should be described not just by vague words but by precise mathematics?' We shall pick up these words again in Chapters 7 and 8.

To Bernstein [78], Bell put it as follows – 'It is very strange in Bohr that, as far as I can see, you don't find any discussion of where the division between his classical apparatus and the quantum system occurs. Mostly you find that there are parables about things like a walking stick – if you hold it closely it is a part of you, and if you hold it loosely it is part of the outside world.'

Bell was in any case very critical of the special role that Bohr (and later von Neumann) assigned to the experimenter – to the setting up of experimental situations and, most of all, to the observing of measurement. In fundamental physical theory, he suggested, 'the concepts of "measurement" or "observation" should not appear at a fundamental level. The theory should of course allow for particular physical set-ups, not very well-defined as a class, having a special relationship to certain not very well-defined sub systems – experimenters. But these concepts appear to me to be too vague to appear at the base of a potentially exact theory.'

These are Bell's criticisms of what I earlier called the major problem of complementarity – the ambiguity in position of the Heisenberg cut, and indeed its very existence. The related minor problem is that, as Bohr's famous discussions with Einstein made clear, measuring instruments were subject to the uncertainty principle, and much of Bohr's analysis of Einstein's thought-experiments depended essentially on that factor. As Jammer [67] comments, 'the double nature of the macroscopic apparatus (on the one hand a classical object and on the other hand obeying quantum mechanical laws) . . . . remained a somewhat questionable or at least obscure feature in Bohr's conception of quantum mechanical measurement'.

Some comments on Bohr and complementarity

I have already reported some views on complementarity, principally those of John Bell (which were taken, incidentally, from papers published in the

1980s). In this section, I shall discuss a very few further comments – selected, of course, from a vast number of books and papers that discuss the topic.

First I return to the book by Schiff [54]. At the beginning of this chapter I mentioned the discussion of complementarity in this book, one of the best written in the period when Bohr's views were wholly pre-eminent, between perhaps 1930 and 1960.

Schiff uses the heading 'Complementarity Principle', a form of words that, without wishing to be carping, I would suggest Bohr never used, and would have been reluctant to use, since it might suggest that complementarity is a physical principle that may be described in a few pithy words. Schiff continues, excellently, that 'This principle states that atomic phenomena cannot be described by classical dynamics; some of the elements that complement each other to make up a complete classical description are actually mutually exclusive, and these complementary elements are all necessary for the description of various aspects of the phenomena.' This is a very good brief account of complementarity – mutual exclusiveness *and* joint completion.

Unfortunately, though, Schiff goes on to relate this difficulty specifically to the physical apparatus and hence, inexorably, to the disturbance interpretation as discussed and criticised above. He then finishes off with the well-considered statement that complementarity 'typifies the fundamental limitations on the classical concept that the behaviour of atomic systems can be described independently of the means by which they are observed'.

This combination of satisfactory and less satisfactory ideas is typical of even the very best accounts of Bohr's work in general texts. While I would by no means claim that Bohr should be immune to criticism, I would suggest that the result of such confusion is that much criticism is based on misunderstanding.

An example of such misunderstanding is the much-quoted text 'There is no quantum world. There is only an abstract physical description. It is wrong to think that the task of physics is to find out how nature is. Physics concerns what we can say about nature.' Many books describe this as a statement of Bohr's views. A recent text [71] calls upon the reader to compare it with a quotation from the well-known positivist A. J. Ayer, and indeed the first three sentences in particular seem very clearly positivistic. The catch is that Bohr did not write this! It was, in fact [67], written by Aage Petersen, Bohr's assistant, in his own attempt to sum up Bohr's philosophy. Many books ignore or obscure this point. In [71], for example, a reference to Petersen's article is indeed given, but the quotation is described as a 'typical Bohr statement'. If this were indeed the case, one wonders why the author of the text did not choose a statement along the same lines from Bohr's voluminous writings. I doubt, in fact, that he could find one that would say *quite* what he would like it to!

I now move back in time to an interesting debate on the Copenhagen interpretation that took place at a crucial time – round about 1960. It was between two highly distinguished philosophers of science, Norwood Russell Hanson and Paul Feyerabend. Hanson wrote perceptively on the philosophy of science [79], but died tragically early piloting his own plane in 1967. Until his death in 1994, Feyerabend was very well known as a guru of California and author of a number of polemical works on philosophy, science and everything else [80, 81]. In the late 1950s, things seemed somewhat different; he was lecturing at Bristol University and writing extremely thoughtful, but unprovocative, articles on quantum philosophy.

Hanson's paper to be discussed here [82], published in 1959, strongly supported Bohr and Copenhagen. He did not regard complementarity as a 'philosophical afterthought', but rather claimed that it had been central to Dirac's great 1928 paper in which he provided a relativistic theory of the electron, incidentally explaining, from a fundamental point of view, the concept of spin, and predicting anti-particles. Hanson's source for this claim was, he reported, Dirac himself. (The two were together at St John's College, Cambridge in the 1950s.) Dirac is said to have asserted that the Copenhagen interpretation 'figured essentially' as 'basic to every operation with the notation'.

I find this claim surprising. For a start, Dirac's biographer [45] has reported Dirac as saying that he 'didn't altogether like [complementarity]' because 'it doesn't provide you with any equations you didn't have before'. (Dirac's own views will be discussed very briefly towards the end of this chapter.) Certainly Dirac must have always retained at the centre of his thinking the kind of relationship between, say, momentum and position which was enshrined in the uncertainty principle, and which lay at the heart of the transformation mathematics invented by Dirac himself and Jordan (mentioned in Chapter 4). I'm not absolutely sure, though, that even these ideas would have played an important role in Dirac's relativistic theory [54, 83]. To suggest that the much more subtle and conceptual arguments of complementarity itself were involved seems far less likely still. I suspect that Dirac, who was much more interested in mathematics than in philosophy, was using the term 'Copenhagen interpretation' rather more loosely than Hanson assumed.

Anyway, Hanson was convinced that all the successes of quantum theory accrued to the account of the Copenhagen interpretation, which has proved itself, he said, 'to work in theory and practice'. If you doubt this, you are instructed to 'ask your nearest synchrotron operator'. Hanson suggested that 'It is arguable that the Copenhagen limitations, far from being the result of philosophical naïveté, are built into the very wave–particle duality micronature has forced upon us, and built also into the symmetry of explanations in terms of that duality. At least it remains clearly to be shown that it is not so . . . Perhaps a Copenhagen-type interpretation is unavoidable if things like wave–particle duality and Lorentz invariance are genuine features of nature.' Thus we are not genuinely 'free to invent and to consider other metaphysical interpretations' because such a possibility 'obscures the historical, conceptual and operationally successful role of the Bohr view . . . *as opposed to other interpretations*'.

In attempting to establish themselves, then, new quantum interpretations must obviously fight with one hand, or even two, behind the back. While Hanson did not actually claim 'that philosophers ought to discontinue all attempts to develop proposals which counter the Copenhagen interpretation,' he did believe that 'until you formulate a new interpretation which works in every particular as well as does the old one, [you should] call your efforts by their proper name, "speculations".' Admittedly, he added, 'This makes them no less worthwhile.'

Hanson was scathing towards 'yesterday's great men' (among whom he included Einstein, Schrödinger, von Laue and de Broglie). Their 'amorphous sighs', he commented 'have not offered one scrap of algebra to back up their grandfatherly advice'. He was even more critical of the proponents of hidden variables, in particular Bohm, whose work is discussed in Chapter 7. Bohm, Hanson said, admitted that 'his interpretation affected no known facts, but only added extra philosophical notions of heuristic value'. Presumably to establish his ability to miss the point in two languages, Hanson finished off by remarking that 'Bohr et Heisenberg n'ont pas besoin de cette hypothèse' (echoing Laplace's reply to Napoleon when he was asked where God appeared in his work).

The judgement of missing the point may doubtless be made more easily today than 45 years ago. Today at least it is clear that the successes of the mathematical content of quantum theory cannot be invoked to support the Copenhagen interpretation, that rivals to Copenhagen must be allowed, and that to deny them any respectability until they catch up with Copenhagen would be to strangle them at, or soon after, birth.

Feyerabend's contribution to the debate was a published version of a lecture given in 1960–61 [64]. Feyerabend was prepared to be a lot more open-minded and forward-looking than Hanson. This does not mean that he merely attacked Bohr. Indeed, he was by no means an opponent of Copenhagen, being 'prepared to defend [it] as a physical hypothesis' and 'to admit that it is superior to a host of alternative interpretations'. Much of his paper was devoted to demonstrating this superiority; Feyerabend claimed to have 'tried to exhibit the advantages of Bohr's point of view', and to have 'defended this point of view against irrelevant attacks'. All this makes interesting and instructive reading.

But Feyerabend was far from swallowing the Copenhagen position whole. He ruled out any suggestion that the framework of complementarity was established permanently and not open to challenge. Rosenfeld [52] had argued that 'any new conception which we need will be obtained by a rational extension of the quantum theory', and that 'new theories of the microcosm will be increasingly indeterministic'. Feyerabend strongly rejected this as a 'dogmatic *philosophical* attitude with respect to fundamentals', which was 'fairly widespread', 'completely unfounded' and 'a very unfortunate feature of contemporary science'. 'Now as ever,' he proclaimed, 'we are free to consider whatever theories we want when attempting to explain the structure of matter.'

Indeed, Feyerabend was prepared to be far more tolerant than Hanson towards speculations that might take a considerable period to mature before they were able to challenge accepted views across the board. In particular, he expressed interest in what he called the 'unpopular and fairly general speculations' of Bohm. Overall, while Hanson's views were looking backward to the days when Bohr ruled supreme, Feyerabend's were healthily looking forward to the open-minded challenge and debate more prevalent today, though Feyerabend may have underestimated the pace of change.

Lastly in this section I want to jump to some 1987 analysis of Murdoch [56], who attempts to do what many would have said was impossible – to axiomatise, or formalise, Bohr's views on measurement. Murdoch starts from the argument that, in quantum physics as in classical, a measurement should correlate the *initial* value of the quantity being measured with the *final* state of the measuring apparatus. He says that he derives this from Bohr's view that 'the observation problem of quantum physics in no way differs from the classical physics approach' [84] and that terminology such as 'creation of physical attributes to atomic objects by measurements' is 'apt to cause confusion' because the words 'attributes' and 'measurements' are used in a way incompatible with common language and practical definition.

It is at least possible to question Murdoch's use of Bohr's remarks. Bohr invented complementarity to *restrict* analysis of measurement to circumstances where a classical prescription could be used. Thus discussion of the value of a quantity *before* it is measured might seem illegitimate – just because it has *not* been measured. It thus seems that Murdoch may be moving beyond the confines of complementarity, and yet still claiming that the classical prescriptions must apply.

Nevertheless, let us take his idea and follow the consequences through. He seems to face a difficulty in that, according to most supporters of Copenhagen, prior to measurement the quantity being measured will not usually possess a unique value. It may appear that Murdoch is assuming what he calls an *intrinsic-values* theory of properties, according to which all observable quantities have definite values at all times. Because the wave-function gives no information about most of these values, and so cannot contain *all* information available about the system it describes, the intrinsic-values theory must be a hidden-variables theory, or, in fact, a Gibbs ensemble theory, as mentioned earlier in this chapter.

However, Murdoch rejects such a theory, and says that Bohr did the same. Rather, Murdoch holds what he calls the *objective-value theory* of measurement, and again he claims that Bohr did too. According to this theory, an observable gains a value when an apparatus appropriate for measuring it is set up. The actual measurement then reveals this pre-existing value. But, in the absence of such an apparatus, no value exists. So, according to Murdoch, 'Bohr can consistently say that successful measurement of an observable reveals its pre-existing value, and also that talk of a pre-existing value is meaningful only in a context in which the measurement has been made. He is, of course, required to deny that we can meaningfully say that the pre-existing value that is revealed by measurement would have been present if no measurement had been made.' Murdoch is able to show that many of the difficulties with Gibbs ensembles which will be discussed in Chapter 6 do not apply to the objective-values theory.

The choice between intrinsic-values theory and objective-values theory is intimately linked to another choice – between what Murdoch calls the *strong meaning condition*, according to which, in order to say that an observable has a given value, a measurement of that observable is required, and the *weak meaning condition*, according to which it is sufficient that a measuring apparatus appropriate for measuring the observable is in place.

Clearly, for consistency the objective-values theory should go along with the weak meaning condition. However, Murdoch is forced to admit that Bohr adopted the strong rather than the weak meaning condition, but regards it as a mistake on Bohr's part.

Overall Murdoch's formalisation of Bohr's views is certainly bold and ingenious. It enables one of the central planks of complementarity (in Murdoch's version, the weak meaning condition) to be coupled to an approach, objective values, which is less vague than that usually attributed to Bohr, but less restrictive than the intrinsic-values theory.

Yet sceptics may ask whether it is not rather fanciful that mere setting-up of an apparatus can 'create' a value for an observable. Does it really do less damage to ordinary language than the idea that the *measurement itself* creates the values, a form of words criticised by both Bohr and Murdoch? Murdoch might reply that the point is not simply whether the observable has a value under these circumstances. It is the *semantic* one – may we avoid ambiguity in referring to the concept of the value of the observable? The sceptic in turn might reply – is it not Murdoch himself who has moved beyond the semantic by discussing what *actually* happens when a particular apparatus is set up?

Overall it may be suggested that the axiomatisation raises as many questions as it answers. The question with which I shall conclude this section is as follows – is this because the axiomatisation is too crude to represent Bohr's subtle and elusive ideas? Or is it that it exposes the lack of real substance in Bohr's position, which may usually be disguised by obscuring rhetoric?

Complementarity in other fields

Bohr was especially keen that complementarity should be regarded not merely as a framework for solving problems of quantum physics, but as providing a lesson for the whole of human knowledge. In the introduction to his 1958 collection of essays [68], he wrote that, while the articles presented the essential aspects of the situation in quantum physics, 'at the same time, [they] stress the points of similarity it exhibits to our position in other fields of knowledge beyond the scope of the mechanical conception of nature'. He stressed that 'We are not dealing here with more or less vague analogies, but with an investigation of the conditions for the proper use of our conceptual means of expression. Such considerations not only aim at making us familiar with the novel situation in physical science, but might on account of the comparatively simple character of atomic problems be helpful in clarifying the conditions for objective description in wider fields.'

Bohr regarded complementarity as providing an 'epistemological lesson' for the whole field of human learning, and beyond physics his own discussions were principally in biology, psychology and social anthropology. In this book it is relevant to treat them only briefly; much more discussion is to be found in books concerned with Bohr's general thought [55, 58].

Bohr was particularly concerned with fundamental biological questions, perhaps because his father, Christian Bohr, professor of physiology in the University of Copenhagen, was himself intensely involved in the debate between mechanical and vitalistic teleological modes of description in biology, between the view that biology must be reduced to physics and chemistry, and the alternative view that ideas such as life and purpose could and should be retained and used. Christian Bohr rejected strict mechanism, but also rejected the idea of an external non-physical entity, a 'vital quality', to be superimposed on a mechanical foundation. Rather, *teleological* accounts of biological processes (accounts discussing 'purpose'), should be regarded as fundamental to the particular discipline, and not contradictory to purely mechanistic descriptions.

In a sense, Niels Bohr may be said to have used complementarity to buttress his father's views. Like his father, he rejected crude vitalism, 'the assumption that a peculiar vital force, unknown to physics, governs all organic life'. Bohr regarded 'life' and 'detailed mechanical analysis' as complementary in the sense that each precluded the other. His reasons for this view, though, developed only gradually.

At its simplest, the point was that detailed mechanical analysis would put an end to life – '[W]e should doubtless kill an animal if we tried to carry the investigation of its organs so far that we could tell the part played by the single atoms in vital functions'. This argument is perhaps not very convincing; it is one of practicality rather than fundamental principle. It could even be argued that medical uses of magnetic resonance that have been developed in recent years *do* investigate the role of individual atoms, without, we hope, damaging the patient.

A more mature view recognises that a living being cannot exist in isolation from its surroundings – breathing and feeding require a constant exchange of atoms and energy between living system and environment. And, to obtain a closer analogy still with quantum processes, the boundary between system and environment must be arbitrary; it is not possible to state categorically whether a given oxygen molecule has, at a given moment, ceased to be part of the environment and become part of the living organism. Yet physical investigation, at its most detailed and rigorous, must be carried out on a uniquely defined system.

Bohr argued then, that – as a matter of principle, not practicality – the presence of life and obtaining a detailed mechanical account of the being are complementary. Mechanical and teleological descriptions are each respectable, but mutually exclusive. The fact of life may be regarded as analogous to the quantum postulate – '[T]he very existence of life must in biology be considered as an elementary fact, just as in atomic physics the existence of the quantum of action [h] has to be taken as a basic fact that cannot be derived from ordinary mechanical physics'.

I now turn to psychology, where again Bohr hoped to shed light on an important debate, that between *free will* and *determinism*, by use of complementarity. Like the debates between wave and particle, and between mechanism and teleology, he held that it was *not* a struggle about the meaning of reality, but a call for clarity in the use of concepts.

From his youth, Bohr had thought deeply about the *subject/object distinction*. A subject may express (free) willingness to perform an action. Equally though the subject may wish to examine her own willingness to perform the action; in this case, though, the subject that expresses willingness becomes part of the object under consideration. As part of an object of study, it may now be appropriate to discuss the willingness to perform the action as derived *deterministically* from previous psychological states.

With the development of complementarity, Bohr was able to describe the two approaches as complementary – so the application of free will and determinism become complementary approaches to psychological problems. Bohr [66] pointed out that 'no sharp separation between object and subject can be maintained, since the perceiving subject also belongs to our mental content', and he suggested that the 'unity of our consciousness' may be analogous to the quantum postulate in atomic physics. As Bohr [68] said, the words 'thoughts' and 'feelings' may be regarded as complementary, and Folse [55] suggests that rational and empirical psychology may similarly be considered as complementary approaches to the same discipline.

There is one more similarity between the biological and psychological applications of complementarity. Just as Bohr opposed the idea of an external 'vitalism' in the first area, so he opposed the idea of a non-physical 'mind' interacting with or directing the physiological being. As in quantum physics and biology, so in psychology, complementarity represents conceptual analysis, *not* dualism.

I now turn to social anthropology, and in particular to an address given by Bohr titled 'Natural philosophy and human cultures' [68] in which only a few hints are given about his views. The complementarity in psychology between thoughts and feelings may be extended to that between *reason* and *instinct*, which may clarify relations between humans and animals, and eventually between different cultures. Humans thrive on reason, on conceptual analysis, and, in so doing, may cut themselves off from the ability of animals to 'utiliz[e] the possibilities of nature for the maintenance and propagation of life'.

When comparing different cultures, civilised man's many abilities based on conceptual reasoning must be balanced against 'the amazing capacity of so-called primitive people to orientate themselves in forests or deserts'. In this address given in the strategic year of 1939, Bohr is keen to point out that 'The main obstacle to an unprejudiced attitude towards the relation between various human cultures is . . . the deep-rooted differences of the backgrounds on which the cultural harmony in different human societies is based and which exclude any simple comparison between such cultures.'

Examination of such different societies, Bohr suggested, shows not superiority and inferiority, but 'that different human cultures are complementary to each other . . . Indeed,' he goes on to suggest, 'each . . . culture represents a harmonious balance of traditional conventions by means of which latent potentialities of human life can unfold themselves in a way which reveals to us new aspects of its unlimited richness and variety.'

Despite the considerable interest of Bohr's ideas on biology, psychology and anthropology, one may legitimately doubt whether they carry total conviction. Certainly they have not generated any real interest in complementarity as a framework to be used across the whole field of human knowledge, as Bohr had hoped. Bell capitalised on this point in his popular lectures. He would mention that Bohr had expected complementarity eventually to be taught in every school, and then ask his audience whether any of them had, in fact, studied it at that level; invariably no hands went up, which, I imagine, came as no surprise to him!

So I was interested in a fairly recent development suggesting that the ideas may have penetrated further than one had suspected – into the field of literary studies. In particular they may suggest a reason why the English poet Philip Larkin preferred not to take part in poetry-readings.

Readers may know that Larkin had a stammer, was deaf, did not much like general company, and disliked the idea of poetry as mass light entertainment, and all this may seem to provide quite enough reason for him to prefer poetry to be read in solitude. Complementarity, though, may provide another, more technical, reason.

The argument is that of Christopher Ricks [85], and concerns in particular Larkin's 1956 poem 'An Arundel tomb'. During a visit to Chichester Cathedral [86], Larkin was moved by a pre-baroque statue of the Earl of Arundel and his wife; he was particularly affected by the couple being carved holding hands. (Only later did Larkin discover that this was actually not an original feature, but a result of re-working in the 1840s following damage in the Reformation and the Civil War.)

Larkin's poem ended as follows:

> Time had transfigured them into
> Untruth. The stone fidelity
> They hardly meant has come to be
> Their final blazon, and to prove
> Our almost-instinct almost true:
> What will survive of us is love.

Ricks was especially impressed by the last line of the poem, the power of which, he suggested, came from the fact that two distinct ideas may be discerned. The classical impersonal view would stress the 'survive'; it makes a general statement about dying and living. The romantic view would stress the 'us', and the poem now becomes a beautiful statement about ourselves. Ricks writes that 'What Larkin achieves is an extraordinary *complementarity* [my italics]; a classical pronouncement is protected against a carven coldness by the ghostly presence of an arching counterthrust, a romantic swell of feeling; and the romantic swell is protected against a melting self-solicitude by the bracing counterthrust of a classical impersonality.'

It seems indeed a perfect example of complementarity – mutual exclusiveness, but also joint completion to give a far more resonant image than either reading individually. And of course the effect would be diluted if not destroyed by the poem being read aloud, when one reading or the other must be selected.

Von Neumann and the Princeton interpretation

It is ironic that, though most physicists traditionally give their allegiance in quantum matters to Bohr, if they ever say what they mean by the Copenhagen interpretation, they often give an alternative set of ideas due almost entirely to John von Neumann [87], a set of ideas often christened the *orthodox interpretation* of quantum theory. (The word 'orthodox' is often applied to Bohr as well, but less often, I think, to his ideas.)

Though ironic, it is perhaps not too surprising. While Bohr's ideas are subtle, and some might say obscure, those of von Neumann are much clearer, and indeed expressed mathematically and axiomatically. Many physicists probably imagine that all von Neumann has done is to put Bohr's ideas into mathematical form. I would agree, though, with Feyerabend [64] that 'when dealing with von Neumann's investigations, we are not dealing with a refinement of Bohr – we are dealing with a completely different approach'.

I call this approach the *Princeton interpretation* in virtue of the fact that von Neumann, and also Eugene Wigner and John Wheeler, other important contributors, spent most of their careers at the Institute for Advanced Study there. (This term, incidentally, is not particularly common.) Another contributor to the approach, as we shall see, has been Dirac, who actually visited Princeton for long periods, but was based, of course, at Cambridge, England.

Occasionally confusion between Copenhagen and Princeton interpretations has abounded. In an important paper of 1970, to be discussed in later chapters, Leslie Ballentine [88] distinguished between what he called Copenhagen and Princeton *schools*, which he took to be comparatively recent formations centred on Rosenfeld and Wigner (both Bohr and von Neumann being dead). His own account of the Copenhagen approach was largely that of von Neumann, and Ballentine criticised this severely.

Shortly afterwards, Stapp [72] wrote his article on the Copenhagen interpretation, stating categorically that the approach of von Neumann was, in most ways, directly opposed to that of Bohr. Ballentine [89] replied that, 'in attempting to save "the Copenhagen interpretation",' Stapp had 'radically revised what was, rightly or wrongly, understood by that term'. He called Stapp's views the 'neo-Copenhagen interpretation', but agreed that his own views and those of Stapp, and hence presumably those of Bohr, were much closer to each other than any of them were to those of von Neumann.

Von Neumann [90], who died at the age of 53 in 1957, was Hungarian and emigrated to the USA in 1930; he was indisputably one of the most brilliant thinkers of this century. He was fundamentally a pure mathematician, but also made very great contributions to computing, where he and Alan Turing share the credit for the stored-memory programmable computer, and to economics, where, as well as producing important models of economic growth, he invented the theory of games – how to behave when one's course of action depends on that of one's competitor and vice versa, with its concepts such as 'zero sum' and 'the prisoner's dilemma'.

In physics, he produced a very important proof in the foundations of statistical mechanics. His important work on the quantum theory was mostly in his famous 1932 book [87], which was published in German, and, frustratingly for John Bell as we shall see, was not translated into English until 1955.

He was dissatisfied with much of the mathematical apparatus used by Dirac, which was highly effective but not very rigorous. Von Neumann's book provided a totally different mathematical foundation for quantum theory (based, I mention only for those who have heard of it before, on the use of Hilbert space). This was highly successful, and, over the 70 years or so since then, von Neumann's has become the standard approach for anyone working on quantum theory and wishing to take the mathematics fairly seriously.

All this praise for von Neumann is a preamble to the fact that, in the same book, he also investigated what he called 'the difficult problems of interpretation, many of which are not even now fully resolved'. And in this area it is fair to say that some of his ideas were rather less satisfactory.

Wave-function collapse

For Bohr, discussion of quantum measurement began with intense and rigorous conceptual analysis. Von Neumann's approach, on the other hand, was formal and mathematical. It may be said that it responded internally to the conceptual problems raised by the approach itself.

A major difference with Bohr was that von Neumann was prepared to treat the measuring device quantum-mechanically, and assign it a wave-function. This avoids preliminary conceptual struggles, but only at the expense of releasing into the world of quantum physicists something that may have looked like a rose, but was, in fact, a thorn, and it's been pricking them ever since. This is the *collapse of wave-function* or, more formally, the *projection postulate*.

To be fair, von Neumann did not invent the idea of the collapse of wave-function. In 1930, Dirac had published a classic text, *The Principles of Quantum Mechanics* [83], in which he used the postulate, as he had indeed done three years earlier at the 1927 Solvay Congress. Even Dirac was almost certainly not the *inventor* of the postulate; many physicists probably developed similar ideas at almost the same time.

Let us, in any case, follow Dirac as he analyses a measurement of energy, say. Standard quantum theory from Chapter 4 will take us so far. We consider a situation where the wave-function of the system just before the measurement is ψ. We take the eigenfunctions of energy as $\phi_1, \phi_2, \ldots$, with corresponding eigenvalues (values of energy) $E_1, E_2, \ldots$

The simplest case is where the wave-function of the system, ψ, is equal to one of the eigenfunctions, say ϕ_2. Then a measurement of energy is *certain* to give the corresponding energy, E_2. The next most simple case is where ψ is the sum of two of the eigenfunctions, say $c(\phi_1 + \phi_2)$, where c is a constant, which must, in fact, be equal to $1/\sqrt{2}$. Then, as stressed in Chapter 4, we do not know which value a measurement of energy will yield. The *probability* of obtaining E_1, the value of energy corresponding to ϕ_1, is $c.^2$, or just $\frac{1}{2}$; so is the *probability* of obtaining E_2, the value corresponding to ϕ_2. (So the probabilities add to 1, as they should.)

But now let us ask what actually happens at the measurement. Standard quantum theory can give no answer. Dirac answered as follows, particularly thinking about what would happen if the measurement were *immediately* repeated. Now if the measurement were one of *position*, it seems clear from relativistic arguments that the second measurement must give the same result as the first; otherwise the particle would have been detected to travel a non-zero distance in zero time. For measurements of other quantities, such as energy, the position is not quite so clear, but Dirac argued, from what he called *physical continuity*, that the same thing should happen; the *same* value should be obtained in an initial measurement *and* in an immediate repetition.

In order that this should happen, he used the projection postulate – at a measurement, the wave-function collapses to that eigenfunction of the quantity measured which corresponds to the value that is actually obtained in the measurement. In the case considered above, if the value E_1 is obtained, then, at the moment of the measurement, the wave-function collapses *from* $c(\phi_1 + \phi_2)$, to ϕ_1. An immediate repetition of the measurement clearly gives E_1 again, as Dirac required. If, though, the initial measurement gives the value E_2, the collapse is to ϕ_2, and an immediate repetition again gives E_2.

For von Neumann as well, the collapse of wave-function seemed little more that a statement of what a measurement is. Yet he admitted that it caused great conceptual problems. In his words, there appears to be 'a peculiar dual role of the quantum mechanical procedure'. In the absence of any measurement, the wave-function develops according to the Schrödinger equation (a process of type 2, as von Neumann called it); at a measurement, it follows the projection postulate (a process of type 1).

There are very considerable mathematical differences between processes of type 1 and type 2. For the purposes of this book, it is easier to describe the differences physically, thermodynamically in fact. Von Neumann himself showed that a process of type 2 is *thermodynamically reversible*; it remains possible at the end of the process to restore the system to its initial state. For a process of type 1, no such possibility exists, and the process is called *thermodynamically irreversible*.

Despite this apparent dichotomy, there have been innumerable attempts to demonstrate that a process of type 1 could be approximately, or in some limit, equivalent to one of type 2. In Chapter 8, we shall meet analysis that admits the difference, but aims to show that the *results* of processes of the two types may be effectively equivalent. None of these attempts are, to my mind, particularly successful.

Indeed, the whole area has become known as the *measurement problem* of quantum theory. It is especially embarrassing because, as Bell in particular stressed, one should really be able to describe the 'measurement' in terms of straightforward quantum-mechanical processes of type 2 of the atoms of the measuring device. So how can the measurement itself be of type 1?

Incidentally, Kragh [45] describes a 1927 discussion between Dirac and Heisenberg about what actually gives rise to the collapse. Dirac said that it is *nature* that makes the choice of measuremental result; once made, the choice is irrevocable, and will affect the entire future state of the world. (Born agreed, and remarked that this was in accordance with the views of von Neumann, whose book had not yet been published.) Heisenberg, however, maintained that behind the collapse, and the choice of which *branch of the wave-function* would be followed, was 'the free-will of the human observer'. These words don't really add much to the mathematics!

Probably, in any case, the reader is itching to reply. Surely, it may be said, I have been making a mountain out of a molehill, creating difficulties about something that is present, and indeed totally clear, in classical physics (or just everyday common sense). Suppose that I toss a coin, but don't look at the result. *In my mind* both possibilities, heads and tails, must exist, with equal probabilities, each equal to $\frac{1}{2}$. I then look at the coin. (I 'perform an observation' or 'make a measurement' if I wish to be erudite.) Of course I see one result or the other, and, in my rather perverse language, it may be said that the wave-function (or the appropriate analogy) collapses (in my mind), so that an immediately repeated measurement (which is just to say that I look at the coin again) will give the same result. Surely, it may be argued, what is happening in quantum theory is just as obvious as this.

Such an argument may itself be given a dignified name – a *knowledge interpretation* of quantum theory, or, even more dignified, an *epistemic interpreation*. It says that the wave-function represents not physical reality, but our knowledge of physical reality, which naturally changes abruptly when we stir ourselves to observe something. I shall discuss this type of interpretation briefly in Chapter 8. I note here, though, that this kind of argument is particularly convincing in the classical case, where *all* physical quantities are assumed to have values at all times, and the observation of measurement has *only* to check up on the value this particular quantity possesses at the appropriate time.

Conventional ideas on quantum theory are not like this – they assume that often a quantity being measured has no unique value prior to the measurement, so the 'observation' cannot *just* observe. The reader – with the bit between the teeth – may wish to reply that surely all that is just so much the worse for these conventional ideas, and perhaps is a little suspicious about what I've said concerning them anyway! Fair enough – and by all means, the wave-function may be supplemented with hidden variables, which may be able to provide values for these other quantities. But – hidden variables have their own problems, as will become particularly clear in Chapter 7. So, for all its own difficulties, the orthodox interpretation of von Neumann must be taken very seriously.

In this context I should point out that what I have given you so far is not much of a measurement – there is no apparatus! In a generally very positive review of Dirac's book, Pauli pointed out [45] that there was 'a certain danger that the theory will escape from reality'. Like Bohr, Pauli wanted good solid classical measuring devices, not just mathematical formulae.

Von Neumann was well able to cope with the measuring apparatus in his mathematical scheme. As I said earlier, he was prepared to write a wave-function for the apparatus; in fact, corresponding to eigenfunctions for the measured property $\phi_1, \phi_2, \ldots$, he had apparatus eigenfunctions $\alpha_1, \alpha_2, \ldots$.

So, if a measured particle has wave-function ψ equal to, say, ϕ_1, before the measurement, in the measurement process a combined wave-function of measured and measuring systems is formed, which is equal to $\phi_1\alpha_1$. The fact that the wave-function of the apparatus is α_1 tells the experimenter that the result of the measurement is the corresponding energy, E_1. A perfect measurement has been performed. (Other measurements may be less than perfect, but we do not need to bother with these. Also note that we don't need to bother with the *initial* wave-function of the apparatus; it's only the *final* wave-function that matters.)

In the more complicated case where the initial wave-function of the observed particle is $c(\phi_1 + \phi_2)$, the *first* stage of the measurement procedure will be the formation of a combined wave-function for measured system and apparatus given by $c(\phi_1\alpha_1 + \phi_2\alpha_2)$. The *second* stage is the collapse *either* to $\phi_1\alpha_1$, and a measuremental result of E_1, *or* to $\phi_2\alpha_2$, and a result of E_2. So the apparatus has been included successfully, but of course the collapse is still as big a problem as ever.

Von Neumann, like Bohr, was worried about the division between measured and measuring systems. For him, a typical measurement was of temperature, and he described it as follows. At one end of a chain we have the body whose temperature is to be measured, at the other is the *ego* of the experimenter. In between are the thermometer, some photons which travel from thermometer to retina, the optic-nerve tract and the brain. Because of what he called *psychophysical parallelism*, incidentally, von Neumann did not feel obliged to treat the brain and, say, the thermometer, differently in his analysis.

He acknowledged that there must be a cut between measuring and measured systems, and also that the position of this cut must be arbitrary – *it must not affect the predicted results.* (At this place in his argument, he was following Bohr, as he himself acknowledged.) The success of his work was that he was able to obtain precisely this result from his mathematics. It doesn't matter, for example, on which side of the cut the optic-nerve tract is included – the predictions are the same.

For all this success, though, we are only returned to the same point, namely, that of the projection postulate. *Where* does collapse take place? Von Neumann gave a clear answer to this question; the final stage of the chain must be entirely different from all previous stages. He said that 'it is inherently entirely correct that the measurement or the related process of the subjective perception is a new entity relative to the physical environment and is not reducible to the latter. Indeed, subjective perception leads us into the intellectual inner life of the individual which is extra-observational by its very nature.' Von Neumann adds that 'experience only makes statements of this type: an observer has made a certain (subjective) observation; and never like this: a physical quantity has a certain value'.

So full responsibility for collapse is put onto the observer – *in the capacity of 'abstract ego'*. Note that, in as direct a comparison as may be made, Bohr would allow measurement to be accomplished by any macroscopic object in a measurement chain. For von Neumann such objects could be treated quantum-mechanically – until one reached the human observer.

The importance of observers has been discussed widely, particularly by one of today's greatest theoretical physicists, John Wheeler, whose life and work have been discussed by Bernstein [78]. If we should say that nothing really *happened* till wave-functions could be collapsed and definite physical results achieved, does that mean that the Universe had to wait for observers to evolve before physics could start? (Astrophysicists may be no more happy with von Neumann than with Bohr!) And who could collapse them – prehistoric man? a child? or, since one is talking in terms of experiments, does it have to be a trained physicist, and if so is a PhD required or just a BSc? In the next chapter we shall meet Schrödinger's famous cat – can it collapse its own wave-function?

Such speculation seems to me almost impossible to return to planet Earth, even in principle. In contrast, Bohr's ideas, with the cut between microscopic and macroscopic, are certainly vague at the moment, but may at least be analysed, elaborated and even tested experimentally, as I shall explain in Chapter 8.

Von Neumann's impossibility proof

Von Neumann's book contained another famous argument, which has turned out to cause just as much confusion as the projection postulate. This was his proof claiming to show that hidden variables are impossible in quantum theory. 'Whether or not an explanation . . ., by means of hidden parameters, is possible for quantum mechanics, is a much discussed question . . .', von Neumann reported, and he claimed to show that 'an introduction of hidden parameters is certainly not possible without a basic change in the present theory'. (The last phrase implies that it is possible, though extremely unlikely, that the experimental results at present explained by quantum theory may at some time in the future be explained by some other theory, *but*, as long as we keep quantum theory, von Neumann argued, we cannot have hidden variables to help us make sense of it.)

While von Neumann's arguments in the previous section were somewhat in discordance with Bohr's views, clearly this proof was greatly to the taste of Bohr and his supporters. If correct, it was the triumphant conclusion of their programme; it showed that the key elements of the programme – the loss of determinism, the renunciation of unique values for all quantities, the resultant special features of measurement – were not just matters of whim or even taste, not just one way of handling matters possibly among several others, but the unchallengeable result of the most rigorous mathematics.

The actual proof is rather long – about 20 pages of von Neumann's book. It is too technical to give much account of in this book, though the actual mathematical steps are not, as these things go, too advanced.

What is tricky, though I can say this more easily with hindsight, is the list of properties that von Neumann assumed for his proposed hidden variables, prior to showing that they could not duplicate the known laws of quantum theory. For his proof to be of total generality, no *unnecessary* restrictions should be put on these hidden variables. Had von Neumann got it right? It was to take 30 years before John Bell answered that question definitively – in the negative. (See Chapter 7.)

There were always a few mavericks who questioned the validity of the proof [67], though, it must be said, not always for reasons that, again with hindsight, seem to go to the nub of the difficulties. But most physicists were only too pleased to accept the word of such a famous mathematician – when, in any case, it told them what they wanted to hear. Cynically, I don't even believe that many of them studied the proof in detail and just missed the problems. I suspect they perhaps glanced through it, or more probably didn't even bother to do that! They believed that everything had come together to support a rather unholy Bohr/von Neumann axis, which should not be questioned, and did not really have to be understood – just accepted. They must be left in peace until Chapter 7.

Wigner and consciousness

Another important contributor to the Princeton interpretation has been Eugene Wigner, who, like von Neumann, was born in Hungary. Wigner emigrated to America in 1933, at the age of 31, and made immense contributions to many areas of theoretical physics, being awarded the 1963 Nobel Prize, specifically for applications of symmetry in fundamental physics. He died in 1995 at the age of 92.

Wigner has also been intensely interested in the type of issues discussed in this book. Many of his papers are included in the collection put together by Wheeler and Wojciech Zurek [91], that illuminates the whole area, and includes most of the important papers published through the years.

The contribution I wish to refer to here is really an extension to von Neumann's idea of the 'abstract ego' collapsing the wave-function. While for von Neumann it all remained rather formal and mathematical, Wigner [92] took it rather more seriously. For him, *consciousness* (or the *mind*) played a more directly physical role, adding an extra term to the mathematical equations, and hence selecting one particular branch of the wave-function and one particular result for the experiment. It thus produced the effect that von Neumann called collapse.

Wigner did not give many details to back up his idea, but he did say enough to irritate John Bell, who made the idea the second of the three 'romantic' world views he disliked. 'As regards mind,' Bell [76] wrote, 'I am fully convinced that it has a central place in the ultimate nature of reality. But I am very doubtful that contemporary physics has reached so deeply down that that idea will soon be professionally fruitful. For our generation I think we can more profitably seek Bohr's necessary "classical terms" in ordinary macroscopic objects, rather than in the mind of the observer.'

6

Einstein's negative views

The 1927 Solvay Congress

In Chapter 4, we left Einstein in early 1926, cautiously optimistic that Schrödinger's scheme might provide what Heisenberg's could not, and did not attempt to – a mathematical description of atomic processes that could also give a physical picture of what was actually occurring. But the demonstration that the two formalisms were equivalent severely dented such optimism.

It was still possible for Einstein to prefer Schrödinger's version as offering more hope for future physical interpretation. Thus it is interesting to review the recommendations Einstein made for Nobel Prizes [1]. In 1928, Einstein suggested that Heisenberg and Schrödinger might share a prize, adding 'With respect to achievement, each one . . . deserves a full Nobel prize although their theories in the main coincide in regard to reality content.' However, Einstein was cautious enough to add that 'it still seems problematic how much will ultimately survive of [their] grandiosely conceived theories', and gave precedence to de Broglie, also mentioning Davisson, Germer, Born and Jordan.

By 1931, more convinced that quantum theory would last (and with de Broglie the 1929 prizewinner), Einstein plumped for Heisenberg and Schrödinger, commenting that 'In my opinion, this theory contains without doubt a piece of the ultimate truth. The achievements of both men are independent of each other, and so significant that it would not be appropriate to divide a Nobel prize between them.' Einstein added 'I would give the prize first to Schrödinger' because 'I assess Schrödinger's achievement as the greater one, since I have the impression that the concepts created by him will carry further than those of Heisenberg', though a footnote said 'This, however, is only my own opinion,

Fig. 6.1. Einstein's thought-experiment at the 1927 Solvay Congress. A beam of electrons reaches a narrow slit S, and, following diffraction, the electrons travel in all possible directions towards the film F.

which may be wrong.' (In the event, Heisenberg received a full prize for 1932, and Schrödinger shared the 1933 prize with Dirac.)

Let us return, though, to 1927, with Schrödinger's own interpretation virtually a non-starter, and Einstein highly antagonistic towards the positivistic ideas of Heisenberg. As I explained in the previous chapter, Bohr regarded complementarity as a bridge between the views of Heisenberg and those of Einstein and Schrödinger. Einstein was not present at its first presentation at Como in September 1927, so there was considerable interest when he and Bohr attended the fifth Solvay Congress in Brussels the following month. Bohr's talk was identical to the one he gave at Como, and it is reported [93] that then 'there took place an extensive general discussion on causality, determinism and probability', those taking part including Lorentz, Born, Pauli, Dirac, Kramers, de Broglie, Heisenberg, Schrödinger, Compton and, of course, Einstein.

When Einstein spoke, he first apologised for not having gone deeply into quantum mechanics. 'Nevertheless', he continued, 'I would like to make some general remarks.' He gave a simple physical example, which we may call a *thought-experiment* or *Gedankenexperiment*, an experiment that need not, and very probably could not, be performed in practice, but provides a useful focus for conceptual discussion.

Einstein considered, as shown in Fig. 6.1, a beam of electrons reaching a screen in which there is a narrow slit. A photographic film in the shape of a hemisphere is placed behind the slit. Because the slit is narrow, the electrons will be strongly diffracted and travel towards the film in all possible directions. While the electrons are travelling towards the film, quantum theory must represent them as a spherical wave moving outwards from the slit; when the wave reaches the film, its intensity at a particular point on the film is a measure of how much exposure will be found at that point. However, once the electrons actually reach the film, each acts individually to produce a highly localised area of exposure.

Einstein considered what he called 'two standpoints regarding [the] domain of validity [of the quantum theory]'. According to what he called interpretation II, which corresponds to Bohr's approach, the theory as so far presented is *complete*; there are no more elements such as hidden variables to be included. One must consider each particle individually, and the value of the wave-function (or, more exactly, the square of its modulus) at a particular point on the film determines the *probability* of that electron reaching that point.

But Einstein is critical of this interpretation because it seems that, whereas, just before the electron reaches the film, it is, in a sense, potentially present over the whole wavefront, at the instant it reaches the screen it acts at one place – not more than one. In Einstein's opinion, this means that the behaviour of the wave, at different points and the same time, is correlated; he considers that this implies 'a contradiction with the relativity postulate', which was always, of course, the gravest of objections for Einstein.

He preferred interpretation I, in which the spherical wave corresponds not to a single electron, but to 'an electron cloud, extended in space', and the theory gives information not about individual processes, but only about 'an ensemble of an infinity of elementary processes'. So the square of the modulus of the wave-function gives the probability that *any* particle of the ensemble arrives at a particular point. Thus there is no need for a special process when the electrons arrive at the film, for, of course, two particles may arrive at different points at the same time. Einstein calls this interpretation 'purely statistical'.

Einstein, it seems, is proposing a hidden-variable theory, in which each electron in the ensemble has a definite, though unknown, position, even before it reaches the film. Indeed, we may naturally assume that he is thinking of a Gibbs ensemble in which all variables have exact values at all times. This kind of interpretation is discussed in the section starting on page 210, but Einstein's own position is actually more problematic than this; it will emerge through the whole chapter, and be discussed in detail towards the end.

In any case, on this occasion at least, the precise details of Einstein's position scarcely influenced Bohr, whose reply (which survives in note form only) [66] began 'I feel myself in a very difficult position because I don't understand what precisely is the point which Einstein wants to [make]. No doubt it is my fault.'

The notes of Bohr's response continue with him saying 'I would put [the] problem in [an]other way . . . I . . . do not know what quantum mechanics is. I think we are dealing with some mathematical methods which are adequate for description of our experiment. Using a rigorous wave theory we are claiming something which the theory cannot possible give. [We must realise] that we are away from that state where we could hope of describing things on classical theories . . .' One would have to describe these remarks as much more a re-statement of his own views than a genuine attempt to respond to those of Einstein. Ominously it seemed that Einstein and Bohr were already talking past each other.

The Bohr–Einstein debate: round 1

There were, though, much more useful meetings between the two outside the conference chamber. They constituted the beginning of the famous *Bohr–Einstein debate*, a fight for the soul of quantum theory, indeed of physics itself. We are fortunate that there is an extensive account of this debate written by one of the participants, Bohr, who initially wrote it for the 1948 Solvay congress, and revised it for an important volume [73] of comments on Einstein's work by various scientists and philosophers, together with autobiographical notes and replies from Einstein himself. This volume was published in 1949 and is often called the Schilpp volume. The account is told, of course, from one side of the debate, but I don't think its handling of what I shall call the first two rounds, the Solvay Congresses of 1927 and 1930, is really open to much dispute.

I would mention that the 1927 discussions were greatly facilitated by the presence of a third physicist, Paul Ehrenfest, who had the greatest respect for both Einstein and Bohr, and was in turn held in high esteem by both. (Sadly, though Ehrenfest had himself made major contributions to physics, in particular to statistical mechanics and quantum theory, he became more and more dissatisfied with his own achievement and committed suicide in 1932.)

In this first round of the Bohr–Einstein debate, Einstein was determined to show that both position and momentum could be measured simultaneously for a particular system; this would constitute a disproof of the orthodox view of the uncertainty principle. It could be regarded as an argument for Einstein's ensemble idea. What would usually be called the uncertainties (or better, indeterminacies) in position and momentum of an individual particle would actually be statistical spreads over the ensemble. Thus, if each member of the ensemble had very nearly the same value of position, for example, there would be a wide spread of values of momentum. Nevertheless each particle would still have a precise value of both position and momentum, and one might hope that thought-experiments could demonstrate this point.

I shall first discuss, as Bohr does in his own account, the fundamental aspects of the uncertainty relations in experiments like that considered by Einstein above. In Fig. 6.2, the only real difference from Einstein's arrangement in Fig. 6.1 is that the slit is wider, and so diffraction of the electron (or photon) beam is mostly restricted to within a range of angles up to a particular angle θ. The

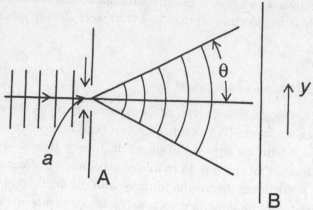

Fig. 6.2. A beam of electrons or photons is incident on a slit A of width a. Following diffraction the beam is mostly restricted to an angular width of θ as it travels to plate B.

electron gains a component of momentum in the y-direction as it passes through the slit; the smaller the slit-width, a, the larger this component of momentum may be, and vice versa. Simple analysis using diffraction theory tells us that the product of these two quantities will be about h, Planck's constant – just as the uncertainty principle demands.

In Fig. 6.3, there is an additional feature, a shutter over the slit, which is opened only for a limited period T. Thus the wave-train proceeding through the slit is of limited extent as shown in the figure, and, from the mathematics of waves, this entails a spread in frequency of the wave and hence a spread in energy of the electron or photon. The larger the value of T, the smaller the spread in energy, and again vice versa, and again detailed analysis shows that the product is at least h, as the time–energy uncertainty principle dictates.

Alternatively we may look at the thought-experiment from the point of view of conservation of momentum and energy. In the case of Fig. 6.2, the uncertainty in the momentum of the particle results from a momentum exchange between the screen and the particle. Because the screen is stationary (to within uncertainty-principle limits), we do not need to consider any transfer of energy. However, in the case of Fig. 6.3, the motion of the shutter must be taken into account. The shorter T, the higher its speed must be, and the greater the possible transfer of energy between shutter and particle. Again, and in both cases, detailed calculations agree with the uncertainty principle.

All this is fairly straightforward, and common ground between the two men. But now Einstein tried to go further, to squeeze more information out of the experiment, more than the uncertainty principle would allow. Could one

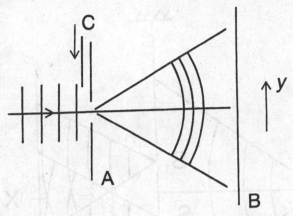

Fig. 6.3. As for Fig. 6.2, with the addition of a shutter C, which is used to open the slit for a limited period T. The train of waves passing towards plate B is thus of limited extent.

not *measure* the momentum and energy transferred between the particle and the screen and shutter by studying the behaviour of the screen and shutter themselves? This would provide *exact* knowledge of the *y*-component of momentum and the energy of the particle, which, together with the *partial* knowledge of its position resulting from the restricted width of the diffraction pattern, and the *partial* knowledge of when it passed through the slit, would defeat the uncertainty principle.

However, as Bohr pointed out in his reply, the previous analysis assumes that the screen remains stationary and that the speed of the shutter has no uncertainty. (Positions of both screen and shutter are 'accurately coordinated with the space–time reference frame', as Bohr put it). This can be the case only if they have infinite mass, so that they can exchange momentum and energy with the particle, without the screen gaining any speed, or that of the shutter being affected. But, of course, under these circumstances, there can be no possibility of *measuring* the effect of the collision on their energy or momentum.

If we *do* want to study experimentally the momentum and energy of the screen and shutter, we must relax the infinite-mass condition, but we must then allow the uncertainty principle to apply to them. There will then be an uncertainty in position of the screen, shutter and, of course, slit (since the slit is a slit *in the screen*). Since the trajectory of the particle is defined relative to the slit, this uncertainty in slit position causes an additional uncertainty in its position. Yet again, as may be shown in detail, one obtains the appropriate relationship between the imprecisions in the values of the momentum of the screen (and hence of the particle) and the position of the particle. In the shutter

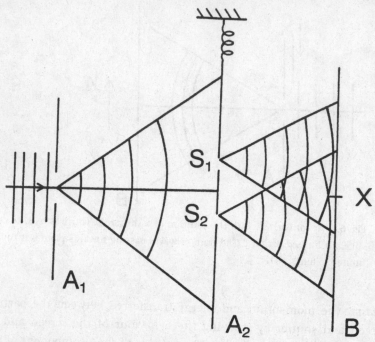

Fig. 6.4. As for Fig. 6.2, with an additional screen A_2 in which are two slits, S_1 and S_2. A_2 is supported by a weak spring.

case, there is a similar relationship between the imprecision in the energy of the shutter (and hence of the particle) and that in the time at which the particle passes through the slit.

(I shall note here, though, that I suspect the reader may be calling foul at this point. Surely, it may be argued, Bohr cannot *prove* or *demonstrate* the significance of the uncertainty relations for one body, the particle, by appealing to the same relations for another body, the screen or shutter! I would certainly agree that one must not assume what one is trying to prove. But actually Bohr was not trying to prove that the orthodox view of the uncertainty principle is correct; it was Einstein who was trying to *deny* this, by showing that both position and momentum (or time and energy) could be measured simultaneously. All Bohr had to do was to show that the orthodox view of the uncertainty principle was consistent, and this he was able to do.)

Now I move to another interesting idea of Einstein – to separate the parts of the apparatus that would perform the two measurements, of position and momentum, respectively. Figure 6.4 shows the original screen, A_1 (with no shutter); mid-way between A_1 and the plate B is a second movable screen, A_2, which is supported by a weak spring. A_2 has two slits in it, S_1 and S_2, and so, if

A_1 and A_2 are both rigid, there will be an interference pattern on B. This pattern is built up, in fact, by individual particles hitting specific points on the plate over a long period, many particles hitting the regions around the maxima of the interference pattern, few the regions around the minima. We shall consider an exceptionally weak beam so that the effect of each electron or photon may be studied individually (and it will take an extremely long time, of course, to build up the interference pattern).

Einstein's suggestion was that the amount and direction of the momentum given to A_2 by a given particle depends on whether that particle passes through S_1 or S_2. If, for example, the particle goes through S_1 and reaches point X on the plate, A_2 must recoil upwards; if it goes through S_2 and reaches the same point, A_2 must recoil downwards. And of course one may argue the other way; the direction of recoil of A_2 tells us which slit the particle passes through. Thus Einstein hoped to maintain a seemingly paradoxical situation – on the one hand, the presence of the interference pattern demonstrating that the electron or photon, in its capacity of wave as it were, has sampled both slits, and, on the other, evidence of which slit it has, in its capacity of particle, actually passed through.

Bohr was able to show, though, that this didn't work. Knowledge of the momentum gained by A_2 to the required precision involves an uncertainty in its position, so the positions of S_1 and S_2 are uncertain to just the extent necessary to smudge out the interference pattern. As Bohr said – 'The point is of great logical consequence, since it is only the circumstance that we are presented with a choice of *either* tracing the path of a particle *or* observing interference effects, which allows us to escape from the paradoxical necessity of concluding that the behaviour of an electron or photon should depend on the presence of a slit in the [screen] through which it could be proved not to pass. We have here to do with a typical example of how the complementary phenomena appear under mutually exclusive experimental arrangements, and are just faced with the impossibility, in the analysis of quantum effects, of drawing any sharp separation between an independent behaviour of atomic objects and their interaction with the measuring instruments which serve to define the conditions under which the phenomena occur.'

Einstein had no choice but to admit that round 1 had gone to Bohr. It was not just that the uncertainty relations survived Einstein's onslaught, but that the clinching argument was exactly the original cornerstone of complementarity. Position and momentum could not be measured simultaneously; wave and particle aspects of light or electrons could not be manifested simultaneously.

Nearly 80 years later, these ideas, and much of Bohr's original viewpoint, have become so well known that they seem almost obvious, practically a priori. And

Bohr's stock, so high for such a long time, has now, at least in some circles, somewhat plummeted. To gain a clearer idea of the sheer brilliance and novelty of his arguments, we turn to a letter sent by Ehrenfest to some colleagues in Holland including Goudsmit and Uhlenbeck [66] – 'Brussels–Solvay was fine! . . . BOHR towering over everybody. At first not understood at all . . ., then step by step defeating everybody. Naturally, once again the awful Bohr incantation terminology. Impossible for anyone else to summarise . . . (Every night at 1 a.m., Bohr came into my room just to say ONE SINGLE WORD to me, until three a.m.) It was delightful for me to be present during the conversation between Bohr and Einstein. Like a game of chess, Einstein all the time with new examples. In a certain sense a sort of Perpetuum Mobile of the second kind to break the UNCERTAINTY RELATION. [This related to attempts to achieve *perpetual motion*, a favourite hobby of would-be-scientists for several centuries.] Bohr from out of philosophical smoke clouds constantly searching for the tools to crush one example after the other. Einstein like a jack-in-the-box; jumping out fresh every morning. Oh, that was priceless. But I am almost without reservation pro Bohr and contra Einstein. His attitude to Bohr is now exactly like the attitude of the defenders of absolute simultaneity towards him . . .'

For Bohr, the only problem in his analysis was the one mentioned in the previous chapter – macroscopic measuring equipment, which Bohr in general stressed must be treated classically, has been made subject to the uncertainty principle. This need not be an absolute contradiction, but it is certainly rather awkward.

Einstein's reaction, expressed in a letter to Sommerfeld [94], was perhaps a little grudging – 'Concerning the "quantum mechanics" I think that as regards ponderable matter it contains just as much truth as the theory of light without quanta. It might be a correct theory of statistical laws, but an insufficient conception of the individual elementary processes.' He was waiting for round 2 of the debate.

Einstein and ensembles

Before we ourselves turn to round 2, I want to discuss Einstein's own ideas in a preliminary and partial way. In his contribution to the 1927 Solvay debate Einstein had appeared to advocate some form of hidden variable (to determine where the particle would hit the film). Indeed, we may usefully think of him as advocating what I call [95] a *Gibbs ensemble*.

In such an interpretation, all variables, position, momentum and so on, have precise values at all times. Only some of these values are related to the wave-function of the ensemble. For example, a wave-function might be an eigenfunction of momentum; that is to say, it corresponds to a particular value of momentum. *All* particles in the ensemble represented by the wave-function then have *that* value of momentum. They also, though, each have a precise value of position, but these values differ for the various particles in the ensemble. All these values are then available to become the results of measurements; measurement, of course, plays no part in creating them. The idea of the Gibbs ensemble is taken from classical physics, and it is discussed in Chapter 2.

I should explain that, in the attempt to interpret quantum theory, a more general type of ensemble is sometimes considered, where it is *not* the case that all variables have values at all times. We shall have cause to discuss this more general type of idea in Chapter 8. But what I call the Gibbs ensemble is so interesting that many authors have discussed it – confusingly giving it many different names. We met it as Murdoch's *intrinsic-values theory* in the previous chapter. Arthur Fine has called it the *complete-values thesis* [96] and a *random-values representation* [94]. Dipankar Home and I have called it a *PIV-ensemble* [97] – an ensemble with pre-assigned initial values before any measurement.

Indeed, for many, such as Popper [69], the Gibbs ensemble *is* just *the* ensemble. Fine too [94] is prepared to use this terminology, preferring the term *statistical interpretation* for the more general ensemble concept. To add to the confusion, Ballentine [88], in what is the strongest claim for ensemble interpretations, in fact rather strongly supporting the Gibbs ensemble, calls *this* idea the *statistical interpretation*.

All this is not intended to confuse the reader, but it does show that considerable care is required in moving to the more general literature; it suggests as well that, where terminology is confused and confusing, so may be the ideas in circulation, and I suspect that this is also the case.

I would suggest that calling *any* interpretation of quantum theory the 'statistical interpretation' is unhelpful. To explain this remark, I would like to clarify the distinction I consider desirable between the words *statistical* and *probabilistic*.

In common parlance, when one counts up numbers of events in different categories, one is doing *statistics*. For example, one may throw a die a number of times, say 100, and count up the numbers of 1s, 2s and so on. It is very likely that each of these numbers will be quite close to 17. From such experimental evidence, or from general ideas (or prejudices?), we may come to the conclusion that each of the numbers has equal *probability* of occurring, which means that they will occur (approximately) equally often. Of course it is assumed in the above that everything is *deterministic*; in principle, if we knew accurately the details of the throw, and the exact strength of the wind at all times, we could predict with certainty which side of the die will show. It is only our lack of knowledge of these details that means we give a probabilistic description. This common kind of situation may usefully be called *statistical*.

The word *probabilistic* may best be used to indicate something different. In quantum theory, conventional or orthodox ideas would say that, if a spin-$\frac{1}{2}$ particle such as an electron has a definite value of its z-component of spin, the x- and y-components just do not have values. It is *not* that half of such electrons (or about half) have x-component equal to $+\hbar/2$, and half to $-\hbar/2$ (which could be described as a *statistical* situation); the values just do not exist. Of course, if a measurement of this component is performed on a number of electrons, we may count the number that give $+\hbar/2$ and the number that give $-\hbar/2$, and we shall expect these numbers to be about equal; that is a good *statistical* situation. But *before* any such measurement, since no pre-measurement values exist, there are not statistics in the sense of counting, and I think it much clearer to use the word *probabilistic* to describe the situation.

Thus *all* interpretations of quantum theory may be termed *statistical* if one is thinking of the results of experiments; indeed one may just say it is quantum theory that is statistical *in that sense*. However, if one thinks of the pre-measurement situation, orthodox interpretations are probabilistic, whereas a Gibbs ensemble interpretation is statistical.

Let us now examine the case that Einstein supported the idea of a Gibbs ensemble. Undoubtedly it is quite strong, and is argued in particular by Ballentine [98], who would indeed claim that his own advocacy of ensembles [88] only follows Einstein's lead. One may start with Einstein's remark quoted at the end of the previous section. He is here saying that the wave-function of quantum theory may be analogous to a wave in the classical theory of light; just as the latter needs to be supplemented with photons, an ensemble of photons being represented by the actual wave, so the wave-function of quantum theory must be supplemented with additional information concerning individual particles, the wave-function representing only an ensemble of such particles.

For the remainder of his life, Einstein made similar comments. In an important paper of 1936 [99], he wrote that: 'The ψ-function does not in any way describe a condition which could be that of a single system; it relates to many systems, to an "ensemble of systems" *in the sense of statistical mechanics* [my italics].' In the Schilpp volume [73] a decade later he repeated the point – '[T]he ψ-function is to be taken as the description, not of a single system, but of an ideal ensemble of systems. In this case one is driven to the conviction that a complete description of a single system should, after all, be possible; but for such complete description there is no room in the conceptual world of statistical quantum theory.'

Given these statements (and many more), it is scarcely possible to dissent from Jammer's comment [67] that 'all through his life Einstein adhered to the statistical [ensemble] interpretation of the existing formalism'. And Ballentine [98] severely criticises Heisenberg for including Einstein in a group that 'expresses . . . its general dissatisfaction with the quantum theory, without making definite counter-proposals, either physical or philosophical in nature'; Ballentine refers to Einstein's ensembles as a 'definite counter-proposal to the Copenhagen interpretation'.

From the above passages, and particularly from the portion I italicised in the 1936 one, it seems at least reasonable to assume that Einstein's ensembles were Gibbs ones – all observable quantities have precise values at all times. In this section and the next I shall assume this to be true, and trace through the consequences. (Thus I leave to the end of this chapter two discordant points. First, it has been claimed, particularly by Fine [94], that Einstein's ensembles were entirely different in nature. Secondly, it would be acknowledged by every-body, including Ballentine and Jammer, that ensembles were only one part of Einstein's overall views on quantum theory.)

If they worked, Gibbs ensembles would, of course, remove a lot of the diffi-culties of orthodox interpretations of quantum theory. The great measurement problem would be removed at a stroke. The existing value of any observable quantity is available to become the result of the measurement of that quan-tity (though, of course, one need not insist that all measurements do give these values exactly). Wave-function collapse is avoided, or, to be more precise, wave-function collapse becomes merely the mental act of separating out those systems which give a particular result in an initial measurement.

So, if they worked, Gibbs ensembles would avoid much conceptual stress; it might be said that they would make advocates of the more orthodox approaches look rather silly! And it might be added that, while many physicists would regard them as ruled out by the uncertainty principle, this is not necessarily the case. The thrust of the ensemble idea is that what in orthodox interpretations would be the uncertainty (or indeterminacy) in momentum or position of an individual particle becomes the range of values of momentum or position for the ensemble, each member of which has a precise value of position and momentum.

Unfortunately, though, Gibbs ensembles don't work. As Fine [94] says, 'the so-called statistical [ensemble] interpretation has been faced with difficulties since 1935, difficulties known to Einstein and simply ignored by him'.

The difficulties with ensembles

The first difficulties with ensembles were pointed out by Schrödinger in 1935 towards the end of a period of intense correspondence with Einstein [94]. Much of this correspondence concerned matters on which they were in general agreement, such as the inadequacy of Copenhagen. They also came up with new ways to demonstrate problems with the conventional ways of looking at quantum theory; Schrödinger introduced his famous *cat*, which I shall discuss later in this chapter. But they also argued against each other's favourite ways of trying to overcome the difficulties, and in the paper [100] that summed up Schrödinger's thoughts, the section titled 'Can one base the theory on ideal ensembles?' must be regarded as criticism of Einstein's approach.

Having explained the classical Gibbs ensemble, Schrödinger suggests that 'At first thought one might well attempt to refer back the always uncertain statements of Q.M. to an ideal ensemble of states, of which a quite specific one applies in any concrete instance – but one does not know which one.' But Schrödinger concludes that 'this won't work'.

One could, for example, set up angular momentum *referred to a specific axis of rotation* so that only those values which were allowed by quantum theory occurred. But as soon as one considers the same motion, but referred to a different axis, values that quantum theory does *not* allow occur. Schrödinger says that 'appeal to the ensemble is no help at all'.

Another example he discusses concerns the simple harmonic oscillator. Let us consider a specific quantum number n, which corresponds to a given energy, E_n, and this energy should be the sum of two terms. The kinetic energy, E_k, should be zero if the particle is not moving, and positive if it is. (I'll call its speed v.) The potential energy, E_p, should be zero when the particle is in its rest position, and increases as x, the displacement from that rest position, itself increases.

Now the point of the Gibbs ensemble is that, for each system, v and x should both take specific values at all times, and hence so should E_k and E_p. Since E_k and E_p should sum to a fixed quantity E_n, and since neither can be negative, E_p, for example, should have a maximum value – when it is equal to E_n, and E_k is zero. This implies that x too should have a sharp maximum. Yet experiment and theory both suggest that there is no cut-off for x. Certainly, once one goes to high values of x, as x increases further the probability of obtaining that particular value of x decreases, but it always remains greater than zero.

Jammer [67] has reported similar calculations on the hydrogen atom, which suggest that, under the Gibbs assumption, almost a quarter of the electrons would have potential energy greater than total energy, and thus negative kinetic energy, which is impossible. He uses such calculations to state that 'the assumption that a particle has simultaneously well-defined values of position and momentum, even though these may be unknown and unobservable, can be rejected independently of the complementarity interpretation'.

A similar problem relates to radioactive decay. Here the decaying particle has positive kinetic energy within the nucleus, but at the boundary of the nucleus there is a region where its potential energy would be higher than its total energy, and so classically the particle may not enter this region. Quantum-mechanically, though, it may *tunnel* through the region to reach another classically accessible region of low potential energy outside the nucleus. On a Gibbs ensemble basis, though, some particles must actually be in the boundary region, where their kinetic energy should be negative.

Other fairly basic properties of quantum theory seem to be at least awkward for the Gibbs ensemble. Let us first consider a spin-$\frac{1}{2}$ particle. Quantum theory tells us that the square of its total spin angular momentum must be $s(s+1)\hbar^2$; s is just $\frac{1}{2}$ here, so we obtain $\frac{3}{4}\hbar^2$. If we perform a measurement of s_x, s_y or s_z, the values obtained must be either $\hbar/2$ or $-\hbar/2$. In Gibbs ensemble terms, that means that, at all times, the values of each of s_x, s_y and s_z must be either $+\hbar/2$ or $-\hbar/2$, and the square of each is always $\hbar^2/4$. Now the square of the total spin angular momentum should just be the sum of $s_x{}^2$, $s_y{}^2$ and $s_z{}^2$, that is $\frac{3}{4}\hbar^2$ as above. This is very nice – Gibbs ensembles seem to be working well.

If we go to a spin-1 particle, the square of the total spin angular momentum is $2\hbar^2$; s_x, s_y, and s_z may be $\hbar$, 0 or $-\hbar$, the squares of each may be $\hbar^2$ or 0, so, if two of $s_x{}^2$, $s_y{}^2$ and $s_z{}^2$ are $\hbar^2$, and the other 0, they add to $2\hbar^2$, as they should.

However, when we move to a spin-$\frac{3}{2}$ particle, our luck runs out. The square of the total spin angular momentum must be $(15/4)\hbar^2$. Each of s_x, s_y and s_z may be $\frac{3}{2}\hbar$, $\hbar/2$, $-\hbar/2$, or $-\frac{3}{2}\hbar$, and there is no way to make the sum of $s_x{}^2$, $s_y{}^2$ and $s_z{}^2$ equal $(15/4)\hbar^2$.

Another important argument against Gibbs ensembles concerns the experiment most characteristic of quantum theory – two-slit interference. Gibbs ideas would appear to indicate that a particle (photon or electron) must go through one slit or the other, since at all times it should have a precise position. But of course the existence of the interference pattern indicates that, in some sense, each particles samples both slits. One might generalise this point by saying that Gibbs ensemble ideas seem powerless to handle any sort of interference effects, yet such effects are at the centre of numerous quantum calculations giving excellent agreement with experiment.

Let us now examine another problem area for Gibbs ensembles, considering an ensemble of spin-$\frac{1}{2}$ particles in an eigenstate of s_z; let us say that they all have s_z equal to $+\hbar/2$. According to the ideas of the Gibbs ensemble, (about) half have s_x equal to $+\hbar/2$, (about) half to $-\hbar/2$, so this gives two sub-ensembles, E_+ and E_-. Now let us consider ensembles of particles in eigenstates of s_x; we may call these D_+ and D_-. We might seek to identify E_+ with D_+, and E_- with D_-, but this cannot be correct because, while all particles in E_+ and E_- have s_z equal to $\hbar/2$, in D_+ and D_- there are (approximately) equal numbers with s_z equal to $+\hbar/2$ and $-\hbar/2$.

This kind of argument is actually quite interesting historically because it seems [67] that it lay behind von Neumann's rejection of hidden variables – his famous proof being an attempt to justify mathematically what he already felt convinced was true. By performing a measurement of s_x on ensemble E above, one might hope to generate sub-ensembles E_+ and E_- with specific values of s_z and s_x. Yet in practice we are not able to do this; ensembles D_+ and D_- are produced with specific values of s_x, but having lost the distinct value of s_z they started off with. This may be regarded as an argument against the existence of Gibbs ensembles.

Like all our arguments, it can, of course, be minimised or even removed by adjusting basic ideas and rules. Some of our arguments turned on the additivity of kinetic energy and potential energy to give total energy. In the spin case, a similar addition rule was imposed for relating total spin angular momentum and its components. The difficulty may then be removed by adapting or suspending these rules.

The interference argument in particular was subject to a debate between Popper [52] – a great believer in Gibbs ensembles, and in general strongly of the view that physicists had created a monster in their interpretation of quantum theory, and Feyerbend [64]. Feyerabend suggested that, to explain interference, ensemble theorists like Popper needed to recognise that, when a second slit is opened, there is 'a new stochastic [probabilistic] process that leads to a new interference pattern. This position is *indeterministic*, as it admits the existence of uncaused individual changes and its indeterminism is about as radical as that of the Copenhagen point of view. It also shares with that point of view its emphasis on the *experimental situation*; predictions are valid only for certain experimental conditions . . . However it differs from the Copenhagen point of view insofar as it works with well-defined states and trajectories.'

In such a concoction, the conservation laws are valid only for averages over ensembles, which would appear to contradict the experiments of Bothe and Geiger, and Compton and Simon, that killed off the BKS theory, as explained in Chapter 4.

Even apart from this, though, it is clear that, when the Gibbs ensemble approach responds to the various challenges made on it, it rapidly loses its beguiling simplicity. One might say that its *chief* attraction is that measurement may merely record, and not, in any sense, create. But as soon as basic conservation laws, which are certainly obeyed in experimental results, are not respected in the pre-measurement values, this feature must be lost. And, with the talk on the 'experimental situation', again the main selling point of the Gibbs ensemble appears to have disappeared. (Incidentally the problems concerning energy balance and conservation may be seen in a different perspective when we move to the work of Bohm on the 'quantum potential' in the next chapter.)

Similar complications occur when we try to rebut von Neumann's position above. Schrödinger indeed had a valid counter-argument; the loss of the precise value of s_z may be a result of interaction with hidden variables in the measuring device. Von Neumann in turn counter-argued with ideas of how to fix the hidden variables of the measuring device as well [101]. His arguments can never be made water-tight. After all, if they could, it would imply a prohibition on hidden variables, and, as I have already said, hidden variables are not prohibited!

But what the arguments *do* achieve is to show how far away from the basic Gibbs ensemble one has to go – again, as Bohr would have stressed, one must include the measuring device as an active participator in the measurement, not just a recorder of a fixed value.

Overall, the Gibbs ensemble idea seems of limited use, unless it is so distorted as practically to lose its own identity and certainly most of its appeal. The reader may even suggest that I have been using a sledge-hammer to crack a nut. I think, though, that the argument is important because it suggests that, to the extent that Einstein's ideas were limited to a Gibbs ensemble, his progress towards genuine answers to quantum dilemmas was marginal.

It is certain, though, that this was not the only string to his bow; it has been suggested [94] that his ensembles were not at all along Gibbs lines, and even that he *intended* his scheme to appear sterile. We shall attempt to gain a deeper understanding of Einstein's ideas at the end of this chapter, building up to this by analysing the further rounds in the Bohr–Einstein debate.

The Bohr–Einstein debate: round 2

The second round of the debate took place at the next Solvay Congress in 1930 – Einstein proposed an ingenious thought-experiment with which he hoped to use his own expertise in relativity to crush Bohr's arguments.

Let us imagine a box containing a certain amount of radiation. In its side is a hole, which may be opened and closed by a shutter. If the shutter were opened for a short interval T, Einstein suggested that it could, in principle of course, be arranged that a single photon would pass through the hole.

The box may be weighed before and after this happens. Now a photon has a particular energy – the product of its frequency and Planck's constant. Einstein's famous relation $E = mc^2$ then tells us that it must have an effective mass, and hence contribute to the weight of the box. Now Einstein said that the box could

Fig. 6.5. Bohr's analysis of Einstein's photon-weighing experiment. A clock C is used to open the shutter S for a period T. The weighing by spring balance of the box B and its contents is examined explicitly in terms of the position of pointer P on scale A. Figure after [67].

be weighed as accurately as one wished, so the uncertainty in the mass of the photon, and hence in its energy, may be zero. But the uncertainty in the time of arrival of the photon is finite, just T, so the product of uncertainties in time and energy may be zero, contrary to the time–energy uncertainty principle!

There is no doubt that Einstein thought he had vanquished Bohr. Rosenfeld [91] wrote much later that 'It was quite a shock for Bohr . . . he did not see the solution at once. During the whole evening he was extremely unhappy, going from one to the other, and trying to persuade them that it couldn't be true, that it would be the end of physics if Einstein were right; but he couldn't produce any refutation. I shall never forget the vision of the two antagonists leaving . . . Einstein a tall majestic figure, walking quietly, with a somewhat ironical smile, and Bohr trotting near him, very excited . . . The next morning came Bohr's triumph.'

For, after a sleepless night, Bohr incontrovertibly *did* triumph, using general relativity *against* Einstein. He started by considering the weighing process explicitly. (Notice how characteristic of Bohr's arguments this is; the word 'weighing' must not be an abstract technical term, but rather one must study the physical process by which the measurement is performed.)

He produced a diagram like Fig. 6.5. The box is weighed using a spring balance, and the result is obtained from the position on the scale of the pointer attached to the box. (There are other forms of balance, but all give rise to similar considerations.)

Now we must accept that the position of the pointer on the scale must have an uncertainty d_y, which is related to that in the momentum of the box, d_p, by the uncertainty principle. To obtain some information about d_p, Bohr imagined the limit of accuracy in the measurement of the mass being d_m. Then d_p could not be larger than the momentum given to d_m in period T, which is just the product of d_m, T and the gravitational constant g.

Now came Bohr's master-stroke. According to general relativity, a clock that is moved in the direction of a gravitational field will change its rate. Thus the uncertainty d_y in the position of the pointer, and hence of the box, gives rise to an uncertainty d_T in the time interval T. The uncertainty in momentum d_p corresponds to an uncertainty in energy d_E, and when the detailed sums are done, one obtains the result that the product of d_T and d_E is at least as great as Planck's constant h – just the time–energy uncertainty principle.

Bohr reports that Einstein himself contributed effectively to the elucidation of the problem, and at one level it must have been as fascinating for Einstein as for Bohr to see how the very simple calculations could express such profound ideas. Yet from another point of view it must have been galling for Einstein to be defeated by Bohr with his own great creation of general relativity. It seems that he accepted that the kinds of argument he used in the first two rounds of the Bohr–Einstein debate would never work; Fine [94], who has had access to much of Einstein's correspondence, says that 'nowhere after 1930 do we find Einstein questioning the general validity of the uncertainty formulas'.

But this was far from indicating an acceptance of Bohr's position. Rather, Einstein realised that more subtlety would be required for the next round of the debate. Before I turn to this, I shall discuss some of the main ideas that are involved.

Determinism, realism, locality

Round 3 of the Bohr–Einstein debate consisted of the famous Einstein–Podolsky–Rosen paper (EPR), and Bohr's reply to it. To help understand these, and to work towards a general assessment of Einstein's ideas on quantum theory, it will be helpful to introduce a few technical terms, and to explain a few I've used before in a little more detail.

Determinism, realism and locality are words that may initially be applied to the Universe; one may have a deterministic or non-deterministic Universe, and so on. Following from that, theories may be, for example, deterministic or non-deterministic, depending on the type of behaviour they describe.

We have already met our first word, *determinism*. It means that, if one can describe completely the state of the Universe at one particular time, it is, in principle, possible to predict precisely its behaviour at all later times. Classical physics was, of course, strictly deterministic, and, as quantum theory developed, one of the greatest surprises was that orthodox interpretations claimed that it was non-deterministic.

A discussion of the second word, *realism*, is rather more tricky – partly because I feel that there is a divergence between the ways it is used by philosophers and by many physicists. Philosophers would traditionally use the word to mean that there is a real world that has certain properties independent of the observer. This does not necessarily need to mean that the observer can interact with this world in a straightforward or classical way; the observer may disturb the system, for example, or, once observation does begin, it may be impossible to define uniquely an observed and an observing system.

A realist theory is one that seeks to express something about this world. It need not find it possible to set out a one-to-one correspondence between elements of the theory and elements of the real world. There may be much mathematical apparatus in the theory that does not relate clearly or unambiguously to anything thought to exist in the world. But there must be elements in the theory that refer, directly or indirectly, to things that have a real existence, and are not just related to the *results* of experiments. Under this loose definition it is possible to claim that Bohr intended to make quantum theory realistic, and to a large extent succeeded. (Whether he quite survived the onslaught of EPR will be discussed shortly.)

Unfortunately – or at least I think it is unfortunate – the same word, realism, has been taken over by at least some physicists to mean a much more definite conception, essentially that of hidden variables. This includes the Gibbs ensemble idea discussed earlier in this chapter, but also the more general idea that observable quantities may not actually possess precise values at all times, but hidden variables dictate what values will be observed *if* any quantity is measured.

This concept, I think, should preferably be called *naive realism*, or possibly *classical realism*, for it is at least close to classical ideas, the Gibbs ensemble being much closer than the more general hidden-variable point of view. Calling it simply realism means that, while Newtonian physics remains, of course, realistic in this sense, Bohr clearly cannot be a realist. It seems to me that two problems are caused by this hijacking of the word. First, it leaves no term to distinguish those like Bohr, who, at the very least, wish to be considered as realists, from those who are out-and-out opponents of realism. Secondly, when different people use the same word in different ways, it is almost bound to lead to confusion.

I now move on to the last of our three words, *locality*. I shall start the discussion by suggesting that, if two events that are to be regarded as *cause* and *effect* take place simultaneously, they would be expected to occur at the same point in space. Let us imagine, for example, a ball travelling towards a window and passing through it. We would certainly expect the window to break (effect) when the ball hits it (cause), so cause and effect, being simultaneous, occur at the same place. We would be most surprised if the window broke when the ball was still a few metres away from it. This idea is called *locality*.

We would often, though, describe two events as cause and effect even when they are not simultaneous, and therefore are not obliged to occur at the same place. Indeed, it may be a more natural choice of definition in the previous case to re-define the cause as the ball being kicked. Obviously now cause and effect are separated in both space and time; again obviously the separation in space s_d will be equal to the product of the speed of the ball and the separation in time s_t. (I am ignoring, here, the fact that, because of gravity, the ball cannot travel exactly in a straight line, and also I am ignoring air resistance; these points, if included, would complicate the analysis slightly, but not change it in principle.)

The speed of the ball will take different values, depending on how hard it is kicked, but, moving to the general case, we may recognise that *any* such particle (or wave, or disturbance) moving from cause to effect has a maximum speed of c, the speed of light, so in *all* cases, s_d must be less than (or, in special cases, equal to) the product of c and s_t. This is the general meaning of *locality*. It may easily be seen that the special case we started off with is just an example of this general result, because if cause and effect are simultaneous, s_t is zero, so s_d must be zero also.

Despite the gesture to relativity in the previous remarks, in some ways they are pre-Newtonian in that they suggest a structure of isolated cause and isolated effect. In classical physics, we have instead the whole of the Universe evolving deterministically. Rather than considering individual causes and effects, we may observe that the equations themselves observe the restrictions just mentioned. Motion of a particle is strictly influenced only by factors at its actual location – a collision with another particle, or a gravitational or electromagnetic field at the position of the particle. Again it is natural to identify as cause and effect events that are not at the same place. We must then introduce the ideas of special relativity; Maxwell's laws of electromagnetism and Einstein's laws of motion will ensure that s_d and s_t are connected in the same way as before.

Actually, even in physicists' most deterministic accounts of experiments, there is one aspect that is *not* treated deterministically – the physicists themselves. They assume that they have free will to set up and interfere with experiments exactly as they wish. If you were to suggest to physicists that the experiments they undertake are not, or not totally, a matter of free will, but, in part at least, are a result of previous deterministic interaction with, among other things, the very equipment they are working with, you would probably be laughed to scorn. (Logically they need not be correct. If experimental result A leads deterministically to the experimenter performing experiment B and achieving result C, the experimenter will say that experiment B always gives result C, but it could be that result A is automatically followed by result C. Don't tell physicists this; it will upset them!)

Anyway, with the usual assumption, the action of the experimenter may be treated as an independent and isolated cause as in what I called the pre-Newtonian approach. The principle of locality then demands what I call here *condition 1* – any action of an experimenter at one particular point in space cannot influence the outcome of a measurement made simultaneously at a different point in space; if the separation in space between action and measurement is s_d, there may be an influence only if outcome follows action by at least s_t, where s_d is c times s_t.

When we turn to quantum theory, condition 1 is still demanded by locality. (This assumes, of course, that the actions of the experimenter are genuinely free. In 1985, John Clauser, Michael Horne and Abner Shimony criticised John Bell for making this assumption, but Bell [76] replied that, in doing so, he was 'just pursuing [his] profession of theoretical physics'.)

In quantum theory, locality demands a second condition as well, since we must at the very least consider the possibility that measuremental results are not pre-determined. (This would, of course, be taken for granted by the orthodox, though it may be disputed by supporters of deterministic hidden variables.) Then *condition 2* of locality says that the result of a measurement of one observable at one point in space cannot influence the result of a second measurement, carried out simultaneously, of another observable at a different point in space. As usual, if the separations in time and space are s_t and s_d, then, for the possibility of a mutual influence between measurement results, s_d must be less than, or at most equal to, the product of c and s_t.

It may be helpful to stress what is and what is not implied by condition 2. Bell's famous example of what is *not* implied is his parable of *Bertlmann's socks*. Bell [76] explained that Reinhold Bertlmann, his friend and collaborator, always liked to wear socks of different colours. If you observed the first one to be pink, you could be sure, without even checking, that the second one was *not* pink. But of course you certainly wouldn't assume that by observing the first sock to be pink, you had, in some mysterious way, forced non-pinkness on the second, and if you observed both simultaneously, you would not feel that the two observations influenced one another. Clearly the pinkness of the first sock and the non-pinkness of the second existed before you started observing; if you like you could describe them as hidden variables – hidden from you until you actually observe them. As Bell said, there was 'no accounting for tastes', but apart from that, 'no mystery'.

The case where there is, or may be, a mystery, is the quantum one where two measurements are performed simultaneously at different places; we may, in an orthodox approach, wish to assume that there are no hidden variables dictating the behaviour at each measurement, and yet the two results are correlated – one influences the other, or, more probably, each influences the other. This is just the kind of situation that Einstein set out to exploit in EPR.

I should note that the laws of relativity do not themselves demand both conditions. Relativity says that a signal may not be sent at a speed greater than c. If condition 1 (Shimony [102] calls it *parameter independence*, by the way) is disobeyed, an experimenter may perform an action that influences an experimental result simultaneously at another point (or, if not simultaneously, too soon to be permissible by the laws of relativity, s_d greater than c times s_t). Effectively a signal may be sent from one point to the other. Thus parameter independence *is* a requirement of special relativity.

However, if condition 2 (Shimony calls it *outcome independence*) is disobeyed, and two measurements are correlated as discussed above, this does *not* give an experimenter the opportunity to send a signal at speed greater than c, because neither measuremental result is itself under the control of the experimenter. Thus outcome independence is *not* a requirement of special relativity.

Nevertheless many scientists, certainly including Einstein, would feel that violation of outcome independence would be against the *spirit* of special relativity, even if not the *letter*. (After all, although *we* cannot use it to send a signal at speed greater than c, it rather implies that Nature itself is doing exactly this.) Einstein and many of his followers would be strong advocates of *both* aspects of locality. (Bohr's view will become clearer in his response to EPR.)

Indeed, Einstein was very keen to uphold determinism, realism *and* locality. Which, if any, had priority, and which, if any, he might have been prepared to abandon to save the others, will be discussed towards the end of this chapter and also in the following one.

Completeness

In contrast to the three words discussed in the previous section, *completeness* is a word applied *only* to theories. For a theory to be complete, 'every element in the physical reality must have a counterpart in the theory', to quote EPR [103]. (The word 'reality' is explicitly in the definition, and I think it would be true that ideas of completeness require some commitment to reality.)

It is immediately clear that several excellent theories are incomplete. Thermodynamics, for example, uses only macroscopic quantities – pressure, volume and temperature for a gas, for example – and declines to consider the actual behaviour of the atoms constituting the substance, but there have been few theories more useful, both in terms of providing understanding of the world and in leading to practical applications (as well, of course, as profiting immensely from practical experience). So, for many purposes, one will not necessarily improve a theory by completing it.

In the case of quantum theory, any classical realist must inevitably conclude that, as it stands at present, the theory is incomplete. It does not allow precise values for all observables – not for both position and momentum, for example. The classical realist believes that such values must exist 'in the physical reality'; they don't have 'a counterpart in the theory', and hence the claim of incompleteness.

This may have been Einstein's personal view, but he realised, of course, that it would cut no ice with Bohr and his friends, who just did not believe that precise values of position and momentum existed 'in the physical reality'. Thus we may interpret Einstein's arguments in the first two rounds of the Bohr–Einstein debate as trying to demonstrate that it is possible to measure simultaneously precise values of position and momentum. If both could be measured simultaneously, one might surmise that they both exist simultaneously in the physical reality, but they do not exist together in the quantum formalism. This implies the incompleteness of quantum theory. However, Einstein failed to achieve this aim.

Following the first two rounds of the Bohr–Einstein debate, Einstein was forced to admit that position and momentum could not be measured simultaneously. However, it was not unreasonable to assume that both quantities might *exist* in the physical reality and in the absence of measurement, but that measurement of one of the quantities disturbed the other so it could not be measured, and so simultaneous measurement was impossible. In the EPR argument, Einstein attempted to demonstrate that a system must at all times have values for both position and momentum.

Einstein actually preferred a different definition of completeness – quantum theory will be incomplete if a system in a given state has more than one wavefunction. Consider, for example, a system with a precise value of momentum; by that token, quantum theory would say that its wave-function should be a particular eigenfunction of momentum. However, if it also has a precise value of position (so quantum theory must be incomplete, as discussed above), its wave-function should be a particular eigenfunction of position. These two wave-functions cannot be the same, so it seems that one system has two different wave-functions.

The Bohr–Einstein debate: round 3; EPR

By 1935, Einstein had been driven out of Germany, and had come to the Institute for Advanced Study at Princeton, where he was to stay for the remainder of his life. He had worked with Boris Podolsky, a young Russian-born physicist, a few years earlier, and, when Podolsky also came to Princeton, contact was quickly re-established. An even younger physicist, Nathan Rosen, started to work at Princeton in 1934, and soon linked up with Einstein. Thus E, P and R came together.

The fundamental idea of the EPR paper undoubtedly came from Einstein; it was a development of his ideas in the first two rounds of the Bohr–Einstein debate. There his thought-experiments had considered measurements of two quantities not permitted joint exact values by quantum theory, but his arguments failed because, in each case, the measurements interfered with each other. In round 1, the measurement of momentum made the position measurement inaccurate; in round 2, the measurement of time was made inaccurate by the simultaneous weighing procedure. In EPR, Einstein's idea – and let me say immediately that it was a brilliant one – enabled him to postulate two measurements that would serve his purpose, but would *not* interfere with each other.

The idea is usually called the EPR *paradox*, and the use of that word has engendered a great deal of heat in its own right. Fine [94], for example, takes pains to establish that Einstein himself frequently used the word. Rosen, though, at a conference [104] held to mark the fiftieth anniversary of EPR, stresses that 'this term is unjustified; there is no paradox'.

I suspect that *this* controversy is unnecessary, arising from ambiguity in defining the word 'paradox'. The literal meaning of the word is something like 'challenging accepted opinion', and the fact that EPR fulfilled this role is clear from the very fact there *was* a conference to discuss it 50 years later.

But the word has come to mean, particularly for mathematicians and hence scientists, a self-contradictory argument – a proof that 2 is equal to 1, for example. Taking this definition, to use the word 'paradox' for EPR would take it as a claim of a fundamental *illogicality* or even *absurdity* in the Copenhagen position, a stronger claim probably than even most of its proponents would wish to make. The confusion over definition has, I imagine, led to much of the acrimony over use of the word. EPR is best described as 'a problem' or 'an example' rather than 'a paradox'.

While the central idea of EPR belonged to Einstein, the selection of wave-functions was performed by Rosen, and the logical analysis by Podolsky, who apparently wrote the final version of the paper after much general discussion. This division of labour caused problems, because Einstein felt that Podolsky, the expert in formal logic, biased the paper too much in that direction. In a letter to Schrödinger reported by Jammer [104], he said that 'It has, however, not really brought out what I actually had in mind, since the principal matter is, so to say, buried under learnedness' [or, in another translation [94], 'smothered by the formalism'].

Because of this, and because Einstein wrote several versions of the scenario on his own later, I won't follow the account given in the actual EPR paper in detail. In fact, the actual thought-experiment is fairly simple. One particle at rest decays into two particles, 1 and 2. From conservation of momentum, these two particles must move off in opposite directions, the magnitudes of their momenta being equal, both p, let us say. This implies that at any moment their positions are related; if you measure the distance particle 1 has travelled, x_1, it is clear that the distance travelled by particle 2, x_2, will be equal and opposite to x_1.

This is true classically. The new point in quantum theory is that, until a measurement is made, according to orthodox views p, x_1 and x_2 have no values. And emphatically one of the particles, particle 2, say, cannot have precise values for both momentum and position. Certainly, if we tried to measure both directly we would run into the problems that Einstein met in the first round of the Bohr–Einstein debate. But the argument of EPR is much more subtle.

Let us imagine that we measure the momentum of particle 1. By our previous argument, this also tells us the momentum of particle 2. However, EPR bring in the postulate of locality, as discussed in a previous section (and use of this postulate is *central* in any argument of EPR type). With this postulate, it is impossible that the measurement on particle 1 could have affected particle 2 (at least instantaneously, and we would assume that particle 2 has a precise value immediately after the measurement on particle 2; we could, after all, measure its momentum at such a time). So if particle 2 now has a precise value of momentum, it must have had it *before* the measurement on particle 1.

Similarly we may imagine ourselves measuring the position of particle 1. Again this tells us instantaneously the position of particle 2. We use *locality* to say that our measurement did not interfere with particle 2, and so particle 2 must have had this value of position *before* the measurement on particle 1.

So before any measurement, and hence, in fact, at any time, particle 2 must have had a precise position *and* a precise momentum. These are, then, both elements of physical reality. Since quantum theory does not allow values for both, it must be *incomplete*. Such, I think, is the original EPR argument, stripped of the mathematics and some rather formal conceptual issues. Note that we do *not* need, even in thought, to perform two measurements on particle 1; as already mentioned, they would interfere with each other if we did. Rather it is the *possibility* that either *could* be performed that leads EPR to deduce that particle 2 has precise values for both quantities.

I will now make two more technical points, the first rather specific and not generally important, but the second very general and of great importance for the rest of the book. The first point is that there is a problem with the way I have described the set-up above in that I implied the source particle had known position and zero momentum. As Dickson [105] has stressed, this very common way of talking about EPR actually contradicts the uncertainty principle for the source. EPR avoid the difficulty by refraining from mentioning the source, and relying purely on the mathematical form of the wave-function of the two particles, without worrying about how it has come about [106].

The second point is that, from a mathematical point of view, the source of the opportunity (for EPR), or the source of the problem (for others) is that the wave-function of the combined system prior to any measurement is *not* a simple product of an expression for each particle. If it *were*, the properties of each particle would be independent, and measurements on one could tell us nothing about the other. Here, though, the wave-function is a sum of many such terms (in fact, an infinite number, each corresponding to a different value of p), and the results of measurements on the two particles must be correlated; we may say that the wave-function of the combined system is *entangled*. This type of wave-function will usually occur when two particles have interacted at one time and then separated. (In our non-mathematical account of EPR, the fact that they were produced by the decay of a single particle is an example of this.) The future behaviour of the particles – as displayed, of course, by measurements on each – is correlated.

The aspect of locality involved in the EPR argument is *outcome independence*, since it relates to correlation of measuremental results. Thus the laws of relativity do not dictate the acceptance of locality here, but, in any case, as I stated before, for Einstein *both* aspects of locality were sacrosanct.

When Einstein discussed experiments of this type in later years, he preferred a more general treatment, as in a paper written in 1948 and included with the Born–Einstein letters [5]. Along the lines of EPR, his system S consisted of two sub-systems S_1 and S_2, separated in space so that Einstein could, as usual, make use of *locality*. S_1 and S_2 have interacted in the past so the combined wave-function is entangled. If one accepts the projection postulate of the previous chapter, different measurements on S_1 will naturally lead to different final wave-functions for S_1, but, because of the entangled nature of the combined wave-function, they also steer S_2 into different wave-functions. But because of locality, these measurements on S_1 cannot change the real physical state of S_2. So we may say that a given system in a given physical state may have a number of different wave-functions – just Einstein's definition of incompleteness.

If EPR showed that quantum theory was incomplete, how did Einstein think it should be completed? He remarked [99] specifically in connection with experiments of this type that 'coordination of the ψ-function to an ensemble of systems eliminates every difficulty'. Particularly if one sticks to the first account above (the one closer to the original EPR), the natural interpretation of that remark is in terms of the Gibbs ensemble, as discussed earlier in this chapter, or, more generally, some sort of hidden-variable theory. This can, of course, eliminate the difficulty in a straightforward way; as with Bertlmann's socks, the correlations between the properties of the two sub-systems are built in from the outset via the hidden variables, and not, in any sense, created by the measurement on one of the sub-systems. There is no doubt that Bell assumed this was Einstein's position, and, regarding himself as a follower of Einstein, made it the starting-point of his own important developments, which are discussed in the next chapter. (For the argument that Bell misunderstood Einstein, see the end of this chapter.)

Lastly in this section, I discuss a variant of the EPR idea introduced by David Bohm [107]. There were technical difficulties with the original EPR experiment [67]; among these difficulties, eigenfunctions of momentum extend over all space, so it was difficult to maintain locality for two such functions. Bohm hoped to avoid such problems.

His thought-experiment consisted of a particle of zero spin decaying into two spin-$\frac{1}{2}$ particles that move in opposite directions along, say, the y-axis. Because of conservation of spin angular momentum, all components of the *total* spin must be zero, but a measurement of, say, the z-component of spin of particle 1 must yield either the result $\hbar/2$ or $-\hbar/2$. This then immediately fixes the z-component of spin of particle 2; it must be $-\hbar/2$ or $+\hbar/2$, respectively, to make the total z-component equal zero. But *locality* will tell us that the measurement on particle 1 cannot have affected particle 2, which must therefore have had this value before the measurement. However, one can repeat the argument for the x-component, or any other component, of the spins, and so conclude that *all* components of each spin have precise values at all times; since quantum theory does not provide these, it must be incomplete.

Bohm's thought-experiment did the same job as EPR, but it avoided the technical problems I mentioned above, and was also rather simpler than EPR itself. While the EPR wave-function is the sum of an infinite number of products of an expression for each particle, in the Bohm case there are only two such products. The Bohm function, in fact, is of the form $(1/\sqrt{2})[\alpha_+(1)\alpha_-(2) - \alpha_-(1)\alpha_+(2)]$, as given in Chapter 4. This indicates that the state with s_z equal to $\hbar/2$ for particle 1 is correlated with that with s_z equal to $-\hbar/2$ for particle 2, and vice versa. If there were just one term, we would have a simple product of a term for each spin, and the behaviour would be uncorrelated. The fact that there are two such terms means that the state is *entangled* – a measurement on either particle selects the value of s_z for *both*.

As written, the wave-function seems to be specific to z-components. Mathematically, though, the same physical state may also be represented by the same mathematical function, but with z replaced by x, y or indeed any other direction in space. This is very much part of the thought-experiment as discussed above; our measurement may be of *any* component of spin.

In an interview [108] in 1986, Bohm confirmed that Einstein 'saw [the scheme] and he thought it was good'. Because of its simplicity, physicists usually discuss Bohm's idea rather than the original version of EPR itself; Bohm's version is sometimes called EPR–Bohm, and often, in fact, just EPR.

Entanglement is a crucial concept, running through from EPR to the work of Bell described in Chapter 7 and on to quantum information theory, the subject of Chapter 9, and I shall take the opportunity now to discuss it a little further.

If the wave-function for the system were $\alpha_+(1)\alpha_-(2)$, we should describe this as non-entangled or *separable*. We may say particle 1 is 'in' state α_+, particle 2 'in' state α_2. This of course gives only partial information about what results we will obtain if we perform measurements on either particle. For example, if we measure s_z for particle 1 we know we will obtain $+\hbar/2$, but if we measure s_x or s_y we have equal probabilities of obtaining $+\hbar/2$ and $-\hbar/2$. But the crucial point is that the results of measurements on the two particles are entirely independent. The result we obtain for a measurement on particle 1 has no bearing on what result we may obtain for a measurement of particle 2.

For an entangled wave-function, as we have said, the situation is entirely different. At least for some possible measurements, results in what we may call for the EPR or EPR–Bohm case the two wings of the experiment are correlated. As we have seen, for EPR–Bohm knowledge of the result of a measurement of s_z in either wing gives complete knowledge of the result of the same measurement in the other wing. It is very difficult to build up any physical picture of two particles in an entangled state (even more difficult, it might be said, than for the states without entanglement we have been considering so far in this book).

The possibility of performing measurements on entangled systems also puts constraints on assumptions that may be made regarding locality, realism and incompleteness. Since a measurement in one wing gives immediate information on the result in the other wing, we are left with two possibilities. Either information has passed instantaneously from one wing to another in violation of locality, or the measurement results existed before any measurements were made, which tells us quantum theory is incomplete and must be completed by some form of realism.

EPR, of course, were happy to assume locality, which meant they deduced incompleteness and a requirement for realism. Both locality and realism were highly congenial to Einstein, of course, so he had no awkward choices. The work of Bell, again based on entangled wave-functions, would later make this position untenable, and for several decades entanglement would be thought of as a major conceptual difficulty. Only much more recently have physicists been prepared to take entanglement as a fact of Nature, or actually a resource, central in the creation of quantum information theory.

Bohr on EPR

Einstein received many letters about EPR 'eagerly pointing out to him just where the argument was wrong. What amused him was that, while all the scientists were quite positive that the argument was wrong, they all gave different reasons for their belief' (as reported by Banesh Hoffmann [67]).

The most important reaction had to be that of Bohr, on whom the initial effect of EPR was quite dramatic. Rosenfeld, who was assisting him at the time, reports that 'This onslaught came down on us as a bolt from the blue. Its effect on Bohr was remarkable' [74]. Clearly the EPR paper had, as Einstein hoped it would, challenged many of Bohr's arguments with great effect. In particular, the *disturbance interpretation*, discussed in detail in the preceding chapter, had to be let go of. As Bohr said in his reply to EPR [109], 'there is . . . no question of a mechanical disturbance of the system under investigation during the last critical stage of the measuring procedure'; a measurement on particle 1 cannot mechanically disturb particle 2.

Rosenfeld says that 'as soon as Bohr had heard of Einstein's argument, everything else was abandoned: we had to clear up such a misunderstanding at once'. Notice how Rosenfeld is not prepared to admit that Einstein actually had a good argument, albeit one that Bohr and Rosenfeld were able to rebut; it could only be a 'misunderstanding'. As misunderstandings go, it must have been rather a profound one, for 'day after day, week after week, the whole argument was scrutinized with the help of simpler and more transparent examples. Einstein's problem was reshaped and its solution reformulated with such precision and clarity that the weakness in the critics' arguments became evident, and their whole argumentation, for all its false brilliance, fell to pieces. "They do it smartly", Bohr commented, "but what counts is to do it right".' Rosenfeld adds that 'When one realises the fundamental nature of the issue at stake, it becomes easier to understand the state of exaltation in which Bohr accomplished this work . . . . It was impressive to watch him thus at the height of his powers, in utmost concentration and unrelenting effort to attain clarity through painstaking scrutiny of every detail . . . He was particularly well served on this occasion by his uncommon ability to go into the opponent's views, dissect his arguments and turn them to the advantage of the truth.'

For all this effort, Rosenfeld is not able to contemplate the possibility that EPR forced some change in Bohr's approach from 'disturbance' to the idea, discussed in the previous chapter, of *wholeness* and *the phenomenon*. 'The refutation of Einstein's criticism', he says 'does not add any new element to the conception of complementarity.'

In fact, the ideas of EPR *as originally presented* seem comparatively easy to dispose of using the precepts of complementarity. For the whole apparatus of Einstein's thought-experiment – the idea that one *might* do various experiments and obtain various results, that one can draw conclusions from each of these possibilities, and put the results together to spread confusion – all this becomes strictly illegitimate.

If one wishes to discuss the measurement of momentum for the first particle, one must do so in the context of an actual measurement. By all means one may then deduce the value of the momentum of the second particle. Indeed, if we wish, we may regard the set-up as an *indirect measurement* of this quantity. (Bohr [109], incidentally, was quite happy with this form of words.) Many classical measurements are of precisely this form; one observes a direct manifestation, not of the particle under investigation, but of another particle or collection of particles with which it has interacted in the earlier stages of the measurement procedure. During this interaction, the particle under investigation and the observed particles have had their properties correlated.

What holds for a classical measurement should also hold for a quantum one, according to complementarity, and *provided that the rules are obeyed*. The rules insist that, if such a measurement is imagined, one *cannot* at the same time contemplate a measurement of the *position* of the first particle. Einstein's scheme is debarred. In the absence of experimental equipment to measure a particular property of a particle, that property simply cannot be discussed.

Having admitted that a measurement on particle 1 causes no mechanical disturbance on particle 2, Bohr continues – 'But even at this stage there is essentially the question of an *influence on the very conditions which define the possible types of predictions regarding the future behaviour of the system* [Bohr's italics]. Since these conditions constitute an inherent element of any phenomenon to which the term "physical reality" can properly be attached, we see that the argumentation of the mentioned authors does not justify their conclusion that quantum-mechanical description is essentially incomplete . . . This description may be characterised as a rational utilization of all possibilities of unambiguous interpretation of measurements, compatible with the finite and uncontrollable interaction between the objects and the measuring instruments in the field of quantum theory.'

Bohr gives a brief account of the ideas of complementarity, along the lines of ours in the previous chapter, and adds that '[these] remarks apply equally well to the special problem treated by Einstein, Podolsky and Rosen . . . which does not actually involve any greater intricacies than the simple examples discussed above'.

While these comments may clear up in a satisfactory way the EPR difficulty *as posed*, from the point of view of complementarity, it cannot have escaped Bohr's attention that the same type of physical situation could be analysed in a way not so susceptible to his reply, because it involves actual rather than hypothetical measurements. One could imagine actually measuring the momentum of particle 1, and then claiming (rightly) to know what result one would obtain in a subsequent immediate measurement of the momentum of particle 2, although one has not interacted with this particle. Or one could just measure the momenta of both particles and exhibit the perfect correlation.

Strictly, one might say that such experiments caused complementarity and Bohr no problems. Provided that the experimental arrangements are allowable from the complementarity point of view, and here they are because both measurements can be carried out simultaneously, we may discuss the experiment as if it were classical, and of course classically the momenta *would* be equal and opposite.

Yet this argument cannot appear convincing. Most physicists will see a measurement on particle 1 as affecting particle 2, and will want to know how this may be explained, or, at the very least, explained away; the response that there is no problem just will not do. Bohr's reply [84] was *wholeness*. In the last chapter I discussed this concept for a simple measured system and measuring device, which may *not* be considered separately, but only as a combined system with shared properties. In the EPR context, this idea is extended to the whole system consisting of *both* particles and any measuring devices involved. Thus it was natural that measurements on the two particles should be correlated, even if, for example, they were carried out simultaneously; they were really two measurements on one coupled system.

One might immediately assume that Bohr must be violating *locality*. Murdoch [56], however, thinks not. He argues that Bohr wishes only to give up the *principle of independent existence*, according to which the real states of spatially separated objects are independent of each other. As we shall see, Einstein was prepared to concede the distinction, though he approved of violating the principle of independent existence no more than of violating locality!

There is another experiment that may be carried out in EPR situations. One may measure the momentum of particle 1 and the position of particle 2. At first sight, this seems exceptionally dangerous for complementarity, because if, as we have done so far, we assume that the measurement on particle 1 gives us a value of the momentum of particle 2, it seems that we have achieved simultaneous exact measurements of the momentum and position of particle 2.

Bohr recognised the problem, and tried valiantly to avoid it [109] by imagining a physical model of the rather mathematical EPR analysis, and arguing that in this experiment simultaneous exact measurement of x_2 and p_1 is impossible (along the lines of the previous rounds of the Bohr–Einstein debate).

With due respect to Bohr, I am unconvinced by this argument. In the original EPR experiment, it does not seem clear that there is a problem in measuring p_1 and x_2 independently, and the point seems clearer still if one moves to the EPR–Bohm set-up and the measurement of s_x for particle 1, say, and s_z for particle 2.

Nevertheless, I think that there is a straightforward answer, which may be seen by returning to a classical analogy. We have already agreed that, in a classical sense, if two systems have interacted, a measurement on the first may give valuable information on the second – or, in fact, may constitute an indirect measurement on the second. But, of course, this can be true only if the second system has been left undisturbed. Once the second system *is* disturbed, any information obtained from measurements on the first becomes irrelevant.

Similarly here in the quantum context, the measurement of p_1 may provide the value of p_2, only until or unless a measurement of x_2 is performed – a good old-fashioned interacting disturbing type of measurement. At this stage, knowledge of p_1 becomes useless for saying anything about p_2. So we know x_2 but not p_2, and the day is saved for complementarity.

I would point out that, as usual, much of the argument of complementarity in the EPR case is semantic or linguistic, or at least may be expressed in such terms. Denial of the principle of independent existence is a command to forget, or at least to modify, many of the ways of using language we learn in childhood. In EPR, a measurement on particle 1 does not cause a mechanical disturbance on particle 2, but it does cause what Fine [94] calls a *semantic disturbance*; what we *say* about particle 1 affects what we can *say* about particle 2.

I now want to ask three questions about Bohr's response. First, did it mark a retreat into positivism? Here there is disagreement. Folse [55] would argue not. Apart from the possible trauma of losing the disturbance interpretation, he says, the essential thrust of Bohr's position was unchanged. Feyerabend

[110], though, who was not ill-disposed towards complementarity, thought that the forced move from the disturbance idea marked a move to a *positivism of higher order*, the acknowledgement that, while one will continue to talk of a quantum-mechanical object, 'such an object is now characterized as a set of (classical) appearances only, without any indication being given as to its nature'.

Fine [94] goes much further. He regards Bohr's response to EPR, particularly the italicised section above ('influence on the very conditions . . .'), as 'virtually textbook neopositivism. For Bohr simply identifies the attribution of properties with the possible types of prediction of future behaviour.' Fine adds 'I think this point needs emphasizing, for many commentators seem inclined to suppose that Bohr's tendency to obscure language is a token of philosophical depth, whereas I find that, as here, where it really matters Bohr invariably lapses into positivist slogans and dogmas.'

My second question [106] is as follows. What was Bohr's actual claim concerning the EPR paper and his own response? Was he arguing that the EPR analysis was flawed in principle, and he was able to refute it on straightforward physical grounds, just as for the previous rounds of the Bohr–Einstein debate? This is what one would assume given the remarks of Rosenfeld and Bohr himself already quoted.

An alternative suggestion is that he was merely showing that the EPR situation could be handled consistently by complementarity, in other words that, if one makes the assumptions involved in complementarity, the conclusions of EPR, inimical of course to complementarity, do not follow. This, as we have seen, is relatively easy to accomplish. Thus any difficulties for complementarity are removed, but of course the argument would by no means have defeated EPR, who do not accept complementarity in the first place!

There may have been an implicit assumption that complementarity had been proved correct by, for example, the first two rounds of the Bohr–Einstein debate. Admittedly we saw that the conclusions drawn from these rounds of the debate could be viewed as positive towards complementarity, but they certainly did not make its adoption inevitable. Indeed, in complete contrast, the work of Bohm, to be discussed in Chapter 7, made it quite clear that complementarity was merely one option. Cushing [111] has noted the unwarranted slide in Bohr's views from the consistency of complementarity to that of its necessity.

I would suggest that Bohr was somewhat stranded between these two options. He was drawn into going further than the strict requirements of complementarity, but was not able to go far enough to deal with the genuinely physical difficulties coming from EPR. For further discussion the reader is referred to a number of recent publications [105, 106, 111–114] on Bohr's response to EPR.

I now pass on to what is, perhaps, the most significant question – did Bohr convince physicists that his answer disposed of the EPR challenge in a convincing way? Here I shall give two rather different answers.

Bohr himself and his supporters certainly regarded his response as a great triumph of *his ideas*. (I think it fair to say that Bohr would never have wanted a *personal* triumph over Einstein.) In the years following 1935, the overwhelming majority of physicists would have supported him. For them, Einstein was merely being reactionary. Bohr had already shown that Einstein's way of looking at things was wrong – it was not surprising that, if you followed it, you ended up with paradoxes! After all, Einstein's writings seemed to require hidden variables, and von Neumann had supposedly shown that these were forbidden. It is not to be supposed that most of these physicists examined Bohr's reply in depth, or, indeed, at all; for them it was enough that Bohr had claimed to demonstrate Einstein's mistakes. (One should not, perhaps, criticise these physicists unduly; they were too busy *using* quantum theory to understand atoms, nuclei, solids and so on to worry about the actual meaning of the theory.)

But when, and to the extent that, physicists did examine the arguments in detail, often their feelings changed, and this has been happening, probably in an increasing way, since the 1960s. Without the disturbance interpretation, Bohr's arguments often seemed unduly abstract, and linguistic rather than physical.

John Bell was a strong supporter of such a reappraisal. In a 1981 paper [76] on EPR and related matters, and having stated that he imagined he understood the position of Einstein (ironically a supposition that has been questioned – see the last section of this chapter), he reports that 'I have very little understanding of the position of . . . Bohr. Yet most contemporary theorists have the impression that Bohr got the better of Einstein in the argument and are under the impression that they themselves share Bohr's views.'

Having quoted Bohr much as we have done, Bell says that 'I do not understand in what sense the word "mechanical" is used, in characterising the disturbances which Bohr does not contemplate, as distinct from those which he does [the ones we have called 'semantic'] . . . Could [Bohr] just mean that different experiments on the first system give different kinds of information about the second? But this was just one of the main points of EPR . . . And then I do not understand the final reference to "uncontrollable interactions between measuring instruments and objects", it seems just to ignore the essential point of EPR that in the absence of action at a distance, only the first system could be supposed disturbed by the first measurement, and yet definite predictions become possible for the second system. Is Bohr just rejecting the premise – "no action at a distance" – rather than refuting the argument?'

Different as it may have seemed in 1935 and 1965, different as it may still largely seem in 2005, I think EPR was a triumph for Einstein. It removed the attractive disturbance interpretation from Bohr, leaving him with a set of ideas that were certainly not illogical, not necessarily positivistic, but on the whole rather unconvincing from a physical point of view. It was a crucial blow, which played a large part in ensuring that eventually physicists would at least contemplate alternatives to Copenhagen.

Schrödinger's cat and Wigner's friend

Following the publication of the EPR paper, between June and October 1935, Einstein and Schrödinger had a period of intense correspondence. (Fine [94] discusses this in detail.) EPR had stimulated Schrödinger to a sustained analysis and criticism of the Copenhagen position on quantum theory, and here he and Einstein were at one. Schrödinger, in particular, developed arguments that both of them saw as demonstrating the violation of locality by Copenhagen. When it came to the solution of the problems, there was rather more disagreement, though all carried out in the friendliest of tones. As already reported, for example, Schrödinger came up with some simple arguments against Einstein's suggestion of ensembles.

Initially merely for his own benefit, Schrödinger wrote up his ideas, and the account was subsequently published [100]. (In this paper, Schrödinger refers to EPR with the remark that 'The appearance of this work motivated the present – shall I say lecture or general confession?') The whole piece makes fascinating reading, but we shall focus on two fairly brief remarks. The first is that Schrödinger extracted the central feature from the EPR paper and gave it the name of *entanglement*. He remarked 'I would call this not *one* but *the* characteristic trait of quantum mechanics, the one that enforces the entire departure from classical thought.' We shall meet entanglement a great deal throughout the rest of the book.

The other novelty came in a single paragraph that found lasting fame – the introduction of the notorious *Schrödinger's cat*. 'One can', Schrödinger says, 'even set up quite ridiculous cases. A cat is penned up in a steel chamber, along with the following diabolical device (which must be secured against direct interference by the cat); in a Geiger counter there is a tiny bit of radioactive substance, *so* small that *perhaps* in the course of one hour one of the atoms decays, but also, with equal probability, perhaps none; if it happens, the counter tube discharges and through a relay releases a hammer which shatters a small flask of hydrocyanic acid. If one has left this entire system to itself for an hour, one would say that the cat still lives *if* meanwhile no atom has decayed. The first atomic decay would have poisoned it. The ψ-function of the entire system would express this by having in it the living and the dead cat (pardon the expression) mixed or smeared out in equal parts.'

One may be relatively happy, Schrödinger is saying, that the wave-function of an atom or an electron consists of a sum of terms corresponding to very different states – for a radioactive atom, for example, one part with the atom surviving, one part with it decayed. Remember that, according to orthodox views, this cannot be taken as meaning that some atoms of an ensemble survive, some have decayed; the smearing, as Schrödinger calls it, applies to each atom.

We are likely to be much less happy that the same applies to a macroscopic object like a cat. Can it really be the case that, as long as no observation is made, or, as Schrödinger says, for as long as the system is left to itself, the cat is somehow smeared out so that it is half living and half dead? At least according to von Neumann, the idea would be that at the end of the hour the observer opens the box, collapses the wave-function, and observes either a dead or a living cat. (So it is really the observer who either kills or saves the cat!)

Dramatic as the conception of Schrödinger's cat certainly is, it is much less of a problem for conventional ideas of quantum theory than EPR. As I have already implied, it is particularly a difficulty for von Neumann's ideas, which stress the human observer as the sole collapser of wave-functions. In contrast, Bohr would never allow a wave-function to be written down for any macroscopic object. For him, the cat, as a classical object, must have a classical state – dead or alive – at all times. (Of course it might be suggested that the cat problem puts added stress on a weak point of Bohr's ideas – the necessity of a cut between quantum and classical – but the *additional* effect it causes is probably not large.)

One may also look at the Schrödinger's-cat argument as a further attempt to demonstrate the incompleteness of quantum theory, thus following EPR. It could be said that there must be a hidden variable to tell us whether each atom really has decayed or not; this will then feed through to a hidden variable telling us whether the cat is dead or alive. Since conventional quantum theory does not contain these hidden variables, it must be incomplete. From this point of view, though, the cat 'paradox' turns out to be a lot less effective than EPR. For the cat, completion of the theory by addition of hidden variables removes all difficulties. Because of the part played by locality in EPR, it is a much more subtle 'paradox', and use of hidden variables only brings out more facets of the problem – as I shall explain in the next chapter.

To finish this section, I shall explain another problem or 'paradox' of quantum interpretation – that of *Wigner's friend* [92]. In structure the problem is analogous to Schrödinger's cat, but with the cat replaced by a human being.

This human being, friend of the actual observer, is not killed as the cat may be, but merely has the opportunity to observe whether an event like the decay of a radioactive atom takes place or not. That being so, the observer has to decide whether the friend collapses the wave-function herself, or remains herself in a state of suspended animation until the actual observer collapses the combined wave-function of atom and friend.

As Wigner says, it would appear solipsistic (self-centred) in the extreme to believe that one's friend does not have the same type of impressions and sensations as oneself and therefore cannot collapse wave-functions; it would imply that the world has no reality, and that the sensations reaching one's own brain are really no more than sensations. His general conclusions would be supported by both Bohr and von Neumann. Wigner used the argument to stress his own view of the importance of the consciousness or mind of living systems.

Einstein's demands for a quantum theory

Earlier in this chapter I reviewed the solution Einstein often proposed to remove the difficulties of quantum theory – ensembles – and provisionally interpreted these as Gibbs ensembles. At that point, though, I warned the reader that such an approach was certainly partial, and possibly misleading. In the last two sections of this chapter, having discussed, in particular, EPR, I want to return to analyse Einstein's beliefs in rather greater depth.

The first question I wish to ask is this: given that Einstein was a believer in locality, realism *and* determinism, were there any of these he would even consider relinquishing, in the light of the success of quantum theory, and in order to protect the others?

I shall start to answer this question by remarking that Einstein's belief in locality has, I think, never even been challenged, and to locality should be added independent existence, as defined in connection with Bohr's response to EPR. In his autobiographical notes in the Schilpp volume [73], he wrote '[O]n one supposition we should, in my opinion, absolutely hold fast: the real factual situation of the system S_2, is independent of what is done with the system S_1, which is spatially separated from the former.' Having discussed briefly the EPR type of problem, and his own conclusions from it, he adds that 'One can escape from this conclusion only by either assuming that the measurement of S_1 (telepathically) changes the real situation of S_2 or by denying independent real situations as such to things which are spatially separated from each other. Both alternatives appear to me entirely unacceptable.' These two possibilities are just denial of locality and independent existence.

Again, at the end of the same volume, specifically discussing EPR, Einstein says that: 'Of the "orthodox" quantum theoreticians whose position I know,

Niels Bohr's seems to me to come nearest to doing justice to the problem', but characterises Bohr as believing that 'if the partial systems A and B form a total system, there is no reason why any mutually independent existence (state of reality) should be ascribed to the partial systems A and B viewed separately, *not even if the partial systems are spatially separated from each other at the particular time under consideration.*' Einstein, in contrast, insisted on independent existence.

As to realism, again there is abundant evidence that Einstein would not have conceived of abandoning this. The quote above talks of 'the real factual situation', and such terminology appears throughout Einstein's work. Quite what he meant by realism may not be quite so clear-cut.

Two important aspects may be identified in realism. The first – and usually it is the one most stressed – is that one's theories should relate to a real external world. The second is that the theories should not explicitly involve the observer; they should give access to an 'observer-independent realm' [94].

My wording is not particularly precise, but it does enable us to make an important point about Einstein. For him the second aspect was extremely important – central, one might say, in his approach to physical theory. He could not accept observers making an appearance in fundamental theory. The certainty of a clash with orthodox interpretations of quantum theory is obvious.

On the first aspect above we may be somewhat less sure. There are many places in his writings where he may easily be interpreted in such a way; in his autobiographical notes, for example, he says that 'Physics is an attempt to grasp reality as it is thought independently of it being observed', which encompasses, to an extent, both aspects of the word. But Fine [94] argues against the rather common idea that Einstein held unswervingly to a real external world that should be mapped in physical theory. He points out that often, when discussing realism, Einstein retreats into what it means for a theory to be realist, rather than advocating anything like correspondence with truth.

For example, Einstein says [73] that 'the "real" in physics is to be taken as a type of program, to which we are, however, not forced to cling *a priori*. No one is likely to be inclined to attempt to give up this program within the realm of the "macroscopic" (location of the mark on the paperstrip "real"). But the "macroscopic" and the "microscopic" are so inter-related that it appears impracticable to give up this program in the "microscopic" alone.' So Einstein clings to realism, even in the atomic realm, but only as a 'programme', not necessarily as a matter of deep conviction.

Fine points to another revealing passage in a history of physics Einstein wrote with Leopold Infeld in 1938 [115]. Following Descartes, they explain how building a theory is like attempting to understand the workings of a closed watch – 'If [a scientist] is ingenious he may form some picture of a mechanism which could be responsible for all the things he observes, but he may never be quite sure his picture is the only one which could explain his observations. He will never be able to compare his picture with the real mechanism, and he cannot even imagine the possibility or the meaning of such a comparison. But he certainly believes that, as his knowledge increases, his picture of reality will become simpler and simpler, and it will explain a wider and wider range of his sensuous impressions. He may also believe in the existence of the ideal limit of knowledge and that it is approached by the human mind. He may call this ideal limit the objective truth.'

One may perhaps say that Einstein was determined that physics should be observer-independent. While not actually objecting to the form of words that his theories may relate to a real external world, in practice one may be a little suspicious of such an idea, and require only that the picture of reality painted by his theories became steadily simpler, in the sense of relying on fewer though more powerful basic statements, and better capable of dealing with experience.

Now I come to determinism. Here there is considerable room for discussion. Perhaps the reader finds this statement unexpected, for among the very most quoted of Einstein's sayings is 'God does not play dice', surely a cry from the heart for determinism!

Yet what at least seems to be opposing evidence may be found in the Born–Einstein letters [5]. In 1953 and 1954, the correspondence took a rather sharper tone than usual, as dissenting opinions on quantum matters rose to the surface. Fortunately, while on a visit to Princeton, Pauli acted as an intermediary. In particular, while agreeing with Born that Einstein had 'got stuck in his metaphysics', Pauli characterised this metaphysics as realistic *not* deterministic. Einstein, Pauli wrote to Born, 'was *not at all* annoyed with you, but only said you were a person who will not listen'. Pauli, indeed, 'was unable to recognise Einstein' when Born discussed his views. It seemed to Pauli that Born 'had erected some dummy Einstein which [he] then knocked down with great pomp. In particular,' Pauli stressed, 'Einstein does not consider the concept of "determinism" to be as fundamental as it is frequently held to be (as he told me emphatically many times) . . . [H]e *disputes* that he uses as criterion for the admissibility of a theory the question: "Is it rigorously deterministic?".' Pauli added that 'Einstein's point of departure is "realistic" not "deterministic".'

Bell in turn stressed these remarks of Pauli, saying that 'It is important to note that to the limited degree to which *determinism* plays a role in the EPR argument, it is not assumed but *inferred*. What is held sacred is the principle of "local causality" – or "no action at a distance".' Bell's own work, which built on EPR, was on exactly these lines.

Yet I don't think one should accept too readily Einstein's willingness to drop determinism. Fine [94] argues that, although Einstein may have put realism first, determinism is always there. He quotes an interesting letter of 1950 from Einstein to Michele Besso [116], in which Einstein says that 'The question of "causality" is not actually central, rather the question of real existents, and the question of whether there are some sort of strictly valid laws (not statistical) for a theoretically represented locality.' Even in a sentence where Einstein is explicitly relegating causality or determinism to a lower place than realism (the question of real existents), strictly valid non-statistical laws squeeze themselves in, and we seem to be back to determinism.

Overall I see no real evidence from Einstein's writings that he was genuinely prepared to sacrifice determinism. We may take it from Pauli's remarks that locality and realism were first and second in Einstein's theoretical demands, but I suspect that determinism remained high on the list as well.

Einstein's ideas for a quantum theory

How was it that Einstein felt able to keep all these conceptual virtues, as he saw them, and yet still aim to improve (or complete) the quantum theory, while most other physicists felt that some, at least, of these demands would have to be abandoned? The answer is that he did not really want to improve the present ideas; he wanted a fresh beginning. In the Schilpp volume [73], he said that 'contemporary quantum theory by means of certain definitely laid down basic concepts, which on the whole have been taken from classical mechanics, constitutes an optimum formulation of the [statistical] connections', but continued that 'I believe, however, that this theory offers no useful point of departure for future developments.'

'Just as . . . one could not go from thermodynamics . . . to the foundations of mechanics' [99], just as Einstein could not have hoped to argue from the perihelion of Mercury to the general theory of relativity, so he felt it useless to hope to argue from the difficulties of the present quantum theory to the theory he envisaged, by adjusting or augmenting the formalism as we have it (completion 'from within', in Fine's phrase [94]).

This helps to explain his unwillingness to expand on the ensemble idea – to explain what kind of ensemble he had in mind, to try to eliminate the problems of ensembles discussed earlier in this chapter. As Guy and Deltete [95] say, Einstein believed that, while ensembles were the only way to solve the internal problems of quantum theory, they were 'inadequate to the task of describing the real physical states of individuals', so developing the interpretation would have been 'a waste of time'. Fine [94] says that ensembles 'provided no more than a setting rhetorically apt for calling attention to the incompleteness of the quantum theory', adding that 'Einstein chose his rhetoric cunningly. For who, learning that a theory is incomplete, could resist the idea that one ought to complete it?' But 'Einstein is thus far less concerned to *show* what ensembles can do than he is to insist what they cannot do' [95]; 'Einstein *wanted* his purely *interpretative* scheme to appear heuristically sterile [sterile for the purpose of intuition]' [94].

For Einstein, the theory had to be completed 'from without'. In his autobiographical notes [73], he asks 'What can be attempted with some hope of success in view of the present situation of physical theory?' His immediate response was to remark that 'it is the experiences with the theory of gravitation [general relativity] which determine my expectations. These equations give, from my point of view, more warrant for the expectation to assert something *precise* than all the other equations of physics.'

Einstein's self-imposed task was to provide a *unified field theory*. He took general relativity as his starting-point, and aimed to describe gravitation and electromagnetism as parts of one unified field. He hoped that, once this was accomplished, suitable approximations would yield quantum theory, as we know it today, as a set of statistical rules giving correct predictions, at least for the range of phenomena studied at present.

It was clear that such a theory must contain a large set of very complicated equations. Here [73] Einstein made completely clear what he regarded as the means of obtaining such equations. 'I have learned something else from the theory of gravitation', he reported; '[n]o ever so inclusive collection of empirical facts can ever lead to the setting up of such complicated equations. A theory can be tested by experience, but there is no way from experience to the setting up of a theory.' The only way could be 'through the discovery of a logically simple mathematical condition which determines the equations completely or (at least) almost completely. Once one has these sufficiently strong formal conditions, one requires only little knowledge of facts for the setting up of a theory.'

Einstein's successful experience with general relativity had led him to a very definite idea of how major theories should be constructed. The process involved little empirical input, but it required the discovery of a logically simple general principle, and willingness to work with vastly difficult equations. The method was not to be successful in this case. Einstein [73] described how he attempted to modify the mathematical structure of general relativity, but had to admit to 'many years of fruitless searching'. Nevertheless he gave a progress report, saying that 'the theory here proposed, according to my view, represents a fair probability of being found valid, if the way to an exhaustive description of physical reality on the basis of [a field model] turns out to be possible at all'.

It is fair to say that most physicists of the time and since have regarded his work as a regrettable waste of time, a hopeless search for the unattainable. Einstein was trying to unify just two fields (or types of interaction) – electromagnetism and gravity. More recently it has been realised that there are two other fundamental fields [117]. The first is the *strong nuclear field*, which causes protons and neutrons to bind together to form nuclei, and, at a more basic level, binds quarks together to form the protons and neutrons themselves. The second is the *weak nuclear field*, which causes radioactive decay. Einstein's most abstruse calculations may be seen to have been attempting to tackle only part of the problem, and actually the more difficult part, since it involved gravity.

Progress towards unification *has* been made, albeit, I would stress, by methods very different from those of Einstein. In 1967, Abdus Salam and Steven Weinberg were able to produce a theory unifying the weak and electromagnetic fields, using a method known to physicists as *gauge theory*; predictions made by this unification – specifically the existence of new elementary particles – have been verified experimentally. Much effort has been applied more recently to including the strong nuclear interaction, and even gravity, in this scheme, but though interesting experimental predictions have been made – such as the proton becoming unstable, with an extremely long lifetime – they have not been verified. And I must stress that, though much effort has also been applied to quantising gravity, only a certain amount of progress has been made. It really does not seem that Einstein was embarked on a profitable line of research!

Before deciding what blame or otherwise should attach to him, let us examine the kind of quantum theory he hoped to extract from his unified field theory in some suitable approximation. It is convenient to start from a minor clash between Bell [76] and Jammer [67]. In a 1964 paper, following a line of argument outlined earlier in this chapter, Bell referred to EPR as follows – 'The paradox . . . was advanced as an argument that quantum mechanics . . . should be supplemented by additional variables.' Jammer rebuked him as follows – 'One of the sources of erroneously listing Einstein among the proponents of hidden variables was probably J. S. Bell's widely read paper . . . Einstein's remarks . . . quoted by Bell . . . are certainly no confession of the belief in the necessity of hidden variables.'

Jammer pointed to the kind of unified field theory Einstein hoped to develop, but Bell replied that such a theory 'in no way excludes belief in "hidden variables". It can be seen rather as a particular conception of these variables.' For Bell, then, whatever Einstein aimed at must have had the form of some of the usual classical variables (as allowed by current quantum theory) *plus* other (hidden) variables, though these hidden variables might themselves be additional classical variables, or possibly more complicated constructions. The

Gibbs ensemble idea we analysed earlier in this chapter certainly comes into this category, the hidden variables just being those classical variables *not* allowed by Copenhagen.

But is this right? Fine [94] argues very cogently that Einstein had no commitment to retaining any classical variables at all. (Thus he did not wish to *add* hidden variables to what Bohr might allow, but to start afresh.) In a letter written to Schrödinger in 1928 [94], he wrote 'Your claim that the concepts, p, q [momentum, position] will have to be given up, if they can only claim such "shaky" meaning seems to me to be fully justified.' The new concepts must be expected to be equivalent (or very nearly equivalent) to our present ones in the region where these present ones work well, but outside this region they may take on a different nature. Thus EPR remains a demonstration that quantum theory is incomplete, but not, as Bell would say, a call for it to be completed by the addition of hidden variables, but for a radical new complete theory using totally new variables to be produced.

If this is the case, it perhaps marks the fundamental divergence between his views and those of Bohr. Both recognised that one could not *both* continue to use classical concepts, *and* also retain all the traditional virtues such as determinism, realism and locality. But Bohr felt it necessary to retain the classical concepts, and had to accept a new way of using them – complementarity, the loss of determinism, and trouble, at the very least, for realism and locality. Einstein was prepared to consider a new range of concepts, and thus aimed to keep determinism and the others. (One might perhaps imagine that *if* he had succeeded in producing a satisfactory unified field theory, and *if* this had been found to reduce to quantum theory in a suitable approximation, he *might* not have been too troubled by non-determinism at that stage.)

Fine [94] gives a nice picture of this using the analogy of a wheel. In the central (classical) core, there is no problem. But for Bohr, if different observers venture out classically along different spokes, the messages they send back cannot be pieced together. Einstein wondered whether the spokes could not be connected if we agreed that they need not be made of the same material as the core. Or, at the very least, could there not be a rim?

Fine's position is certainly plausible. Yet I am inclined to draw a different conclusion from it. Fine regards it as demonstrating Einstein's superiority to Bohr – '[T]he tale of Einstein grown conservative in his later years is here seen to embody a truth dramatically reversed. For it is Bohr who emerges the conservative, unwilling (or unable?) to contemplate the overthrow of the system of classical concepts and defending it by recourse to . . . conceptual necessities and *a priori* arguments . . . Whereas, with regard to the use of classical concepts, Einstein's analytical method kept him ever open-minded.'

My view differs somewhat. I believe that Bohr achieved a great deal – he achieved, in fact, precisely what he required, a way of connecting the idea of measurement with the abstract mathematics produced by Heisenberg and Schrödinger in an unambiguous way. It is merely a great pity that complementarity, instead of being thought of as a powerful scheme, extremely useful in managing and coordinating ideas of quantum theory, came to be regarded as a set of ideas of great philosophical depth, which no physicist who wished to be taken seriously could question or even examine.

In contrast, it was greatly to Einstein's credit that he continued to discuss and question the Copenhagen synthesis. Yet for all that, his main achievement – EPR – was a negative one. It undoubtedly wounded Bohr, perhaps mortally. But one still has to say that Einstein's *positive* achievement in the area was zero. While he did keep opposition to Bohr just about alive, for all his talk of ensembles, for all his years of work on unified field theories, he created no genuine results. To that extent I have to fall in behind most of the physicists of the last 70 years. It is not necessary to discuss fading mental powers, but I would suggest that the magnificent triumph of general relativity led him to an unnecessarily restricted view of how theoretical advance *must* be made. For there *were* important advances to be made, which were made, eventually, by Bohm and Bell, as we shall see in the next chapter.

The reader may think I am being harsh. Did Einstein not, after all his undoubted achievements, have a right to choose his own approach? Of course (and Fine suggests that, despite the failure, he was 'genuuinely content to have made the effort'). It's possible to wish, though, that he had spent perhaps just a little time exploring other approaches – the completion from within as well as from without.

Am I not, though, being hypocritical? If Einstein *had* succeeded, would I not be extolling the virtues of refusing to follow fashion, of paddling your own canoe, or agonising over important problems rather than obtaining easy solutions to trivia, and so on? I certainly would, but the hard fact is that he did not succeed, and Einstein, like less gifted scientists, must be prepared to bow to that judgement, particularly when there do seem genuine reasons for thinking his work was always unlikely to be fruitful.

7

Bohm, Bell and experimental philosophy

David Bohm: life and times

In what is perhaps Virginia Woolf's most famous novel, much of the narrative describes intentions, hopes, plans for the visit *To the Lighthouse*, a visit forestalled by bad weather. There follows a section 'Time passes' in which all is war, death, decay, desperation; only the poets thrive. Then life returns slowly and timorously; the visit is at last made, in sombre reflective mood.

Between perhaps 1930 and 1952, the study of the meaning of quantum theory went through its own period of emptiness. Von Neumann had done most to cause it. Einstein could not disturb it . . . When interest did creep back, it was a result of the work of David Bohm.

Bohm had already experienced a chequered career. He was born in the USA in 1917, and rapidly built up an exceedingly high reputation as a theoretical physicist. After initial collaboration with Robert Oppenheimer, he specialised in the physics of *plasmas* – gases in which, because of high temperature or low pressure, atoms are *ionised*, or broken up into negatively charged electrons and positively charged *ions*. Plasmas [118] are important in astrophysics and also in the effort to achieve *controlled nuclear fusion*. In nuclear fusion, small nuclei fuse together to form one large one, with the emission of energy; high temperatures are required, and so one is forced to deal with plasmas. This is also exactly the process that causes stars to radiate energy. (More recently, readers may have met the suggestion that *cold fusion* may be achievable – in a test-tube! Unfortunately it seems unlikely that these hopes will be realised [119].)

Bohm was appointed at Princeton in 1946, and began to think about metals, in which free electrons provide the electrical conduction. Bohm realised that, because the electrons *were* free, a metal *was* a plasma, and he was able to apply his considerable expertise on gaseous plasmas to clearing up long-standing conceptual problems in the theory of metals. This was work of immense importance, which has been enormously influential in the subsequent development of the physics of solids [50, 120]. Thinking ahead to Bohm's later work, we may say that he was treating the electrons *as a whole*, rather than as individual particles.

Also in this period he became interested in the structure of quantum theory. He wrote a textbook [107] that quickly became recognised as highly authoritative and comprehensive. This book was written largely along orthodox lines [67] – there is, for example a proof or 'proof' that quantum theory is inconsistent with hidden variables. But after writing this book, and a discussion with Einstein, Bohm became dissatisfied with the views of Copenhagen.

(Coincidentally the same thing happened to another famous physicist – Alfred Landé. In Chapter 4 I mentioned briefly his important contributions to atomic physics in the 1920s. In 1951 he wrote a textbook [121] on quantum theory, also supporting an orthodox position, but almost immediately he too changed his mind and became an ardent and prolific critic of Copenhagen [67], although I do not go into his views in this book.)

At this stage in his career Bohm was beset by political problems. Early participation in moderately left-wing activities had led to his being called before the House Un-American Activities Committee for questioning in 1949. (David Pines [122], an important collaborator of Bohm at the time, says that the committee, which included Richard Nixon, was really only out to get Oppenheimer through Bohm and others.)

Bohm opted for silence under the fifth amendment of the US constitution. Princeton responded by refusing to consider him for a tenured post, and paying him for the last year of his contract only on condition that he did not enter the campus. Despite his considerable achievements he was blocked from all employment in the USA (including the fusion project, on which his expertise could have saved millions of dollars [122]).

Bohm worked in Brazil from 1952 to 1955, in Israel until 1957, and then at Bristol, until he became professor at Birkbeck College in London; here he stayed until his death in 1992.

His main work in the last 40 years of his life began with his alternative approach to quantum theory of 1952, which developed into a wide-ranging study of the nature of science and its relation with human knowledge and language [123–126].

The breadth of his interests and his appeal is demonstrated by the titles of a collection of essays published in his honour in 1987 [127]. Not surprisingly they involve science, both orthodox and unorthodox, and mathematics, but also work that would most naturally be classified as philosophy, psychology, neuroscience, biology, linguistics and the theory of art. Bohm also became close to the Indian philosopher J. Krisnamurti, and Pines [122] suggests that many expected Bohm to become his successor.

I shall say a very little more about Bohm's later writings towards the end of this chapter.

De Broglie and pilot-waves

Before describing Bohm's 1952 theory, I shall return briefly to the ideas of Louis de Broglie [67, 128]; I referred to these briefly in Chapter 4, describing them as the only real opposition at the time to those of Bohr.

Actually de Broglie put forward two rather distinct theories. The first, the *double-solution interpretation* of 1926–7, proposed, in addition to the usual wave-function, a second function, which differs from the wave-function only in a localised region. We may describe this region as containing the *particle*, which may be said to move under *guidance* from the actual wave-function.

De Broglie was unable to complete the working out of this theory, and at the 1927 Solvay Congress he presented a simpler version of his idea – the *pilot-wave theory*. There is now no second function, but one still thinks of a particle existing in a given region; the probability of its position taking particular values is obtained from the probability density in the standard way, and its motion is also determined by the wave-function. It is clear that de Broglie's ideas follow in a rather straightforward way from his original suggestion on the wave nature of the electron described in Chapter 4. We may say that, while for Bohr and complementarity it was wave *or* particle, for de Broglie it was wave *and* particle.

De Broglie's ideas received short shrift from Bohr's supporters. In particular [67, 128], Pauli argued that they could not handle the kind of situation where an atom collides with an electron, and gains from it enough energy to jump from one energy-level to another. According to Pauli's analysis, if one used de Broglie's ideas it seemed that, contrary to experiment, electron and atom could not separate with well-defined energies. De Broglie was not able to reply convincingly, and soon gave up his suggestion. In a major account of Bohm's theory [128], Peter Holland says that de Broglie's idea remained 'largely an unfulfilled programme', but that 'Bohm's model is essentially de Broglie's pilot-wave theory carried to its logical conclusion.' In particular, while de Broglie considered only single particles, Bohm was able to handle the many-particle case.

Bohm theory – the minimalist approach

It was in 1952 that Bohm turned independently to the same type of idea [129]. The titles of his papers refer to *hidden variables*, the position of the particle being the hidden variable, since it does not appear in the wave-function. As Bell [76] points out, it is actually the position that shows up immediately in a measurement, the wave-function appearing only in the statistics of many such results. Thus the fact that the former rather than the latter is the 'hidden variable' is, according to Bell, 'a piece of historical silliness'.

Since Bohm's own treatment was rather technical mathematically, I shall start with a simplified approach due to Bell [76]. (Holland [128] calls it the *minimalist approach*, though he regards it as unsatisfactory because it leaves out some important aspects of the theory.)

In this approach, one defines the wave-function in the usual way. But in addition there is a particle with a precise position and a precise momentum at all times – hence it follows a specific trajectory as time advances. I would emphasise that this position and momentum exist *entirely independently of any measurement*.

Particle position and trajectory are related to the wave-function. It is postulated that the positions of a large number of particles will differ, and their distribution will give precisely the probability density, which, as explained in Chapter 4, is essentially the square of the wave-function. Thus we make contact with conventional ideas, but from a very different perspective. The conventional view would be that one has a distribution of positions only *after* measurements of position have been made. Before any measurements any distribution is *potential*, and can be made actual only by measurement. At the time of the measurement, and in terms of the definitions of the previous chapter, the situation changes from *probabilistic* to *statistical*. But in the de Broglie–Bohm theory, it is *statistical* all the time; there is a distribution of actual values.

We use probability because of our own ignorance, not because of any indeterminacy in the properties of the particle; these properties themselves are well defined at all times. This is analogous to the state of affairs for Boltzmann statistics, for which we use statistics because we do not know, and do not want to know, the position and velocity of each particle. Nevertheless we take it for granted that each particle does actually have a precise position and velocity at all times.

To obtain the trajectory, we define the speed of the particle as the ratio of *probability current* to probability density. (The probability current is a name for the flow of probability, and, like the probability density, it may be obtained mathematically from the wave-function.) Technically, this definition ensures that the total probability remains equal to one – there would be something badly wrong if it didn't, so this just ensures consistency.

Though the particle *has* a position at all times, it is important to realise that a measurement of position does *not* just record this value. Rather the measurement procedure must be described as a physical interaction between measured particle and measuring system, during which the value of the position changes deterministically. For Bell, this demonstrates the truth of Bohr's teaching that '[e]xperimental results are products of the complete set-up, "system" plus "apparatus", and should not be regarded as "measurements" of pre-existing properties of the "system" alone.'

I would emphasise that, in the measurement procedure, there is no requirement to give the apparatus or the observer a special status, and there is no process of wave-function collapse lying outside the usual equations describing the evolution of the system.

The exciting thing is that, totally contrary to the claims of Bohr, von Neumann and their supporters, this fairly simple set of ideas reproduces the usual results of quantum theory, yet it remains totally realist – the particle has a precise position and momentum at all times – and deterministic (or causal). For Holland [128], as possibly for Einstein according to the discussion in the previous chapter, the important point is not the determinism – which is almost an unexpected bonus – but the realism – '[T]he important issue [that Bohm opposes] is not so much the denial of causality . . . but the claim that no model at all can be constructed of an individual system.' The interpretation is often called the *causal interpretation*, or the *quantum theory of motion*. The latter is the title of Holland's book [128]; it makes the point that, in the theory, one may discuss the motion of individual particles in trajectories.

(I have said that the theory is realist and deterministic. The reader may well ask – is it local? No, it is not, and this will be hugely important later in this chapter.)

In the de Broglie–Bohm theory, 'wave and particle are on an equal footing', as Holland says, in that 'they both exist'. Yet he proceeds to say that the wave is the 'senior partner' because the particle equation can be deduced from the wave equation, and not vice versa. The wave actually plays a dual role in the theory; it determines not just the probability of the particle's position, but also its motion.

Bohm theory – the quantum potential

Having introduced Bell's minimalist approach to the de Broglie–Bohm theory, I now turn to Bohm's original version. This certainly yields a far deeper physical appreciation of what is happening, but at the cost of using quite sophisticated mathematical techniques. The *Hamilton–Jacobi method* is such a technique used in classical mechanics as an important alternative to specifically Newtonian methods – leading to the same results but often giving additional power and insight. Bohm was able to link the Schrödinger equation of quantum theory to the Hamilton–Jacobi equation of classical mechanics.

Specifically, the Schrödinger equation was decomposed into two equations. The first just ensured that total probability was conserved. The second was the interesting one; it had the structure of the Hamilton–Jacobi equation, but with an extra term for the quantum case. Because this term is an addition to the classical potential energy, it is called the *quantum potential* (and yet another common name for the method is the *quantum potential method*).

The actual mathematics is well beyond the scope of this book, but since the quantum potential is the key to understanding what is happening, I shall say a few words about it. (Everything else, including particle position and trajectory, works exactly as in the minimalist approach of the previous section.)

It is important to realise that the quantum potential does *not* just depend on position like a classical potential would; rather it depends on the wave-function itself – in a rather complicated way. It is from the point of view of the moving particle that the effect of the quantum potential just adds to that of the classical potential.

The presence of this term is a demonstration that the theory has a specifically quantum nature. The de Broglie–Bohm approach is sometimes criticised (sometimes praised!) for attempting to restore classical physics, but, as Holland [128] says, it is not 'an attempt at squeezing quantum mechanics into a classical language'. Rather it 'provides an appropriate context in which to discuss classical and quantum mechanics in the same language, as instances of a general theory of waves and rays, and not a means of reducing one to the other'. Holland comments that 'The resulting theory stands in a clear and obvious relation to its classical counterpart. The principal feature it shares with the classical paradigm is that the individuality of experience is comprehensible.' The theory does not merely imply 'an *extension* of classical notions, but requires the development of a new physical intuition. Bohm locates the novelty of quantum mechanics not in its statistical or discrete aspects, but in a new physical conception of the state of a system . . . that manifests itself in the motion of particles through a new type of potential, the quantum potential.'

A rather strange point about the quantum potential is that its effect does not fall off with distance, as would be the case for any classical potential. The force between two particles may be large even when they are a very large distance apart. This is one aspect of the *non-locality* of the theory. More fundamentally, we may deduce, directly from the equations of motion, that the behaviour of one particle depends on the positions of all the others at that time. Clearly this behaviour is highly non-classical. I would also mention that, if it were possible to *control* the hidden variables (which does seem a very remote possibility), we could violate *parameter independence* (as defined in the previous chapter); this would enable signals to be sent at speeds greater than that of light, in fact instantaneously. Clearly this prediction, while not necessarily wrong, is not very pleasant. We return to a full treatment of non-locality later in this chapter.

The existence of the quantum potential helps to explain such phenomena as interference and tunnelling, as I shall describe later in this chapter. Holland discusses two ways of thinking about the combined effect of classical and quantum potentials. One may say that the classical potential has non-classical effects 'via the mediating role of the quantum potential'. Alternatively, as originally suggested by Bohm and Hiley, we may interpret the quantum potential as an *information potential*; particles move under their 'own' energy, but are guided by the quantum potential, 'much as a ship on automatic pilot may be steered under the influence of radar waves of considerably less energy than the ship's power source'. (Again this idea goes right back to de Broglie's *pilot-wave*.)

While Bohm's interpretation agreed with the accepted theory in regions where the latter was empirically correct, he [129] suggested areas where his new ideas permitted 'modification of the mathematical formulation which could not even be described in terms of the usual interpretation' because they involved the hidden variables themselves. This would entail detailed adjustments of his equations, which he claimed would only cause differences from current theory for distances of about 10^{-15} m, where current theory was unsatisfactory anyway.

Lastly in this section I return to what the reader may have forgotten, or perhaps may have suspected I had forgotten – Pauli's objections to de Broglie's original ideas, which were taken to be so damning that they blocked their study for a quarter of a century. Bohm was able, in fact, to dispose of these arguments fairly easily, using two distinct arguments. First he showed that Pauli had discussed the example using too simplistic a model for the incident particle; he had assumed that it had a precise momentum, and therefore was of infinite extent. Secondly, the apparatus really needed to be supplemented with a device to measure energy; with this in place, a specific result would be obtained, as experiment demanded. The fact that Bohm was able to answer Pauli's challenge so easily shows that the opposition to de Broglie had been based little on direct argument, much more on prejudice.

Responses to Bohm

Now let us turn to the responses to Bohm's 1952 papers. As would be expected, the leading proponents of the orthodox position were highly antagonistic. Heisenberg [130] criticised the particle trajectories as 'superfluous "ideological superstructure"' – superfluous because the interpretation led to the same results as the orthodox one. He regarded Bohm's tentative suggestion that some modification of quantum theory permitting experimental testing of the causal interpretation could be made as 'akin to the strange hope that sometimes $2 \times 2 = 5$'.

Heisenberg had a much better point in being concerned that symmetry between position and momentum, the establishment of which by Dirac and Jordan was one of the major glories of the mathematics of the quantum theory, is lost by Bohm's elevating position to a privileged position. It must be said, though, that glorious as the structure may indeed be from the aesthetic point of view and in the absence of measurement, it is exactly this feature that causes so much trouble when one *does* measure something. Position does have a central role in measurement – all measurements are ultimately measurements of position – and its central place in Bohm's theory recognises that fact. (Holland remarks that it is by no means unreasonable that *physical asymmetry* should accompany *mathematical symmetry*.)

Pauli [131] too swooped down on Bohm's theory, as he had on de Broglie's version a quarter of a century before. Born [5] actually wrote to Einstein that 'Pauli has come up with an idea . . . which slays Bohm not only philosophically but physically as well.' This appears to have been wishful thinking. Pauli described Bohm's work as 'artificial metaphysics', but his physical arguments were surprisingly limp. Like Heisenberg, he did not believe that Bohm could modify the quantum formalism without contradicting established experimental results, and he asked for an explanation of the relationship between probability density and wave-function. Bohm and his associate, Jean-Pierre Vigier, were actually able to provide the latter, but even without this reply, one could hardly say that Bohm had been 'slain'.

Nevertheless the overwhelming majority of physicists ignored Bohm, for several reasons. First, of course, they were totally convinced that Bohr had given all the correct answers long ago, and that any attempt to provide a more physically accessible model was doomed to failure. And Bohm's approach used hidden variables, which von Neumann was believed to have shown were untenable long before. I should mention that Bohm produced a counter-argument to von Neumann, specifically requiring hidden variables for the measuring apparatus as well as the observed system. This argument was not, however, particularly convincing, and actually, as Bell later showed, it was wrong.

Secondly, Bohm made no concessions in his account of the theory; it was detailed and technical. Now, of course, physicists *will* take time to read detailed and technical papers – it's an important part of their job – but *only* if they are convinced it is worth making the effort. In this case, most were convinced or practically convinced that Bohm must be wrong; it seemed much easier to suppose that an error had crept into his analysis than that the great von Neumann had got things totally wrong. And even if Bohm *did* happen to be correct, most physicists would take the line of Heisenberg and Pauli; anybody quite happy with Bohr's approach would inevitably find Bohm's strange and uncongenial, and, most important, unnecessary, since it merely reproduced the known and accepted results of orthodox quantum theory.

When Bohm suggested areas of potential divergence from current theory, he fared no better. He was indeed in something of a 'catch-22' situation. To the extent that his interpretation *agreed* with current ideas, it was pointless – *armchair philosophy* was a popular description. To the extent that it *disagreed*, it must surely be wrong, since it seemed difficult to restrict such discrepancy to the region where current theory was not fully established. (Bell was to avoid this catch, as we shall later see.)

The support of one person – Einstein – could have made a considerable difference. Bohm referred to Einstein's criticisms of the usual interpretation of quantum theory in the very first paragraph of his first paper, and thanked him for 'several interesting and stimulating discussions' at the end. But Einstein was unimpressed. 'Have you heard,' he wrote to Born [5], 'that Bohm believes (as de Broglie did, by the way, 25 years ago) that he is able to interpret the quantum theory in deterministic terms?' He adds '[T]hat way seems too cheap to me.' In his 1971 commentary on this letter, Born is somewhat surprised, feeling that 'this theory was quite in line with his own ideas'. Bell quotes both Einstein and Born and is also surprised.

From our perspective of the last chapter, Einstein's views may seem less surprising. His idea of what was required was so high-minded that Bohm's relatively simple ideas were bound to seem naive, trivial, 'cheap' in comparison. Nevertheless I do find it sad that Einstein was not flexible enough at this point to acknowledge that approaches totally different from his own did deserve attention.

Bohm [127] was disappointed by Einstein's rejection – '[I]t was important that the whole idea did not appeal to Einstein, probably mainly because it involved the new feature of non-locality . . . I felt this response of Einstein was particularly unfortunate . . . as it certainly "put off" some of those who might otherwise have been interested in the approach.' (Incidentally, while Bohm's suggestion that it was non-locality that offended Einstein is entirely logical, at the very least we may note that it was not the point he made to Born.)

Holland [128] suggests that Einstein's attitude may have been a 'tactical mistake', that 'whatever he perceived as its drawbacks, some model such as that of de Broglie and Bohm was better than none at all in countering the prevailing vagueness of interpretation, at least as a makeshift before a more satisfactory foundation could be found'. But Holland doubts that Einstein's advocacy would, in fact, have made much of a difference, because Einstein's own favoured options were universally disregarded.

Anyway, as Bohm [127] says, 'These proposals did not actually "catch on" among physicists . . . Because the response to these ideas was so limited, and because I did not see clearly, at the time, how to proceed further, my interests began to turn in other directions' (some of these directions being discussed extremely briefly later in this chapter).

This state of affairs lasted, according to Holland, for about 25 years – '[T]he theory did not enter the mainstream of physics either as a research topic or in textbooks . . . [N]o development or application of the . . . theory was made.' Indeed, in the 1971 commentary referred to above, Born comments that 'Today one hardly ever hears about this attempt of Bohm's, or similar ones by de Broglie.' The one important exception was the work of John Bell.

Bell and the de Broglie–Bohm theory

So at last we meet John Bell formally, though I have already reported many of his comments on other ideas and interpretations – apart from the crucial nature of his own contributions to the subject, he played a very important role through his clear and penetrating discussion of nearly every development.

Bell [132–135] was born in Belfast in 1928. Since his family was not affluent, he was unable to attend one of the leading grammar schools, but instead went to Belfast Technical College, and when he entered Queen's University it was initially as a technician not as a student, but with the support of the physics department at Queen's and with the award of scholarships, he was eventually able to study for degrees in physics and mathematical physics.

Bell worked at Harwell from 1949, initially on the design of accelerators. In 1953 he was given leave of absence to work for a year under Peierls at Birmingham University, eventually obtaining a PhD. Peierls was greatly impressed with Bell's ability, and arranged for him to transfer to the newly formed Theoretical Physics Division at Harwell, where he became an expert in the field of elementary particles. As we have seen in Chapter 5 and shall see again in Chapter 8, Peierls was a strong supporter of Copenhagen, so he had little time for Bell's discordant views, but nevertheless remained a great supporter of the man himself.

In 1960 Bell moved to CERN, the European Centre for Nuclear Research, where he worked till his death in 1990. (So he died two years before Bohm, though he was eleven years the younger.) Bell became a leading member of the theoretical-physics division at CERN, and an international authority on accelerator and elementary-particle physics. During working hours he used quantum theory in a thoroughly pragmatic way, and, of course, extremely successfully.

It was outside working hours, so to speak, that Bell became involved in the problems discussed in this book. What he jokingly referred to as a hobby was to bring him fame both inside and outside the community of physicists.

Bohm's theory played a large part in developing Bell's interest. From his student days, Bell had thought deeply about the meaning of quantum theory and the possibility of hidden variables. He had heard of von Neumann's impossibility proof, but at this stage the book itself had not been translated into English and Bell did not know German, so he took the word of Born [136] in a well-known popular treatment of quantum theory that hidden variables were indeed impossible.

'But in 1952', Bell says, 'I saw the impossible done . . . in papers by David Bohm.' For Bell the fact that Bohm's description was deterministic was of interest, but he says that '[m]ore importantly, in my opinion, the subjectivity of the orthodox version, the necessary reference to the "observer", could be eliminated.'

'But why then', Bell asks, 'had Born not told me of this "pilot wave"? If only to point out what was wrong with it? Why did von Neumann not consider it? More extraordinarily, why did people go on producing "impossibility proofs" after 1952, and as recently as 1978? When even Pauli, Rosenfeld and Heisenberg, could produce no more devastating criticism of Bohm's version than to brand it as "metaphysical" and "ideological"? Why is the pilot wave picture ignored in text books? Should it not be taught, not as the only way, but as an antidote to the prevailing complacency? To *show us that vagueness, subjectivity and indeterminism, are not forced on us by experimental facts, but by deliberate theoretical choice* [my italics]?'

Bell thus attempted to turn the tables on Heisenberg and the others. They were not entitled, he implied, to assume a kind of intellectual high ground, and to cast Bohm as a vulgar interloper trawling dubious ideology. Bohm had convincingly demonstrated that their own grandiose metaphysical claims were bogus. Complementarity had to meet the challenge from de Broglie–Bohm, and perhaps other interpretations, head-on, not by assuming the role of guardian of established truth.

Bohm indeed suggested [127] that 'if de Broglie's ideas had won the day at the Solvay congress of 1927, they might have become the accepted interpretation; then, if someone had come along to propose the current interpretation, one could have said that since, after all, it gave no new experimental results, there would be no point in considering it seriously. In other words I felt that the adoption of the current interpretation was a somewhat fortuitous affair.'

Bell liked the de Broglie–Bohm theory so much that he made it the second of the three *unromantic world views* he approved of. 'This idea seems to me so natural and simple, to resolve the wave–particle dilemma in such a clear and positive way,' he wrote, 'that it is a great mystery to me that it was so generally ignored.'

To say that Bell liked the scheme is certainly *not* to say that he considered it the final answer. 'It is easy to find good reasons for disliking the de Broglie–Bohm picture', he said. 'Neither de Broglie nor Bohm liked it very much.' Bohm [127], for instance, wrote that 'I saw clearly at the time [1952] that the causal interpretation was not entirely satisfactory. I felt that the insight it afforded was an important reason why it should be considered, at least as a supplement to the usual interpretation. To have some kind of intuitive model was better, in my view, than to have none.'

Holland [128] describes the de Broglie–Bohm approach as a 'physicist's theory', a down-to-earth set of ideas providing a complete specification of the state of an individual particle, and open to improvement and modification. This, I think, was Bell's point of view as well.

Bell and the impossibility proofs

Bell made Bohm's work the starting-point for his own two magnificent contributions to the foundations of quantum theory, both written in 1964. The two papers are superb, not just because of the extreme importance of their arguments, but also for their clarity and simplicity, and also because they include the suggestion of a crucial experiment. While Bohm's work was technical and complicated enough to be almost universally ignored, Bell cut his arguments, and those of others he discussed, to the bone, so as to *demand* attention. While Bohm made only vague suggestions of areas of physics where there *might* conceivably be differences between his own predictions and those of current theory, Bell was able to pinpoint a specific type of experiment where it seemed quite plausible that quantum theory might be violated.

The first of Bell's papers, the earliest ideas of which, Bell says, dated from as early as 1952, dealt specifically with the possibility of hidden variables in quantum theory, and the impossibility proofs — that of von Neumann, and those produced by later physicists. Bell started by remarking that Bohm's work had led to 'the realization that von Neumann's proof was of limited relevance . . . gaining ground . . . [though remaining] far from universal'. But, he said, there was no adequate published account of what was wrong. (In a footnote, Bell specifically criticised Bohm's discussion of this point. 'It seems', he said, 'to lack clarity, or else accuracy' Bohm had implied that it was the presence of hidden variables in the measuring apparatus that allowed circumvention of the impossibility proof. Bell replied that such hidden variables were actually neither required, nor sufficient, to obtain this result.)

'Like all authors of non-commissioned reviews', Bell added, '[the writer] thinks he can restate the position with such clarity and simplicity that all previous discussions will be eclipsed.'

Bell immediately proceeded to set up a hidden-variable model so simple that, unlike that of Bohm, physicists could neither ignore it, nor claim that there 'must' be a mistake in it. Bell considered a spin-$\frac{1}{2}$ particle such as an electron, and discussed what were essentially measurements of different components of spin angular momentum. If one wished to measure only the z-component, it would, of course, be easy to set up a hidden-variable scheme that would agree with experiment. If λ is the hidden variable, one can just give appropriate numbers of particles the value of λ equal to $+\frac{1}{2}$ and $-\frac{1}{2}$; then, if λ is $\frac{1}{2}$, the measurement gives the answer $\hbar/2$, and if λ is $-\frac{1}{2}$, it gives $\hbar/2$.

Bell's problem was much more difficult, because his model had to give results in agreement with quantum theory for measurement of *any* component of spin, not just that along the z-axis. Nevertheless, Bell was able to come up with a relatively simple model that did exactly what he required. His values of λ were in a *continuous range* between $-\frac{1}{2}$ and $+\frac{1}{2}$, and he produced an expression that gave the correct probabilities of obtaining $+\hbar/2$ or $\hbar/2$ for a measurement of *any* component of spin, in terms, of course, of λ.

The formula was very contrived, and Bell emphasised that he had no intention of putting his model forward as a conceivable 'real' structure. It was, in any case, lacking, in that it did not say what happened to the hidden variable at the time of the measurement. But this was all irrelevant. What Bell had shown was that a hidden-variable theory agreeing with quantum results *was* possible. It was clear that von Neumann had been wrong; the log-jam that had been preventing progress, or even much discussion, in the area for more than 30 years had been removed!

This could only be half the story, of course. Bell now had to show exactly what was wrong with the von Neumann proof, and with the later ones claiming to show the same result. But the important point was that, with a working hidden-variable model under his belt, so to speak, he could try every assumption of von Neumann against the model; presumably at least one assumption must be wrong. What had seemed eminently reasonable in the abstract must prove to be obviously incorrect in a particular context. (Later Bell was to use the de Broglie–Bohm theory as a working hidden-variable model in exactly the same way; his moral was — always test your reasoning against simple cases.)

Such proved to be the case. One of the basic assumptions of von Neumann turned out to be totally defective. It concerned two quantities, O and P say, and their sum, $O + P$. Let us consider measuring each of these quantities for an ensemble of systems, and calculate the average of each set of results. Then von

Neumann's infamous assumption was that the average value of the results for $O + P$ should equal the sum of the average values of the results for O and P individually. The assumption certainly seems very reasonable (even obvious?), and it is well known to be true for the usual quantum-mechanical states. Surely it must be true for the postulated hidden-variable states (states characterised by particular values of the hidden variables) as well?

But Bell showed why not. First he pointed out that it was actually rather a freak that the result *did* apply to the quantum-mechanical states. Suppose, for example, O and P are the x- and y-components of spin, s_x and s_y. Then we require different apparatuses – Stern–Gerlach devices with fields in different directions – to measure s_x, s_y and $s_x + s_y$. (The last may be measured by a device with field mid-way between the x- and y-axes.) It certainly is not the case — there is absolutely no reason why it should be — that, for any particular state, the result of a measurement of the third should be the sum of the results of measuring the first two. It is rather a strange feature that, as I have said, for quantum-mechanical states the *average* values over an ensemble *do* add up in this way. And Bell stressed that there is certainly no reason to extend this feature to the hypothetical hidden-variable states, as von Neumann postulated. Without this postulate, the von Neumann proof collapses.

Bell was easily able to show that the postulate does *not* apply to the hidden-variable states in his own model, or in that of de Broglie and Bohm.

Bell dealt with other so-called impossibility proofs in similar fashion. One proof, for example, had an innocent-looking assumption that measuring quantity O must give the same result whether you measure it together with quantity P or with quantity Q. Again it seems reasonable enough, but it is not true for the de Broglie–Bohm theory. Neither is there any reason to expect it to be true; as Bell pointed out, measuring O and P requires a different experimental set-up from that for measuring O and Q. Again, without this assumption, the proof collapses. (Models where the result of measuring O depends on whether you measure P or Q with it, that is, depends on the context of measurement, are called *contextual*. So any hidden-variable interpretation of quantum theory must be contextual.)

So Bell had confirmed that hidden variables *could* exist. This had to be a blow for the followers of Bohr, and potentially it could be a support for Einstein's position; and Bell would have welcomed both these points. But he then picked up the point that Einstein would definitely *not* have liked, the non-locality of the Bohm model. This had been clearly pointed out by Bohm himself, and Bell merely spelled out the argument mathematically for the simple case of two spin-$\frac{1}{2}$ particles. 'The trajectory of 1 . . . depends in a complicated way on the trajectory and wave function of 2', he concluded, adding that 'the

Einstein–Podolsky–Rosen paradox is resolved in the way which Einstein would have liked least'.

Bell concluded his paper by wondering whether this non-locality might be a property of *all* hidden-variable theories agreeing with quantum theory, not just de Broglie–Bohm. At this point events took a strange turn. By a chapter of accidents [67], this paper, which had been submitted for publication in September 1964, was not published until 1966. (A revised version of the paper was mis-filed by the journal, while a request from the journal for this version went astray.) By this time, Bell had published another paper in a different journal, and in this paper, which was submitted in November 1964, and published amazingly in 1964 itself, he answered his own question – in the positive.

Bell and non-locality

To discuss non-locality in quantum theory, Bell generalised the EPR–Bohm thought-experiment. He used the form of an impossibility argument, assuming locality, and arguing to hidden variables, and hence to a mathematical condition on the results of the experiment that is *not* obeyed by quantum theory.

The first short part of the paper just took the straightforward EPR–Bohm set-up, with measurements along a particular axis in each wing of the experiment. If the measurement on the first spin gives the result $\hbar/2$, then, according to quantum theory, the measurement on the second must give the result $-\hbar/2$. Bell says: 'Now we make the hypothesis, and it seems one at least worth considering, that if the two measurements are made at places remote from one another the orientation of one magnet does not influence the result obtained with the other.' This is an assumption of *locality* (more specifically of *parameter independence*, as it was defined in the previous chapter, and to help to justify it, Bell gives a reference to Einstein's views stated in the Schilpp volume [73].

Since, Bell goes on to say, we may predict the result of the second measurement from that of the first, and the measurements may be along *any* axis, the results of the measurements must be pre-determined. (He is also using *outcome independence*, since he is not allowing the result of the one measurement to influence the result of the other; correlation then requires predetermination.) Since the wave-function does not provide this predetermination, it must be achieved, Bell says, by 'a more complete specification of the state', that is to say, deterministic hidden variables.

Although this initial argument is simple, indeed practically a re-run of the EPR–Bohm argument itself, Bell regarded it as important. Twenty years later, he wrote that in EPR it is locality that is assumed, not determinism, which is not assumed but inferred. He pointed to 'a widespread and erroneous conviction that for Einstein determinism was always *the* sacred principle'. In a footnote he comments – 'And his followers. My own first paper on this subject starts with a summary *from* locality to deterministic hidden variables. But the commentators have almost universally reported that it begins with deterministic hidden variables.'

Despite Bell's fairly casual manner of introducing locality in the paper, there is no doubt that it was something he wished to hold to. Since his main proof was to establish a discordance between locality and quantum theory, he felt it quite possible that it was quantum theory that was wrong. Realism, in the form of hidden variables, was also very important to him, determinism much less so.

Now I shall proceed with the main part of the proof. We assume hidden variables λ; this may indicate a single parameter or a set of them. As Bell says, we might well think it natural to have two sets of hidden variables, one set relating to each particle, but this is just a particular example of his scheme.

The experimental set-up is more general than that used by EPR. The motion of the particles is the same, but the measurements in each wing of the experiment may be of the component of spin angular momentum *along any direction*, whereas for EPR all measurements of spin were along the z- or the x-axis.

Bell now spells out the effect of his crucial locality assumption. The result of the measurement on the particle in the first wing must depend *only* on the component of spin being measured in this wing (that is to say, the direction of the Stern–Gerlach field in this wing) and the value or values of the λ for this particular pair of particles. It must *not* depend on the field direction in the second wing. Similarly the result in the second wing must depend *only* on the field direction in this wing and the values of the λ, *not* on the field direction in the first wing.

Bell first shows that there are easy ways of fixing up a single λ to provide agreement with the quantum prediction for the cases usually discussed until that time, with both fields along the z-axis (the original EPR), or with one along the z-axis and the other along the x-axis. Also agreement with quantum theory may easily be obtained in the general case *if* the result in one wing of the experiment could depend on *both* field directions – though this, of course, is exactly what Bell had excluded.

But, and this was Bell's important result, in the case of general field directions, and with locality obeyed, it was impossible to choose values of the λ so as to duplicate the quantum-mechanical results. To put the idea in reverse, as Bell says, 'a grossly non-local structure . . . is characteristic . . . of any . . . theory which reproduces exactly the quantum mechanical predictions.'

Bell's argument was simple but clever. It involved analysing three experiments, so, counting both wings in each, *six* field directions, but these are not all different – each occurs in two different experiments. Bell produced a fairly intricate relationship between the results of the three experiments which local hidden-variable theories must obey, but was able to show that quantum theory disobeyed it.

It is important to note that, while what we called the initial part of Bell's argument started from quantum theory and argued to non-locality, it is actually just the second and main part which is used from now on. This is the only part tested in experiments; violation of the relationship mentioned would be a disproof of the combination of locality and realism. Note that this statement as such makes no reference to quantum theory, which only enters the argument when one says that, if quantum theory is correct, the relationship *is* violated, so we may not have local realism.

I stress Bell's concluding remarks – 'The example considered above', he says, 'has the advantage that it requires little imagination to envisage the measurements involved actually being made.' Indeed, even apart from their great intrinsic interest and importance, Bell's results had made contact with experiment. Bohm had been able to make only the vaguest of suggestions of where his approach *might* differ from quantum theory, and he had been dismissed as an *armchair philosopher*. In contrast, Bell could point to a specific type of experiment where there seemed a reasonable possibility that quantum theory *might* be wrong – since it disobeyed the demands of local hidden-variable theories. The particular experiments, it turned out, had not been performed, so it was not possible to argue that Bell's suggestion had already proved unacceptable. It was a clear opening for experimentalists, and they were not reluctant to take up the challenge. It may be said that Bell had founded *experimental philosophy*.

Indeed these two papers of Bell changed markedly the climate of opinion towards research into the foundations of quantum theory. Much of the indifference had been swept away by Bell's demolition of von Neumann. Perhaps more significantly, Bell had provided the opportunity for a direct experimental test of what had been for several centuries the central belief concerning Nature: that it consisted of individual systems with well-defined properties acting locally. Of course it could be said that those following Bohr over the previous three decades had implicitly assumed that this belief was wrong, but, as I have said, the possibility of an experimental test puts matters on an entirely different plane. Henry Stapp [137] was prepared to call this result 'the most profound discovery of science'. David Mermin [138] reports a comment that 'Anybody who's not bothered by Bell's theorem has to have rocks in his head.'

Bell's own research described in this section led to an ever-increasing flood of published papers, and slowly new areas of research, both theoretical and experimental, were opened up. A few of these topics will be described in the following chapter. In the following sections, though, I introduce the most important experimental work on Bell's ideas themselves.

The early period of experimental philosophy

The mathematical condition obeyed by local hidden variables that Bell produced in his original paper was not very convenient for experimental purposes, and over the next few years many more suitable expressions were produced by Bell and others. These proofs used four experiments with *four* different field directions each being used twice. The best-known theoretical result was produced by John Clauser, Michael Horne, Abner Shimony and Richard Holt [139]. It was of the form of an inequality to be obeyed by the results of the four experiments according to any local hidden-variable theory; a certain quantity must not be greater than 2, whereas for quantum theory it takes values as high as 2.7. The condition is known as the CHSH–Bell inequality, or usually just the Bell inequality.

This was just part of a considerable amount of discussion about the possible experimental test of Bell's ideas [67]. Various types of experiment were considered, but it turned out that by far the most accessible at the time involved a pair of photons emitted in swift succession (in a so-called *cascade*) from an excited state of a calcium atom as it drops back to its ground state. While the EPR–Bohm thought-experiment uses a pair of spin-$\frac{1}{2}$ particles such as electrons, and the components of their spins are measured, in the cascade type of experiment it is the *polarisation* of each photon that is measured. The polarisation of an electromagnetic wave was discussed in Chapter 2, and the idea extends to photons. Relative to any chosen axis, a photon may be found to have horizontal or vertical polarisation, but the wave-function is *entangled* so that the behaviour of the two photons is not independent; a measurement on one gives information about both, and we are in a typical EPR-type situation.

A detailed experimental suggestion was made in 1969 [139]. Each photon would pass into a *polariser* or *linear polarisation filter*. Photons with polarisation parallel to a given axis pass through, and on to a detector; those with polarisation orthogonal to the axis are absorbed. This is a *one-channel analyser*, since only one beam of photons emerges, which is a considerable handicap, particularly when coupled with the other major problem – the efficiencies of the photon detectors are very low, less than 20%. So it happens that many pairs of photons polarised parallel to the given axis will *not* be detected. When one photon is detected, we don't know whether its partner had parallel polarisation but failed to be detected, or orthogonal and was absorbed. And of course pairs of orthogonally polarised photons will not show up at all. To start to cope with this, the experiment must be performed three times – once with each polariser in turn removed, and once with *both* removed. A new, rather more complicated, inequality could be produced for this experiment.

Unfortunately, though, detection efficiencies were too low to check whether the inequality was obeyed or not. At this stage, an auxiliary assumption is required [139, 140] – the probability of detection of a pair of photons is independent of the presence and orientation of the polarisers. With this assumption, the proposed experiments could at last give a clear indication of whether the constraints of local hidden-variable theories were obeyed.

During the 1970s, four such experiments were performed [67, 91, 104]. Three of these, two led by Clauser [141, 142] and one by Ed Fry [143], produced results that disobeyed the (new form of) Bell inequality and so pronounced *against* local hidden-variable theories; in fact they agreed well with quantum theory. The fourth [144] did precisely the opposite; its results satisfied the Bell inequalities. Overall the majority of physicists who expected quantum theory to be vindicated were probably quite satisfied – three out of four was not bad when experiments were very tricky. However, when one took into account the auxiliary assumption, it could hardly be said that the results were conclusive. A detailed summary was written by Clauser and Shimony [145].

Indeed, it was clear that there were two problems with the experiments. These became known as *loopholes*. The first is the detector loophole. As we have said, the efficiency of the photon detectors was very low. Various additional assumptions were required to produce inequalities that could be tested experimentally. We have already met one. Another often used [140] was that, when photons are travelling to a detector, inserting a polariser in the path may reduce the probability of detection but cannot increase it.

The importance of these assumptions, and the problems they caused for any interpretation of the data, were pointed out particularly by a group of scientists I shall call, a little loosely, the *realists*. This group, which has included Franco Selleri, Emilio Santos, Tom Marshall and, more recently, Caroline Thompson, believe very strongly in the traditional virtue of local realism and thus that, once all loopholes are removed, the experiments must show that quantum theory is defective.

At first sight one might feel that the auxiliary assumptions made to connect theory and experiment are very reasonable, and that the realists are merely being awkward in dwelling on them. However, this would not be fair. In his work on the von Neumann theorem, Bell had shown that assumptions that seem sensible enough in the abstract may easily be seen to be totally unreasonable in the context of a given model.

Essentially the assumptions imply the fact that, in the analysis of the experiments, a *fair-sampling assumption* is being made; one assumes that the photons that reach the detector and are then in fact detected are a fair sample of all the photons initially produced. This is again a natural enough assumption, but unfortunately there is no reason to think that it is necessarily correct. The whole point of a hidden-variable model is that different members of an ensemble have different properties, and one should not impose any arbitrary restrictions on these differences.

In fact, though, the realists received little support from Bell. Although he may genuinely have hoped that quantum theory might be violated and local realism upheld, he wrote [76] in 1981 'But the experimental situation is not very encouraging . . . . It is true that practical experiments fall far short of the ideal, because of counter inefficiencies, or analyzer inefficiencies, or geometric imperfections, and so on. It is only with added assumptions, or conventional allowance for inefficiencies and extrapolation from the real to the ideal, that one can say the inequality is violated. Although there is an escape route there, it is hard for me to believe that quantum mechanics works so nicely for inefficient practical set-ups and yet is going to fail badly when sufficient refinements are made.'

The second loophole was called the locality loophole; it had been highlighted by Bell in his 1964 paper – 'Conceivably [the quantum-mechanical predictions] might apply only to experiments in which the settings of the instruments are made sufficiently in advance to allow them to reach some mutual rapport by exchange of signals with velocity less than or equal to that of light. In that connection, experiments . . . in which the settings are changed during the flight of the particles are crucial.'

Indeed, till now the directions of the polarisers had been decided well in advance of the production of the photon pair. Could not a message be sent from the polarisers to the photon source so that the photons would somehow be emitted with given values of the hidden variables? Or from one polariser to the other so that they could conspire to trick the experimenter?

By the end of the 1970s, Alain Aspect at Orsay, and his collaborators, Philippe Grangier, Gérard Roger and Jean Dalibard, felt that 'technological progress (mostly in lasers) [was] sufficient to allow a new generation of experiments [146, 147]'. They planned that these experiments would take very major steps towards closing the locality loophole, and make at least a little progress with the detector loophole. In their 1982 experiments, lasers were used to excite the atoms, giving a high rate of cascade processes and hence high accuracy.

Experiments that previously had taken many hours could now be performed in minutes. (Actually the most recent of the previous four experiments had used lasers, but Aspect was able to take advantage of further progress in laser technology.)

Another important improvement was that *two-channel analysers* were used. The polarisers transmitted photons with parallel polarisation, and reflected (rather than blocked) the orthogonal ones. So two beams were produced in each wing of the experiment, and coincidences between photons in each beam in the first wing and those in each in the second could be studied – a four-fold coincidence counting system. Clearly one is much closer to the original EPR–Bohm idea, and the necessity for awkward assumptions has been reduced but not, of course, eliminated.

To attempt to close the locality loophole, Aspect arranged that, in some of the experiments, photons could be *switched* between polarisers at two different angles in each wing. Switching took place so fast, in comparison with photon-transit times, that there could be no possibility of photons, polarisers and detectors conspiring in advance so as to produce the quantum-mechanical results. In practice the switching could not be made random, as would have been desirable, but the generators controlling the processes ran at different frequencies in the two wings, so one would expect them to be uncorrelated.

I shall now briefly describe the results [146, 147]. Since the case with switching is necessarily less accurate, let us first consider the case without the switches. The results may be described in terms of the angle, θ, between the polarisers in the two wings. A certain quantity S should always lie between -2 and $+2$ for local hidden-variable theories. Quantum theory, however, predicts that S should be greater than 2 for a wide range of θ (roughly $0°$–$30°$), and it should reach a maximum of 2.70 ± 0.05 at $\theta = 22.5°$. The experimental results gave S very close to the quantum predictions throughout, and, at $\theta = 22.5°$, S was found to be equal to 2.697 ± 0.015. Though switching makes the results less accurate, the general nature of the results was not changed at all when switching was applied.

Experimental philosophy: recent work

In this section we sketch some of the developments in connection with Bell's theorem, experimental and theoretical, since Aspect's experiments, including efforts to close the loopholes [148].

While it was fairly generally accepted after Aspect's experiments of 1982 that the locality loophole had been closed, four years later Anton Zeilinger [149] pointed out a problem. There was a possible awkward coincidence involving the time between each beam switching to a particular polariser and the time taken for a signal to travel between the two switches. For particular cases it seemed that the switching might be in effect irrelevant. It would be another 13 years before this particular problem would be solved.

Progress has also been slow on closing the detector loophole. Philip Pearle [150] showed that it would be possible to contradict the usual Bell inequality conclusively if the detector efficiency were as high as 84%, provided, of course, that quantum theory was obeyed. Philippe Eberhard [151] was able to show how to reduce this to 67% by considering a range of quantum states, but even this is far higher than that of any detectors used in these experiments until very recently.

The major advance in the late 1980s was the use of parametric down-conversion for producing pairs of entangled photons very much more copiously than had been possible with previous methods. In this method, pioneered by Carroll Alley and Yanhua Shih [152], and by Zhe-Yu Ou and Leonard Mandel [153], photons from a high-frequency laser source are incident on a crystal such as BBO, β-barium oxide. This crystal has the property that there is a term in the dipole moment induced in the material that is proportional to the square of the electric field rather than the field itself.

The result is that an incident photon produces two photons, the output beams being called the signal and idler beams. For conservation of energy, the sum of the energies of photons in these two beams must be equal to that of the inci-dent photon. Thus there must be the same relationship between the frequencies of the various photons, and both output frequencies must be lower than the input frequency; thus one speaks of 'down-conversion'. Momentum must also be conserved in the process. The beams are then reflected onto a beam-splitter from opposite sides and thence to detectors. (The beam-splitter allows the beams to interfere; for a little more detail on beam-splitters, see the account of the Mach–Zehnder interferometer later in this chapter.)

The early work used what is called a type-I scheme, in which both output photons have the same polarisation. The conceptual difficulty in this case is that the two photons in the output are actually in a product state (a separa-ble or non-entangled state) rather than an entangled state. One must approxi-mate this state to an entangled state by considering only pairs of photons that take different exits from the beam-splitter, and hence travel away from each other.

The increased intensity of the source means that Bell's inequalities may be violated by several standard deviations, but, because one is using only half the pairs of photons actually produced, one is very vulnerable to questions about fair sampling, and certainly has made no progress towards closing the detector loophole.

In 1995, Paul Kwiat and co-workers [154] in Zeilinger's group then at Innsbruck took a major step forwards by solving the problems involved in using a type-II scheme, in which the output photons are of different polarisations and a genuinely entangled state is produced, so there is no requirement to discard any part of the ensemble of pairs. The source was many times brighter than those produced by type-I schemes, and so has considerable application in quantum information theory, for which see Chapter 9. The experiment enabled Bell's inequality to be violated by about 100 standard deviations, and also provided an improved detector efficiency; it was thus a step towards closing the detector loophole.

Then, in 1998, Gregor Weihs and colleagues [155], again in Zeilinger's group, performed the major work that finally closed the locality loophole. They enforced the condition of locality by maintaining sufficient distance between the measurement stations in each wing of the experiment, and by achieving completely independent data registration in each wing. As these authors said, it had taken more than 15 years of major improvements in technology to allow significant improvement on the work of Aspect.

Weihs and his co-workers insisted that, if one defined an individual measurement as lasting from the first moment when the choice of measurement direction in a particular wing of the experiment might be influenced till the actual registration of the photon, then a measurement in one wing of the apparatus must be performed quickly enough that no relevant information can travel to the other wing of the experiment before the appropriate measurement is completed in that wing.

Selection of measurement direction must also be completely unpredictable, so the measurement direction must be chosen by an actual random-number generator rather than a pseudo-random-number generator. Any common activity in the two wings was avoided; individual events were to be registered independently in each wing, and compared only after the experiment had been terminated. The latter point required independent and highly accurate time bases in each wing.

In the experiments, the observers were separated by 400 m, so the time available for a measurement was about 1.3 μs (1.3×10^{-6} s). Polarised entangled photon pairs were sent to each wing of the experiment through optical fibres, which differed in length by less than 1 m, so the photons were registered simultaneously within 5 ns (5×10^{-9} s).

In each wing the direction of the local polarisation measurement was switched by extremely fast electro-optic modulators controlled by a physical random-number generator sampled at a frequency of 500 MHz (500 μs^{-1}). Thus the total time taken by the measurement was about 100 ns (10^{-7} s), much less than the 1.3×10^{-6} s allowed.

Individual data registration in each wing was ensured by the provision of an independent time-interval analyser and atomic clock in each wing. These were synchronised only once before each experimental cycle to a precision of 3 ns (3×10^{-9} s). In each wing information on times of detection was stored in its own personal computer, and the files were analysed for coincidences only long after measurements had been completed. In this experiment not only was the locality loophole closed convincingly, but also Bell's inequalities were violated by about 30 standard deviations.

During the 1990s, experiments on other types of Bell's inequalities were performed. In 1990, John Rarity and Paul Tapster [156] were able to demonstrate the violation of a Bell's inequality for an entangled photon pair based not on different directions of polarisation but on phase and momentum. This idea had been independently suggested by Horne, Shimony and Zeilinger [157]. James Franson [158] had suggested experiments using entanglement in energy and time, and experiments [159] have been carried out along these lines, although there has also been an argument [160] that there may actually be a local hidden-variable explanation of these particular experiments.

There has also been an interesting suggestion by Peter Samuelsson, Eugene Sukhorukov and Markus Büttiker [161] of studying entanglement and violation of Bell's inequalities in mesoscopic conductors, conductors that are between microscopic and macroscopic in size. A superconductor would generate entangled states by injecting pairs of electrons into different leads of a normal conductor.

An interesting set of experiments has been carried out by Wolfgang Tittel and co-workers [162–164] in Gisin's group in Geneva. They studied Bell's inequalities for entangled photons separated by more than 10 km. The source for the photons was at a telecommunications station in Geneva, and the photons were sent along installed standard telecom fibre under Lake Geneva to analysers in Bellevue and Bernex, distances of 4.5 and 7.3 km, respectively, while Bellevue and Bernex are themselves 10.9 km apart. Compared with experiments in the laboratory, absorption in the wires meant that numbers of coincidences decreased by a factor of about ten, and fringe visibility was also lower by about 8%. Nevertheless the experiments were still able to demonstrate violations of the Bell's inequality by about 25 standard deviations.

These experiments were important for two reasons. First, they demonstrated that quantum correlations are indeed maintained over quite large distances and times. Two alternative suggestions had been made. One was that the correlations might be described by environment-induced decoherence along lines suggested by Hans-Dieter Zeh [165] and discussed in detail by Wojciech Zurek [166]. The other is that there might be spontaneous collapse, as suggested by Wendell Furry [167], Philip Pearle [168] and especially by Gian-Carlo Ghirardi, Alberto Rimini and Tullio Weber (GRW) [169]. Both these ideas are extremely important and are discussed at length in the next chapter, but the Geneva work makes it clear that they do not spoil the predicted quantum effects over these distances.

The second reason why the experiments are important is that they lead directly to the study of quantum cryptography, one of the main branches of quantum information theory, which specifically requires the maintenance of quantum correlations over long distances. As will be discussed in Chapter 9, much of the theoretical and experimental work up to the 1990s on the rather abstract topic of Bell's inequalities was to move seamlessly on to the study of the very much more applied topic of quantum information theory.

We now return to the detector loophole. Great efforts have been made to close this. Many suggestions of how to do so have been made, using a variety of strategies. One suggestion [170] is to use non-maximally entangled photons for which the requirements on the detectors are somewhat lower. Santos and collaborators [171] have suggested experiments with an extra measurement, that of a recoil atom, which provides additional information on each pair of entangled photons so that no additional assumptions are required. Santos has also related Bell and entropy inequalities [172], and proposed an optical test of Bell's inequalities without any fair-sampling assumption using a 'restricted but appealing' family of local hidden-variable models [173].

Most of the strategies, however, use much more massive particles for which the detection rates can be far higher than for photons. One idea [174, 175] is to analyse correlated spins in pairs of atoms obtained by dissociation of a beam of *dimers* containing two atoms of sodium or mercury. Several proposals use pairs of atoms entangled by use of a cavity filled with black-body radiation [176–179], or a classical radiation field [180].

A lot of attention has been paid to the possibility of demonstrating Bell's inequalities with relatively massive elementary particles, namely protons and neutrons, but especially K-mesons (kaons) and B-mesons, again with the hope of eventually closing the detector loophole.

The analysis of early experiments using protons [181] was not conclusive. A recent experiment using neutrons in the neutron interferometer is discussed in the section of the following chapter on neutron interferometry. In recent years a great deal of work has been done on the behaviour of kaons in particular in Bell-type experiments [182–186], though it is still being debated [187, 188] even whether the Bell inequalities have been violated convincingly. Kaons are probably the elementary particles with the most bizarre properties [189], including the fact that particle and anti-particle differ only over the timescale of the strong interaction, roughly 10^{-23} s, so there are several interesting features of their behaviour that may be used to provide entanglement.

Incidentally, one of the main participants in this work has been Reinhold Bertlmann, whom we met in connection with Bell's story of his socks in Chapter 6. Bertlmann had an important collaboration with Bell [190, 191] in the 1980s in the area of particle physics. More recently Bertlmann has been able to combine his expertise in this area of physics with his great interest in the ideas and discoveries of Bell himself.

Following a brief summary of the state of affairs by Aspect [192] in 1999, in 2001 there was announced the first claim of an experiment succeeding in closing the detector loophole. This was performed by Mary Rowe and collaborators [193] in David Wineland's group at the National Institute of Standards and Technology in Boulder, Colorado.

In this experiment, pairs of beryllium ions are initially confined in an ion trap. (More will be said about ion traps in the next chapter.) Two laser beams are then used to stimulate a transition that entangles the ions. The resulting wave-function is a superposition of two states: in the first, both ions are in what we may call level A; and in the second, both are in level B.

Whether any ion is in level A or level B is determined by probing the ion with circularly polarised light from a 'detection' laser beam. During this pulse, ions in level A scatter many photons, and about 64 are detected by a photomultiplier; ions in level B scatter very few photons. Thus statistics showing that Bell's inequality is violated by about eight standard deviations may be built up. What is more significant is that a measurement outcome is recorded for every experiment, so no fair-sampling assumption is required and the detector loophole is closed.

In the same edition of *Nature*, Grangier [194] provided a crisp summary of the progress that had been made, and also made the point that the hard-won experience built up studying Bell's inequalities was exactly what was needed for quantum information processing.

More recently Francesco De Martini's group in Rome has demonstrated considerable progress [195] along an alternative route to closing the detector loophole using a source of entanglement of high brilliance, and again aiming at measuring all entangled pairs.

On the down side there has been a suggestion [196] that there are further loopholes. These are a memory loophole, if the values of relevant hidden variables for a particular pair of tested particles depend on previous measurement choices and outcomes; and a simultaneous/measurement loophole, if measurements in both wings of the apparatus are carried out simultaneously. It seems, though, that any effect of these further loopholes would be small.

On the up side there have been suggestions [197, 198] for tests using homodyne detection and continuous variables that may be able to produce results free of *all* loopholes simultaneously. This would be the end-point of a process of steadily greater technical sophistication over 40 years.

Bell's theorem without inequalities: Greenberger, Horne, Zeilinger

Bell's theorem tells us that the results of quantum theory may not be simulated by any theory including locality and hidden variables. Bell's original proof and all our discussion of theory and experiment so far have been centred around inequalities. In the next two sections I shall explain two striking arguments that lead to the same conclusion without the use of inequalities.

The first is the (Dan) Greenberger–(Michael) Horne–(Anton) Zeilinger argument, always known as GHZ [199–201]. To discuss this, I shall have to go a little bit deeper into the structure of operators, eigenstates and eigenfunctions, which was introduced in Chapter 4. The discussion will be restricted to spin-states.

In Chapter 4, we said that, if a spin-$\frac{1}{2}$ system is in the particular state $s_z = \frac{1}{2}\rangle$, then a measurement of the observable s_z is bound to give the value $+\hbar/2$. Similarly, if it is in the state $|s_z = -\frac{1}{2}\rangle$, a measurement of s_z is bound to give the result $-\hbar/2$.

This is connected to the fact that we may define an operator $\hat{s}_z$ that acts on $|s_z = \frac{1}{2}\rangle$ and $|s_z = -\frac{1}{2}\rangle$, in the following way: $\hat{s}_z|s_z = \frac{1}{2}\rangle = +(\hbar/2)|s_z = \frac{1}{2}\rangle$ and $\hat{s}_z|s_z = -\frac{1}{2}\rangle = -(\hbar/2)|s_z = -\frac{1}{2}\rangle$. It should be remembered that $\hat{s}_z$ *is* an operator, not a normal number or physical quantity. If it acts on a general state of the system (a general linear combination of $|s_z = +\frac{1}{2}\rangle$ and $|s_z = -\frac{1}{2}\rangle$, which, of course, quantum theory allows), it will *not* give a constant times the state itself.

The fact that operator $\hat{s}_z$ acting on $|s_z = +\frac{1}{2}\rangle$ gives a constant, equal to $\hbar/2$ times $|s_z = +\frac{1}{2}\rangle$ tells us that $|s_z = +\frac{1}{2}\rangle$ is an eigenstate of $\hat{s}_z$ with eigenvalue $\hbar/2$, and that a measurement of observable s_z on a system whose state is given by $|s_z = +\frac{1}{2}\rangle$ is bound to give the answer $\hbar/2$. Similarly $|s_z = -\frac{1}{2}\rangle$ is also an eigenstate of $\hat{s}_z$ with eigenvalue $-\hbar/2$; a measurement of s_z on a system with state given by $|s_z = -\frac{1}{2}\rangle$ is bound to give the answer $-\hbar/2$.

(I imagine that the reader may feel that everything I have said is rather abstract, and might wish I would write down explicit expressions for the various operators and eigenstates. This can be done, operators being represented by matrices, eigenstates by vectors. Then the eigenvalue relations work out nicely. Unfortunately it's beyond the scope of this book, but in any case it is probably more meaningful in the long run, if a little less obvious in the short run, just to use the more symbolic relations as I have given them.)

Analogous to operator $\hat{s}_z$ are operators $\hat{s}_x$ and $\hat{s}_y$. These operators have of course their own eigenstates, which are different from those of $\hat{s}_z$. In the GHZ argument, the operators play the central part. It is essential to realise at all times that, since they *are* operators, in most cases they do not commute; in other words, for example, $\hat{s}_x\hat{s}_y$ is *not* equal to $\hat{s}_y\hat{s}_x$. (This is rather like saying that putting on your socks and then putting on your shoes is completely different from, and much more sensible than, putting on your shoes and then putting on your socks. Putting on either, of course, is an operation.) The fact that $\hat{s}_x$ and $\hat{s}_y$ do not commute is intimately connected to the fact that they do not have common eigenstates.

For the GHZ argument, it is handy, though not strictly necessary, to work not with $\hat{s}_x$, $\hat{s}_y$ and $\hat{s}_z$, but with related operators $\hat{\sigma}_x$, $\hat{\sigma}_y$, and $\hat{\sigma}_z$, $\hat{\sigma}_x$ is given by $(2/\hbar)\hat{s}_z$, and so on. Then, for example, we have the nice relations $\hat{\sigma}_z|s_z=\frac{1}{2}=|s_z=\frac{1}{2}\rangle$ and $\hat{\sigma}_z|s_z=-\frac{1}{2}=-|s_z=-\frac{1}{2}\rangle$. Thus the appropriate eigenvalues are ± 1.

We have now actually reached the GHZ argument itself. It concerns three spin-$\frac{1}{2}$ particles, a, b and c, which have interacted, and then fly to detectors in the horizontal plane. Thus they are in an entangled state. Rather than just $\hat{\sigma}_x$, we have a quantity for each spin, $\hat{\sigma}_x^a$, $\hat{\sigma}_x^b$ and $\hat{\sigma}_x^c$, and similarly for the y- and z-components. Operators for components of σ for *different* spins always commute, so that, for example, $\hat{\sigma}_x^a\hat{\sigma}_y^b=\hat{\sigma}_y^b\hat{\sigma}_x^a$, but those for the *same* spin do not, or in other words $\hat{\sigma}_x^a\hat{\sigma}_y^a$ is not equal to $\hat{\sigma}_y^a\hat{\sigma}_x^a$.

GHZ consider the three compound operators $\hat{\sigma}_x^a\hat{\sigma}_y^b\hat{\sigma}_y^c$, $\hat{\sigma}_y^a\hat{\sigma}_x^b\hat{\sigma}_y^c$ and $\hat{\sigma}_y^a\hat{\sigma}_y^b\hat{\sigma}_x^c$. Since each compound operator contains individual operators for each spin, at first sight it would seem almost certain that these compound operators will *not* commute. However, GHZ have chosen them cleverly enough that in fact they *do* commute. (We may say that the various components of the non-commutativity are designed to cancel out.)

In turn this implies that the three compound operators will have simultaneous eigenstates. There will be eight of these; for each of these the eigenvalues of the three compound operators will be a particular combination of ± 1, ± 1 and ± 1, and we may restrict our argument to the case in which all three are $+1$. It is natural to wonder what the eigenstate is for this case. We can write it as

$$\frac{1}{\sqrt{2}}\{|s_z^a=\tfrac{1}{2}, s_z^b=\tfrac{1}{2}, s_z^c=\tfrac{1}{2}\rangle - |s_z^a=-\tfrac{1}{2}, s_z^b=-\tfrac{1}{2}, s_z^c=-\tfrac{1}{2}\rangle\}$$

though actually we do not need to do so. As we have said, the argument concentrates on the operators. Note, though, that the state is highly entangled.

Since the operators for components of different spins commute, we may measure simultaneously the x-component of one spin and the y-components of the other two. If we consider the three compound operators above, then, since the spin-state in each is an eigenstate of the appropriate operator with eigenvalue unity, the product of the three measurements in each case must itself be unity. This in turn means that, among the measurements on each individual spin, there must be an even number of results of −1 and an odd number of results of +1.

Thus, if two of these results are known, we can predict with certainty the result of the third. We see a very clear analogy with EPR. The three spins are in an entangled state and have separated; measurements on two of the spins also provide knowledge about the third. If locality is assumed, all three results are pre-determined, so quantum theory is incomplete as EPR argued.

Thus, as Mermin [201] in particular says, *each* spin must have instructions for the result to be obtained for a measurement with each field-setting. A little thought convinces one that there are just four distinct ways in which this can be achieved. These are as follows: (a) each spin gives +1 whatever the setting; (b) one spin gives +1, the others −1 whatever the setting; (c) each spin gives +1 for the x-component and −1 for the y-component; and (d) one spin gives +1 for the x-component and −1 for the y-component, while the other two spins do exactly the reverse.

All these cases lead to an even number of negative results for measurements of σ_x on the three spins, or in other words the EPR-type analysis including locality leads to the definite conclusion that $\hat{\sigma}_x^a \hat{\sigma}_x^b \hat{\sigma}_x^c$ is equal to +1. However, quantum theory tells us something very different. It is indeed the case that the GHZ state is an eigenfunction of $\hat{\sigma}_x^a \hat{\sigma}_x^b \hat{\sigma}_x^c$, but its eigenvalue is not +1 as the EPR-type analysis above shows, but −1.

In fact it is possible to show that

$$(\hat{\sigma}_x^a \hat{\sigma}_y^b \hat{\sigma}_y^c)(\hat{\sigma}_y^a \hat{\sigma}_x^b \hat{\sigma}_y^c)(\hat{\sigma}_y^a \hat{\sigma}_y^b \hat{\sigma}_x^c) = -\hat{\sigma}_x^a \hat{\sigma}_x^b \hat{\sigma}_x^c.$$

Note that, in working out the above, one needs to be very careful with non-commutativity. If one were to ignore this and just treat the σ's as physical quantities rather than operators, one could use $\hat{\sigma}_y^2 = 1$ for a, b and c, and show that the left-hand side of the equation is equal to the right-hand side but with a plus sign rather than the minus sign. Taking into account the rules of non-commutation gives the minus sign.

So the GHZ state is an eigenstate of each operator on the left-hand side of the equation I just wrote down with eigenvalue +1, and also of the operator on the right-hand side with eigenvalue −1. Thus the conclusion is that the GHZ argument is a total contradiction of the possibility of retaining the

Fig. 7.1. The Mach–Zehnder interferometer. Light from source S has two paths through the apparatus available to it, labelled 1 and 2. The diagram shows the possible paths to detectors D_1 and D_2, but, as explained in the text, with both paths of exactly the same length, all photons are detected at D_1.

correctness of quantum theory in all cases together with the ideas of hidden variables and locality. It is obtained with an exact relationship rather than with inequalities.

An experimental observation of the GHZ state has been made by Dirk (Dik) Bouwmeester and co-workers [202] in Zeilinger's group. (Bouwmeester is now at the University of California at Santa Barbara.) The experiment was based on techniques developed for work in the area of quantum information theory by the same group, demonstrations of quantum teleportation [203], which is discussed in Chapter 8, and also another interesting technique of entanglement swapping [204]. Again this shows the intimate connection between studies in the area of Bell's theorem and those in connection with quantum information theory.

This experiment, like many in the area and indeed elsewhere, uses a variation of the Mach–Zehnder interferometer, an instrument invented over 100 years ago for optical measurements. The original form of the interferometer is shown in Fig. 7.1.

Light from a source S is incident on glass slab A, the top surface of which is half-silvered, so that half the light incident on it is reflected and half transmitted. This device is often called a *beam-splitter* for obvious reasons. Thus beams of light of the same intensity proceed along paths 1 and 2, and are reflected at mirrors on the lower side of glass slab B_1 and the upper side of slab B_2, eventually recombining at slab C, the lower side of which is again half-silvered. As shown in the diagram, two paths away from slab C are available for the light, ending at detectors D_1 and D_2.

If the two paths between A and C are of exactly the same length, all the light travels to detector D_1, none to D_2. The key to understanding this is the fact that the phase of a wave of light changes by π, equivalent to half a period, when the light is reflected at a medium, if but only if the medium has a higher index of refraction than the medium the light is in. Thus there is a phase change when a beam of light in air is reflected at a mirror surface, but not the other way round.

For light proceeding to D_1, both beams of light recombining have had two phase changes; therefore the two are in phase and there is constructive interference. However, for light proceeding to D_2, light that has travelled along path 1 has undergone two phase changes, but that which has travelled along path 2 has undergone just one. Therefore the two components are out of phase, and there is destructive interference, no light travelling to detector D_2.

When used for its original purpose, material such as a type of glass may be inserted in one or other beam, and its optical properties investigated by measuring the changes in the sizes of the signals at the two detectors. More recently, with the understanding that light may be sent through as individual photons, and that particles such as electrons also have a wavelike nature, a great variety of experiments has been performed, all of which are variations on the same basic idea.

In the Bouwmeester experiment, shown in Fig. 7.2, the source produces a pulse of pairs of polarisation-entangled ultra-violet photons. The idea of the experiment is to transform two pairs of these photons into three entangled photons in a GHZ state and a fourth independent photon called the trigger photon. The GHZ entanglement is observed only if both the trigger photon and the three entangled photons are actually detected, so detection plays the double role of projecting the photons into the GHZ state and performing a measurement on the state.

In the apparatus we see that two polarising beam-splitters reflect vertically polarised photons and transmit horizontally polarised photons. The third beam-splitter acts independently of the polarisation of the photons. We do not go into the details of the optical manipulations, but merely state that, in a small number of cases, two pairs of photons are generated by a single pulse, and all four photons are eventually simultaneously detected, one by each of detectors D_1, D_2, D_3 and T. The first three detections correspond to the GHZ state and the fourth is the trigger photon.

Fig. 7.2. A sketch of the experimental set-up used by Bouwmeester and his collaborators [202] to demonstrate GHZ entanglement. With one photon registered at the trigger detector T, the three photons registered at D_1, D_2 and D_3 demonstrate the required correlations (from [202]).

As we saw, the GHZ state must be a superposition of two states, and a fair amount of checking has to be performed to ensure that there is indeed a superposition, not just a statistical mixture of the two states.

The possibility of one pulse yielding this result is very low, of order 10^{-10}, but with the rate of pulses being about 7.6×10^7 s^{-1}, one event was registered every 150 s, just sufficient for the experiment to be performed.

Bell's theorem without inequalities: Hardy

We now turn to a second means of establishing Bell's theorem without inequalities, that produced by Lucien Hardy in 1992 [205, 206]. His method was based on a thought-experiment of Avshalom Elitzur and Lev Vaidman [207], which is often called the bomb problem. This thought-experiment uses what is called *interaction-free measurement*, and I will now describe it.

Let us imagine we have a stock of bombs, each of which happens to operate via a sensor that explodes the bomb if any light at all reaches it. However, in a consignment of bombs, some sensors are defective in such a way that light may pass right through them without causing the bombs to explode, and without affecting the light itself in any way. It is required to produce a stock of bombs in good working order.

Shining light at each bomb will not work. Those bombs which do not explode will have been found to be defective; those that were originally good will have exploded and thus be good no longer. Classically the problem is insoluble. With a Mach–Zehnder interferometer, however, we may take advantage of interaction-free measurement.

Before considering the experiment itself, let us return to Fig. 7.1 but imagine a non-transmitting object inserted in path 1. When a photon is sent through the interferometer, there is a probability of $\frac{1}{2}$ that it will be absorbed or scattered by this object. However, there is also a probability of $\frac{1}{2}$ that it will proceed along path 2. In this case we have only one path reaching beam-splitter C, so the photon has equal probabilities of leaving this beam-splitter along either path, and thus equal probabilities of $\frac{1}{4}$ of being detected at either D_1 or D_2.

We start the experiment exactly as in Fig. 7.1; photons are sent through and all are detected at D_1. The bomb is now inserted in path 1, and photons are sent through the interferometer one by one. If the bomb is good, there is a probability of $\frac{1}{4}$ that it will explode when a photon is sent through, and probabilities of $\frac{1}{4}$ that either of the detectors will click. If it is defective, however, we are back to the situation of Fig. 7.1, and D_1 must click. So D_1 clicking tells us nothing, but D_2 clicking tells us that the bomb is good and has not been exploded.

We therefore send photons through until either the bomb explodes or D_2 clicks. If neither of these things has happened after a large number of trials, but D_1 has clicked each time, we conclude that the bomb is defective. If it has exploded, clearly it was good but is good no longer. However, if D_2 has clicked, we may add this bomb to our stockpile of good but unexploded bombs. For a good bomb, the probability of its exploding at any trial is double that of D_2 clicking, so the overall probability of failure, the probability that the bomb explodes at some stage, is $\frac{2}{3}$, and the probability of success is $\frac{1}{3}$.

The bomb has been studied without interacting with it. (Interaction would, of course, be disastrous for the project.) Hence the idea is called interaction-free measurement, a topic that is of considerable interest in its own right, but here we study it as the basis of Hardy's experiment.

As shown in Fig. 7.3, Hardy used two Mach–Zehnder interferometers, one for electrons and one for positrons, arranged so that the paths intersect. Each interferometer has a single input mode $s^\pm$, two possible paths through the interferometer, $u^\pm$ and $v^\pm$, and two possible output modes, $c^\pm$ leading to detectors $C^\pm$ and $d^\pm$ leading to detectors $D^\pm$. Each interferometer contains two

Fig. 7.3. Hardy's thought-experiment to demonstrate Bell's theorem without inequalities. Two Mach–Zehnder interferometers, one for positrons and one for electrons, are arranged so that, if a positron takes path u^+ and an electron takes path u^-, they meet at P and annihilate one another (from [205]).

beam-splitters, $BS1^\pm$ and $BS2^\pm$, and a crucial point is that the latter may be removed when required.

If each interferometer were isolated, arguments analogous to our previous ones show that all electrons and positrons would leave the interferometers along paths $c^\pm$, and only detectors $C^\pm$ would click. However, the overlapping of paths is designed to give a possibility of electrons and positrons in the two interferometers meeting and annihilating each other. We will imagine that a positron and an electron are created simultaneously and fed into the respective interferometers. If the positron is imagined to take path u^+, and the electron to take path u^-, they meet at P and annihilate each other with probability unity. This will happen in a quarter of all events.

Since that is the fate of the particles if paths u^+ and u^- are taken, we may analyse the other experimental possibilities by saying that there are three components of the state of the combined system, each of which will occur with equal weight. These correspond to the paths being (a) u^+ and v^-, (b) v^+ and u^-; (c) v^+ and v^-.

We now need to analyse what happens as the particles move to the various detectors. We have said that either or both of the beam-splitters BS2$^\pm$ may be removed, so there are four possibilities: (i) both are absent; (ii) BS2$^+$ is present and BS2$^-$ absent; (iii) BS1$^+$ is present and BS2$^-$ absent; and (iv) both are present.

This all looks much more complicated that it actually is! The easiest case to consider is (i) because, with both beam-splitters absent, particles moving along u$^+$ proceed to register at C$^\pm$, and those moving along v$^\pm$ proceed to register at D$^\pm$. Thus there may occur clicks at C$^+$ and D$^-$; or at D$^+$ and C$^-$; or at D$^+$ and D$^-$. Each of these events occurs with probability $\frac{1}{4}$, the remaining probability, also of $\frac{1}{4}$, corresponds to the mutual annihilation and hence no clicks at all. Note that clicking of C$^+$ and C$^-$ is *not* permitted.

Possibilities (ii), (iii) and (iv) may be analysed along the same lines, though they are a little more complicated because we have to take into account interference at the beam-splitters that are present. We will not give the details; the most interesting point is that in cases (ii) and (iii) simultaneous clicking of D$^+$ and D$^-$ does not occur. (For case (iv) all possible events occur, though not with equal probabilities.)

We must now attempt to model the situation using local hidden variables. We let the state of each positron–electron pair be described by hidden variables before it is sent through the system. The values of the hidden variables may take different values each time the experiment is repeated.

When we analyse the result of each run of the experiment, the only requirement is that the result of a measurement on the electron does not depend on whether BS2$^+$ is present, and similarly the result of an experiment on the positron does not depend on whether BS2$^-$ is present. Note that this restriction is directly analogous to the one imposed by Bell in his initial proof of his theorem.

Hardy was now able to show that the set of constraints we mentioned in the treatment of the problem using quantum theory, in other words the events that did not occur for cases (i), (ii) and (iii), could not be modelled using our local hidden-variable approach. Again we have a clear contradiction between quantum theory and local realism, and again it is without inequalities.

Subsequently Hardy [208] was able to adapt the proof to rule out local realism for all entangled states, except, rather strangely, maximally entangled states such as the EPR state, again without inequalities.

Tests of this later proof of Hardy's have been performed by De Martini and co-workers [209, 210, 195]. In these experiments, entangled photons were produced by parametric down-conversion using a BBO crystal. Their polarisations were then independently rotated by variable amounts, ψ_1 and ψ_2, and the two photons were then injected into two input modes of a polarising beam-splitter. The polarisation of each photon was then rotated by angles θ_1 and θ_2, and the photons then passed through further polarising beam-splitters followed by detectors.

The arrangement was extremely flexible because ψ^1, ψ^2, θ_1 and θ_2 could be varied independently, so a complete picture of the behaviour of the system could be built up. The dictates of local realism could be analysed, along the lines of the arguments of Hardy just given, and these requirements were shown to be violated by the experimental results.

What do the experiments mean?

Many experiments testing Bell's inequalities as well as the versions of Bell's theorem without inequalities have now been performed. Apart from one of the very earliest experiments, all found violations of local realism, often by many standard deviations; all have results close to the predictions of quantum theory. While no individual experiment yet performed is totally free of loopholes, both the locality loophole and the detector loophole have been closed in individual experiments.

The realists – Selleri, Santos, Marshall, Thompson – deserve credit for having drawn attention to the detector loophole. For this group, their position on the Bell tests was only one component of their views, though an important one. Their arguments were spelled out most clearly by Selleri in his 1990 book [140], *Quantum Paradoxes and Physical Reality*. Selleri was a strong advocate not only of realism, but also of comprehensibility, the belief that it is possible to form mental images corresponding to atomic objects and atomic events.

He argued, not for strict determinism, but for what he called *causality*; he believed that there was a connection between present and future, but it was probabilistic in nature. Locality was also a crucial assumption, but Selleri talked of *probabilistic Einstein locality*, based on (his definition of) causality, not strict determinism.

While the realists acknowledged that Bohm's work demonstrated the inadequacies of the Bohr–von Neumann position, they did not welcome the fact that it was non-local, so it was the dualistic position of Einstein and de Broglie that was their favoured starting point. This has real waves and real particles for photons and electrons alike. In interference experiments, one may imagine a *full wave* containing the photon or electron, and an *empty wave* or *ghost wave* propelled by a *phantom field* in Einstein's terms.

Interesting experiments have been performed, using both neutron interferometry, which is discussed in the following chapter, and optical photon techniques.

Leonard Mandel and collaborators [211] have used the latter to try to ascertain whether the empty wave affects experimental results by causing stimulated emission or modifying interference patterns. There has, though, been no encouragement for the realist position.

Though Thompson [212], in particular, continues to contest the findings of the Bell experiments carried out so far, in particular arguing that the handling of the experimental data involves systematic distortion [213], the room for the realist position does appear to have diminished over the last few years, and for the rest of this section it will be taken for granted that the Bell experiments have disproved local realism.

While much of our discussion has been of experiments and closing loopholes, there has also been a great deal of conceptual and theoretical work in the area of Bell's inequalities. This has included detailed analysis of concepts such as locality and realism, and their places in the Bell type of analysis, and extensions from deterministic to stochastic hidden variables. Much of the fundamental work is included in the proceedings of two fairly early conferences [211, 214], while a much more recent conference [215] specifically on all aspects of John Bell's work is also of considerable interest. Many books [216–222] also focus on Bell and the results of his ideas.

It might be said that Einstein had been, so to speak, hoist with his own petard. He had invoked hidden variables in order to avoid the apparent non-locality of EPR, only to find that, if there are hidden variables, they must themselves be non-local. There was irony for Bell as well. He regarded himself as a follower of Einstein [76]; he reported to Bernstein [78] his feeling that 'Einstein's intellectual superiority over Bohr, in this instance, was enormous; a vast gulf between the man who saw clearly what was needed, and the obscurantist.' Yet, as he continued, 'Einstein's idea doesn't work. The reasonable thing just doesn't work'; and Bell is often known, in the words of John Gribbin, as 'the man who proved Einstein was wrong' [223] (though Robert Romer retorted [224] that he was actually 'the man who proved Einstein was right').

Of course, as we said in Chapter 6, the argument of EPR led formally to the conclusion that one must have either hidden variables (HV) or non-locality (NL) (or both). All that was ruled out was the assumption of no hidden variables (NHV) together with locality (L). So EPR allowed (i) HV and L; or (ii) HV and NL; or (iii) NHV and NL.

Since Bell theory and experiment ruled out only (i), leaving (ii) and (iii), the strictly logical conclusion of EPR was certainly not proved incorrect. But of course it must be admitted that Einstein's personal preference was undoubtedly for (i), and it is in that context, and that context only, that it may be said he was proved wrong.

Thus while Bell [76] writes that 'It would be wrong to say "Bohr wins again", the argument was not known to the opponents of [EPR]', he is forced to add 'But certainly Einstein could no longer write so easily, speaking of local causality " . . . I still cannot find the fact anywhere which would make it appear likely that the requirement will have to be abandoned".'

However, there was an important limitation on the kind of locality required by hidden variables according to Bell's theorem. In Chapter 6 we explained the ideas of parameter and output independence, which were introduced by Shimony [102] and Jarrett [214] specifically for this situation. (The actual terminology was that of Shimony.)

The important conclusion of these authors was that the Bell-type non-locality constituted breakdown only of output independence, *not* of parameter independence. In other words, experimental outcomes in different wings of the experiment may be connected non-locally, but a measurement setting at one location can have no immediate effect at another point. Thus we cannot use the non-locality to send messages faster than light; there is a 'no-signalling' theorem, and Shimony talks of 'peaceful coexistence' between quantum theory and special relativity.

It should be mentioned that the remarks of the previous paragraph apply if there are no hidden variables, or if hidden variables may exist but are out of our control, so we can only study averages across all possible values of the hidden variables. The latter is, of course, what we normally assume for the Bohm case. The mere assumption of hidden variables then makes no difference to our ability to send signals.

It seems extremely unlikely, but, if we were ever able to gain control of the hidden variables, we would actually be able to harness the determinism inherent in the Bohm system. We may say that, at the level of the hidden variables, parameter independence is violated, and, in fact, output independence is upheld [225].

Let us now return to Einstein and contemplate what he might have tried in order to come to terms with Bell's theorem. As we have seen in the previous chapter and this one, the arguments of EPR put him under no pressure to renounce any of his desiderata — locality, realism and determinism. (The only pressure he was under was not to appear too stubborn towards the likes of Bohr, Born and Pauli, and this pressure he was largely able to resist.)

Bell's theorem made the situation entirely different. Essentially Einstein would have to choose which of locality and realism would have to be sacrificed in order to protect the other. We discussed remarks of Fine [94] in Chapter 6, and Howard [226] and Cushing [225] have commented on these. Cushing argues that, with parameter independence maintained, Einstein might have felt that physics could function successfully even allowing this measure of non-locality, and might have plumped for retaining realism.

With a theory such as that of Bohm, this could also allow him to keep determinism, which would be a bonus, although not in my opinion as important for Einstein as realism or locality. But as we have said, Bohm's theory allows a breakdown of parameter independence *in principle*, and altogether it seems far from obvious which way Einstein would have moved.

Bohm: the later years

For the last two sections of this chapter we return to Bohm himself and the de Broglie–Bohm theory, and survey more recent events and developments. From 1952, while Bohm collaborated with younger scientists in the development of the causal interpretation, much of his own efforts were devoted to broader and deeper analyses of the methods of science, and the scope and the limitations of the knowledge we may gain from it.

In his 1957 book, *Causality and Chance in Modern Physics* [123], he began by discussing the *philosophy of mechanism* as it developed during the nineteenth century. Next he described his own causal interpretation, though regarding it only as 'a rather schematic one which simplifies what is basically a very complex process by representing it in terms of the concepts of waves and particles in interaction'. Finally he outlined a more general concept of natural law *beyond mechanism*, recognising 'the infinite richness of the real relationships existing in natural processes, and of our need to express partial aspects of these infinitely rich relationships in terms of finite laws based on experiments and observations done up to a particular period of time, which can reflect adequately only a limited part of the infinite totality that exists in nature'.

Bohm's most famous book, published in 1980, is *Wholeness and the Implicate Order* [124]. Bohm felt that notions implying the undivided *wholeness* of the world would be more fruitful than the present fragmentation. (Notice an acknowledged relationship to Bohr's ideas at this point.) *Language* was also important for Bohm; he proposed a new mode of using existing language – the *rheomode* (flowing mode) that would prevent us from discussing observed fact in terms of separately static things. *Order* is also of crucial importance. In the *implicate order*, space and time are no longer dominant, and 'an entirely different set of basic connections is possible'; the ordinary notions re-appear in the *explicate* or *unfolded order*, but implicate order is primary, explicate order secondary.

Bohm's last book, *Thought As a System* [126], was published posthumously in 1994. It deals with thought and knowledge, dismissing the idea that thought processes may report neutrally on an objective world.

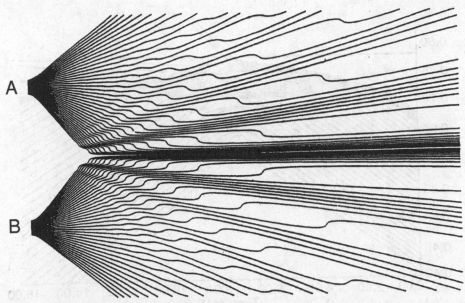

Fig. 7.4. Two-slit interference according to the Bohm theory. Particles emerge from slits A and B and subsequently move along trajectories, bunching around the maxima of the usual interference pattern. From [227] by kind permission.

Development of the de Broglie–Bohm theory

I now turn to the development of the de Broglie–Bohm theory over the last 50 years [125, 128]. In fact it is fair to say that, for the first 25, interest was minimal. When it did increase in the late 1970s, it was not clear whether credit should go chiefly to Bell, who *had* advertised the theory consistently, or to the advent of computer graphics, which showed the theory working in practice, rather than merely as somewhat esoteric equations on paper.

Holland [128] says that 'serious interest was re-kindled when the trajectories corresponding to the two-slit experiment were explicitly displayed' in a 1979 paper of Chris Philippidis, Chris Dewdney and Basil Hiley [227]. (Apart from these three, of whom Hiley worked closely with Bohm for many years, others centrally involved in the elaboration of the theory have included Holland, Tassos Kyprianidis, Pan Kaloyerou and also Jean-Pierre Vigier, who has worked on such theories right back to an initial collaboration with de Broglie.)

The two-slit interference case is shown in Fig. 7.4. The particles emerge from the slits at the left. Each trajectory has kinks in it, due to interference with the *empty wave* passing through the other slit. (Whenever a particle goes through one slit, that is to say a full wave, it must be accompanied by an empty wave

Fig. 7.5. Quantum tunnelling according to the Bohm theory. Particles are aimed at the barrier. Their behaviour as a function of time (which is on the horizontal axis) is shown. Each follows a distinct trajectory. Some particles tunnel through the barrier; others are reflected, with or without initially entering the barrier region. From [228] by kind permission.

passing through the other slit.) Alternatively we may say that the path of the particle is directed by the quantum potential. The net effect is that particles bunch together around what become the maxima of the interference pattern, and are pulled away from the minima; thus the usual interference pattern is obtained.

What is new and striking is that, in complete contrast, of course, to the orthodox approach, particles *do* have continuous trajectories, and in particular each particle goes through one slit or the other. Note that, because of the quantum potential, the trajectory is not straight even in what might be described as a field-free region away from the slits.

Another interesting computation is shown in Fig. 7.5. It shows *quantum tunnelling*; whereas in classical physics, particles reaching a potential barrier higher than their own kinetic energy *must* be reflected, quantum theory predicts a possibility of tunnelling through. The effect is the basis of radioactive decay. In orthodox quantum theory one handles the situation purely in wave terms. Incoming and reflected waves are matched to a decaying wave-function in the barrier region, and to a transmitted wave beyond; this is an elementary calculation for all students of quantum theory.

In Bohm theory [228], one works with particle trajectories. In Fig. 7.5, particles are aimed at the barrier. Some are slowed down in the barrier but eventually tunnel through. Others enter the barrier, and, after following sometimes complicated paths inside it, are reflected. Others are turned around and so reflected by the effect of the quantum potential without even reaching the barrier. Again the clearness of the picture provided is in marked contrast to the orthodox approach.

I shall now briefly refer to the hydrogen atom [128]. An interesting point here is that, with a return to particle trajectories in the Bohm theory, one might expect some relationship with the Bohr orbits of the pre-1925 old quantum theory. Such orbits do exist in the Bohm theory (although more complicated trajectories are also allowed). The simple orbits differ from the Bohr case in that the radius may take any value, and it is the *speed* of the particle that is quantised. Rather as in the Bohr case, an integer number of de Broglie wavelengths fits into a quantum orbit.

The most striking feature of the Bohm interpretation is that, in its ground-state, the electron is at rest; the force resulting from the quantum potential just balances that from the classical potential. At first sight this is a surprising result, but it does certainly explain very nicely the stability of the atom – which, it will be remembered, was one of the very big problems at the beginning of the quantum theory. If the electron is at rest, it is not accelerating, so it should not be expected to radiate energy and spiral into the nucleus.

The last elementary application of the Bohm interpretation that I shall deal with is the *delayed-choice experiment* [76, 128], a clever idea of Wheeler. This is a normal two-slit experiment, except that counters may be swung very quickly into each path, the time taken being much less than that for the electrons to pass from slit to screen. When the counters are absent, the electrons act in a wavelike way, we may say, and experience both slits and both paths. But when they are present, the electrons act as particles, and apparently pass through only one slit and travel along only one path. Yet this decision – whether to pass through one slit or both, whether to travel along one path or both – must be taken *after* the electron has travelled much of the way from slits to screen. A paradox, Wheeler would say – 'no phenomenon is a phenomenon until it is an observed phenomenon'.

In Bohm theory, the experiment presents no problem at all. Both particle and wave always exist, and will behave in an appropriate fashion whether or not the counters are present. The *earlier* behaviour of wave or particle is certainly not affected by the *later* insertion or non-insertion of the counters.

I have described a few elementary applications of the theory. As the theory came of age in the 1980s, though, a great range of more advanced topics was handled [125, 128]. *Many-body systems* were investigated in detail. *Spin* – for a long time a problem for the theory – has been considered successfully. A little progress has been made in the areas of *relativistic quantum theory* and *quantum field theory*. It must be admitted though that, at least at present, Bohm theory has a long way to go. It cannot, for example, handle pair creation, namely the simultaneous creation of an electron and a positron from radiation, which is a central type of event in relativistic quantum theory.

So it seems that the de Broglie–Bohm theory may be a worthy competitor for orthodox interpretations, though, as Holland [128] says, 'It is not presented as a conceptually closed edifice offering the final word on quantum mechanics, and its originators never intended it to be this. Rather, it is a view worth developing for the insight it provides, and as a clue for possible future avenues of enquiry.'

Finally let us return to the question of whether Bohm theory may contradict or transcend orthodox interpretations of quantum theory. While the two cannot actually contradict each other, since the mathematical formalisms are identical, there may be areas where the orthodox interpretations simply give no information, but Bohm theory may be more forthcoming.

An important example [128, 225] is that of *tunnelling time*. Although, as I have said, calculation of tunnelling probabilities is absolutely standard in orthodox quantum theory, calculation of the time taken for the particle to move through the barrier is not possible. This is fundamentally because, as mentioned in Chapter 4, in non-relativistic quantum theory there is no operator for time analogous to those, for example, for position and momentum.

With its explicit trajectories Bohm theory may be able to make an important contribution here, going beyond what is possible for orthodox interpretations, or indeed if one restricts oneself just to the formalism of quantum theory. Direct comparison with experiment is also conceivable. Richard Leavens [229] in particular has analysed the possibilities, and a recent volume [230] contains many approaches to studying time in quantum theory, including those based on Bohmian mechanics.

8

A round-up of recent developments

Everett and relative states

In this chapter I shall discuss briefly a number of the interesting developments – interpretational, theoretical, experimental – that have taken place in the foundations of quantum theory over the last few decades, where appropriate relating them to the work of Bohr or Einstein. While the topics considered will range well beyond the specific areas studied by John Bell, I think it is fair to say that it was the interest stimulated by his ideas that led to nearly all the work described.

This could not, though, apply to the very first ideas I discuss, which date from as early as 1957, when a PhD student at Princeton University, Hugh Everett, wrote a thesis titled *The Theory of the Universal Wave Function* [231]. A short version of this was published [232], and it was followed by a brief positive assessment [233] of Everett's ideas by John Wheeler, who had guided and encouraged him. (Sixteen years later, both of Everett's papers, and a number of related ones, were collected in a single volume [234].)

I shall now sketch his ideas. Till now, we have allowed wave-functions for microscopic systems, such as atoms or electrons, to be sums of wave-functions for highly distinct states, so that the corresponding properties of the systems – position, momentum and so on – do not usually have precise values, at least not till a measurement is made. When a perfect measurement *is* made, though, we have taken it for granted that a unique result is obtained by the *measuring apparatus*, and the *mind* of the observer will correspondingly be left in a unique state, which will be registered permanently in the *memory* of the observer. (The memory will play an interesting role in some of the ideas to be discussed.) This process ending in a unique result is just von Neumann's collapse, which has led to so many difficult questions in the preceding chapters – 'how does the collapse take place?', 'how does one *define* a measurement from a fundamental point of view?' and so on.

Everett solved these problems at a stroke – no collapse takes place! 'The whole issue of the transition from "possible" to "actual" is taken care of in the theory in a very simple way', he writes; 'there is no such transition' [231]. Let us consider an example. Suppose that a measurement of the z-component of spin of a spin-$\frac{1}{2}$ particle is made and (in conventional terms) there are equal probabilities of obtaining the results $+\hbar/2$ and $-\hbar/2$. According to von Neumann, the final state of the total system (spin plus measuring device plus mind of the observer) must correspond *either* to the z-component of spin being $+\hbar/2$, measuring device registering that fact, and mind of the observer being aware of it, *or* to the same with $-\hbar/2$ replacing $+\hbar/2$. According to Everett, on the other hand, the final wave-function is a mathematical sum of *both* these possibilities. In particular, the mind itself appears to be somehow split between totally different states.

Requiring no collapse means that we do not need von Neumann's two distinct types of quantum-mechanical process; we do not need to pick out certain sub-systems as 'measurement devices' and treat them according to special rules. We do not use the idea of measurement selecting a particular *branch* or *component* of the evolving wave-function; we have the *universal wave-function* of the title of [231], and we will claim to be working deterministically.

We may sum up Everett's vision by saying that the quantum state is a real physical state and a complete description of reality. It always evolves unitarily. Everett tells us to take the quantum formalism absolutely literally.

Wheeler [233] emphasises that the theory of measurement becomes a straight-forward and uncontroversial example of correlation between sub-systems. In each branch of the wave-function, the various states of different sub-systems, such as observed and observing systems, are correlated. Everett says that 'There does not, in general, exist anything like a single state for one subsystem of a composite system . . . One can arbitrarily choose a state for one subsystem, and be led to a relative state for the remainder.' He thus gives [232] the title of '"Relative state" formulation of quantum mechanics'.

So Everett avoids many of von Neumann's problems, but seemingly only at the expense of failing drastically to ensure the *aim* of von Neumann's procedure – that, following a measurement, the apparatus is left in a distinct state, and, in particular, the mind of the observer recognises a distinct result for the measurement.

Everett himself does little to resolve the dilemma. Much of the content of his papers is mathematical, and Squires [235] fairly characterises him as not so much interpreting quantum theory as asking for the formalism without collapse to be taken seriously; this formalism will still require interpretation. Bell [76] too says that 'it becomes obscure to me that any physical interpretation has either emerged from, or been imposed on, the mathematics'.

Much of Everett's own efforts were directed towards demonstrating consistency *within any branch* of the wave-function. If a second measurement of the same quantity as the first one is made, for example, then in each branch the second observer will obtain a result consistent with the first. Each branch evolves separately, in fact, just like the single branch left after a von Neumann collapse; in an Everett interpretation, the different branches do not usually interfere with each other.

This was an exceptionally important result, but, as I have said, it does not explain the strange point that different branches, containing different experimental results and different states of the mind of the observer, co-exist inside the universal wave-function. (Actually I should say, in the interests of historical accuracy, that I have gone somewhat further than Everett himself, who did talk of splitting of 'observer states', but did not mention minds explicitly. I have been somewhat closer to a paper by Leon Cooper and Deborah Van Vechten, which is also included in [234]. I might also mention, incidentally, looking ahead to later in this chapter, that Everett certainly never talked about splitting of 'worlds'.)

Recently Squires [235] has attempted to interpret, or at least take seriously, Everett's original position. He argues that the presence of many branches in the wave-function may be quite reasonable. Being part of the wave-function, the observer 'cannot stand outside the world to observe it'; the observer who is conscious of a particular experimental result 'is totally unaware of the other parts of the wavefunction' because '[i]n the act of making a conscious observation, "I" become that part of the wavefunction corresponding to a particular observed value.' Squires calls this the *many-views interpretation* of quantum theory. David Albert and Barry Loewer [236] call their similar idea the *many-minds interpretation*, and Michael Lockwood [237, 238], Henry Stapp [239], Jeffrey Barrett [240] and Matthew Donald [241] have also discussed the connection between the mind and the quantum. Readers may find themselves (like the author) appreciating the argument logically, but unsure whether they have fully assimilated the idea; others may, perhaps, be in two minds as to whether they like it or not.

De Witt and many worlds

In the years following 1957, Everett's work received very little attention; Jammer [67] mentions a comment that it was 'one of the best kept secrets in this century'. Then in 1967 Bryce de Witt wrote an article drawing attention to Everett's work. (In 1970 and 1971 he wrote two further articles, which are included in [234].) Undoubtedly these articles popularised, even glamorised, the ideas, the rather elusive concepts of Everett being supplemented (or replaced) by ideas that, though exceptionally strange, are much more definite. Unfortunately, though, many [76, 125, 242, 243] feel that de Witt's ideas are so different from those of Everett that they need to be considered as an entirely separate development.

For Everett the various branches of the wave-function co-exist (in some way). For de Witt, at the moment of the measurement there is a splitting into two different worlds – one for each distinct measuremental result. For the example discussed before, after the measurement there exist two worlds; they are identical, except that in one of them the result of the experiment is $+\hbar/2$, and in the other it is $-\hbar/2$. (Observed system, observing system and mind of the observer all agree on the matter in *each* world.)

That of course is just for one measurement. We must imagine that at *every* measurement there is a similar splitting, a tree-like structure emerging. It will be clear that, as more and more splittings take place, the number of different worlds must become enormous. At any moment, many of the worlds may differ only very slightly from the one we are in, but others, which split from ours long ago, much may be altogether different. We have the *many-worlds interpretation* (MWI) or the *many-universes interpretation*.

De Witt [234] says that 'I still recall vividly the shock I experienced on first encountering this multiworld concept. The idea of 10^{100+} slightly imperfect copies of oneself all constantly splitting into further copies, which ultimately become unrecognizable, is not easy to reconcile with common sense. Here is schizophrenia with a vengeance.'

Can this vision – which at first sight seems closer to science fiction than to science – really be true? Many reject it just because of its strangeness, while others may be attracted to it for the same reason! Scientists are often abjured to use the most economical hypothesis, and supporters of the MWI claim that theirs is the simplest interpretation, the one most *economical of assumptions* (on which there will be more in the following section); opponents retort that it is very *expensive on universes*!

I would mention one further point of de Witt. Till now I have assumed that world-splitting takes place *at a measurement*. Certainly many of his statements would agree with this – but not all. In two successive sentences he says that 'The universe is constantly splitting into a stupendous number of branches, all resulting from the measurement like interactions between its myriads of components [splitting at measurement?]. Moreover, every quantum transition taking place on every galaxy, in every remote corner of the universe is splitting our local world into myriads of copies of itself [splitting at *any* interaction?].'

(Bell [76] suggests that de Witt (and Everett) intend splitting at any interaction, but that the ambiguity may be deliberate; 'I suspect,' he says, 'that Everett and de Witt wrote as if instrument readings were fundamental only in order to be intelligible to specialists in quantum measurement theory'.) I shall come back to this point as well in the following section.

I would mention here that, if splitting is restricted to measurements, different worlds will always differ macroscopically, because results of measurements will differ. But if splitting may occur at any interaction, worlds may differ only microscopically – for example, a photon taking two different paths. Interference may then take place, and the two worlds come together or *fuse*. This process is the opposite of splitting, and when it occurs the total number of worlds decreases rather than increasing.

I should point out that, though the MWI does seem bizarre, it has some definite advantages. I have stressed that neither Bohr nor von Neumann can cope with cosmology; they cannot consider a wave-function of the Universe. Bohr specifically works with a quantum experiment set up and analysed by an external classical observer; von Neumann too requires collapse by an external observer. Since astrophysics clearly has no external observer, many astrophysicists, such as Stephen Hawking, prefer to talk in terms of the MWI, which requires neither such an observer nor a classical region, and they are thus able to discuss the wave-function of the Universe [244].

Another advantage relates to the so-called *cosmological anthropic principle*. It is well known [245–247] that, if many of the constants of physics (charges and masses of elementary particles and so on) had values only very slightly different from their actual ones, life would not have been possible in the Universe. It seems that the existence of life is just an amazing accident. One way round this, of course, is to suggest that the Universe has been specifically designed for our benefit, though, for whatever reason, I don't think even the majority of believing scientists choose this position. If that is the case, John Polkinghorne, an eminent elementary-particle theorist who became an Anglican priest, is a notable exception [248].

Another possibility is just the MWI. This tells us that many worlds exist, and it is at least a possibility (though not obvious) that the values of the physical constants mentioned above result from quantum processes in the early days of these worlds, so they may differ in the different worlds. In at least some of these worlds the values of the constants will be expected to support life; it must be one of these particular worlds that we live in, and we will be entirely oblivious to the vast number where life would be impossible.

A rather similar possibility [249, 250] is that the Universe goes through a long succession of sequences in which there is a Big Bang, followed by an expansion and then a contraction back to a point again. It is again at least possible to conjecture that in different periods of expansion and contraction the physical constants may differ. As with the MWI, if there are enough sequences, some will be expected to support life.

In recent years, the main proponent of the MWI has been David Deutsch [53]. (I would mention that Wheeler's advocacy of the MWI has become, at most, lukewarm; he [53] believes it carries 'too much metaphysical baggage', and has returned to supporting Bohr. I also mention that Deutsch calls the MWI the Everett interpretation; others often conflate Everett and de Witt.)

Deutsch [251, 252] has introduced two new points into the discussion. First, rather than believing that the number of worlds constantly increases, he believes that there is at all times the same large number, probably an infinite number, of worlds. At the beginning of the Universe, all the worlds are identical, but as measurements (or 'choices' or 'decisions', as Deutsch calls them) are made, these universes divide into different groups corresponding to different outcomes.

The other, more controversial point concerns predictions of the MWI that differ from those of accepted interpretations. De Witt took it for granted that the MWI 'leads to experimental predictions identical with those of the Copenhagen view'. He continued 'This, of course, is its major weakness. Like the . . . Bohr theory it can never receive operational support in the laboratory.'

Deutsch disagrees, though the scientific equipment required to perform his crucial experiment is itself somewhat bizarre! In the experiment, an ordinary interference experiment takes place, so that at one time there are two states of the system, but subsequently, if things are undisturbed, they combine to re-form a single state. A superbrain with a memory at the quantum level observes the system and, according to the MWI, is itself split into two copies in different worlds. Now Deutsch's experiment 'hinges on observing an interference phenomenon inside the mind of this artificial observer. This can be done . . . by his trying to remember various things so that he can conduct an experiment on his own brain while it's working . . . And what he tries to observe is an interference phenomenon between different states of his own brain. In other words, he tries to observe the effect of different internal states of his brain in different universes interacting with each other.'

While there are two states of the system and two worlds, the observer must write down that 'I am hereby observing one and only one of the two possibilities.' Subsequently interference does take place, both between the states of the system under investigation and between states of the brain in the two worlds. The results should demonstrate that both states of the microscopic system, both brain states, did exist before the interference reduces them to one. This, as I said, is according to the MWI.

On an orthodox interpretation, though, as soon as information enters the brain of the observer, the system under investigation collapses from a combination of two states to a single one, to match a single brain state and, of course, a single world. No interference can subsequently take place. Thus an orthodox interpretation and the MWI must lead to different experimental predictions.

The 'only' difficulty in performing this experiment is producing such a super-brain, capable of working at the quantum level. (Our own brains work in practice at the classical level, even though there may be quantum effects that remain outside our control.) Deutsch suggests that such an experiment might be possible within decades.

These arguments of Deutsch, support for the MWI and the idea of the super-brain, were components of his visionary work on quantum computation. This is discussed in detail in Chapter 9 on quantum information theory.

Discussion of the many-worlds interpretation

Now I turn to what other people have thought about the MWI. First I tackle the claim mentioned before, that of its conceptual simplicity. Deutsch says, for example, that it is the 'simplest interpretation of quantum theory'. There are two aspects of this claim. The first is just that one has removed the von Neumann collapse with all its problems, and gone back to the Schrödinger equation pure and simple, with no additions whatsoever.

This may well be true of Everett and relative states. It really cannot be said to be so for de Witt and the MWI. In the first place there are world-splittings, new worlds are formed; these are actual physical events, not just algebraical manipulations, and by no stretch of the imagination can they be called 'nothing'.

But in addition we have to decide *when* a splitting takes place. Let us first suppose that splitting takes place at a measurement. In this case we are back precisely at the von Neumann difficulty – how does one define a measurement in fundamental terms, and when precisely does it take place? (Bell, who put this argument against von Neumann many times, put it forcefully against the MWI in [53].)

I might mention that de Witt [234] does have as one of his chief postulates his *postulate of complexity*, according to which the world is decomposable into systems and apparatuses, but it may be felt that this merely recognises the problem rather than actually solving it.

The alternative possibility that splitting takes place at any interaction may seem more hopeful, but it really seems to meet the same problem – how strong must the interaction between particles be, how close may they come before a splitting occurs? And what property will characterise the various branches of the wave-function after the splitting – in the case of spins, will they have precise values of s_z or s_y or some other quantity? (As Bell says, it is not clear from the mathematics that there may not be branches for s_z *and* s_y *and* any other quantity we choose!)

The second aspect of the claimed simplicity of the MWI is given by what de Witt calls a metatheorem – 'The mathematical formalism of the quantum theory is capable of yielding its own interpretation.' If it is to justify this, he says, the MWI must answer the following question: 'How can the conventional probability interpretation of quantum mechanics emerge from the formalism itself?'

I would remind you that, in quantum theory, if, prior to a measurement of energy, the wave-function of a system is given by $c_1\phi_1 + c_2\phi_2$ (where, for simplicity, c_1 and c_2 are mathematically real quantities), the probabilities of obtaining E_1 and E_2, the values of energy corresponding to ϕ_1 and ϕ_2, are $c_1{}^2$ and $c_2{}^2$, respectively. This is called the Born rule. In orthodox interpretations, this is a postulate. To satisfy de Witt's metatheorem, MWI enthusiasts must prove it.

During the early 1970s Everett, de Witt himself and Neil Graham [234] all made valiant attempts to do so. They used the idea of repeated measurements of the same quantity on an ensemble of identical systems in identical states. They thus built up statistics in one particular branch, and there is a claim that the desired probability rule holds in the overwhelming majority of worlds. There may be *maverick worlds* in which the rule is broken, but, as Graham says, we must assume that our own world is a typical one.

As early as 1973, though, Ballentine [253] published a paper severely criticising these 'proofs'. This was merely the start of a debate that has continued for the succeeding 30 years, perhaps diverting attention from more important questions such as the general adequacy of the MWI. The position of Ballentine and fellow critics has been that the 'proofs' illegitimately smuggle in just the argument that leads to the result. Supporters of the MWI, on the other hand, have suggested that the argument is legitimate, and required for mathematical consistency.

Ballentine admits that some of the mathematical results obtained by the MWI advocates are indeed interesting. Since, though, much of the mathematical nature of the MWI and the orthodox interpretation is identical (in one case one must sum over co-existing branches, in the other over probabilities of occurrence of unique branches) it is not clear that the maths necessarily accrues to the account of the MWI rather than the orthodox, or indeed any other, interpretation. Ballentine concludes that, for deriving the statistical postulate of quantum theory, '[t]he many-universes interpretation is neither necessary nor sufficient.'

Bell [76] says that, insofar as de Witt's metatheorem is true, 'it is true also in the pilot-wave [de Broglie–Bohm] theory . . . Everett has to attach weights to the different branches of his multiple universe, and . . . does so [as in our example above]. Everett and de Witt seem to regard this choice as inevitable. I am unable to see why, although of course it is a perfectly reasonable choice with several nice properties.'

In 1986, Deutsch [53] conceded that 'Everett was slightly wrong. I think that even in his interpretation, one requires a little bit of extra information . . . to arrive at the interpretation. But not much – very much less than in the conventional interpretation . . . It is the little piece of mathematics which provides the connection between the wave-function . . . and the concept of the many parallel universes. But I do agree with Everett so far as to say that his is the *simplest possible* addition to the purely instrumental quantum theory.'

I shall now turn to Bell's views on the MWI itself. I must start by saying that he didn't like it [53] – 'I have strong feelings against it . . . It's extremely bizarre, and for me that would already be enough reason to dislike it. The idea that there are all of these other universes which we can't see is hard to swallow. But there are also technical problems with it which people gloss over or don't even realise when they study it. [I have touched on some of these problems above.] . . . I believe that the many-universes interpretation is a kind of heuristic [intuitive], simplified theory, which people have done on the backs of envelopes but haven't really thought through. When you do try to think it through it is *not* coherent.'

Bell [76], in fact, made the MWI the last of the *romantic worlds* of quantum theory that he had no time for. He hated the idea that journalists may pick up slogans like 'All possible worlds are actual worlds' and repeat that as 'what scientists believe'.

Yet despite Bell's distaste for the MWI, he did make a novel suggestion about manipulating it into something more to his interest, if not approval. The manipulations were, though, quite severe; he admits that readers of Everett and de Witt might not immediately recognise his formulation. ('I am not sure', he also admits, 'that my present understanding coincides with that of de Witt, or with that of Everett, or that a simultaneous coincidence with either would be possible.')

First Bell sweeps away the many worlds themselves. '[I]t seems to me', he says, 'that this multiplication of universes is extravagant, and serves no real purpose in the theory, and can simply be dropped without repercussions.' Elsewhere he says 'I do not myself see that anything useful is achieved by the assumed existence of the other branches of which I am not aware. But let he who finds this assumption inspiring make it.'

For Bell, what *is* important is what I have so far hardly mentioned – Everett's emphasis on memory contents as 'the essential material of physics'. An observer in a particular world has a memory of a given sequence of past events. Bell admits that, for Everett, these events have actually taken place; Everett tries, Bell says, 'to associate each particular branch at the present time in a tree-like structure, in such a way that each representative of an observer has lived through the particular past that he remembers'. But Bell does not think Everett succeeds, and, in any case, thinks the attempt is against the true spirit of his approach. *All* Bell requires is memory contents. 'We have no access to the past', he says, 'but only present memories. A present memory of a correct experiment having been performed should be associated with a present memory of a correct result being obtained. If physical theory can account for such correlations in present memories it has done enough – at least in the spirit of Everett.'

Bell now relates the Everett picture to that of de Broglie and Bohm. At first sight there doesn't seem much similarity, but Bell says that both use an exact and complete Schrödinger equation, and add in additional structure. For de Broglie–Bohm, this structure is the continuous trajectories, for Everett (in Bell's view) just memory contents. Bell thinks that the lesson of Everett for de Broglie–Bohm is that, *if we like*, we may leave out the trajectories, and *merely* use the positions of the particles at successive instants, randomised so that there is no suggestion of continuity. 'For we have no access to the past', Bell says, 'but only to memories, and these memories are just part of the instantaneous configuration of the world.' So in Bell's interpretation of the Everett theory 'there is no association of the particular present with any particular past'.

'Everett's replacement of the past by memories . . . cannot be refuted', Bell says, but he still does not like it. Apart from technical difficulties with relativity, he would like to take more seriously the past of the world, and of himself – '[I]f such a theory were taken seriously it would hardly be possible to take anything else seriously.'

On a more general note, I mention that this is not the last time in this chapter that the suggestion is made that an interpretation thought to be incoherent may be sharpened up – and then looks rather like de Broglie–Bohm.

Having said that Bell did not like the MWI, I should mention his one caveat – he thought it might have something useful to say about EPR. There have been many suggestions that adoption of MWI-like interpretations may remove much of the problem from EPR and Bell non-locality. Squires [235], for example, speaking specifically of the relative-states formalism, says that 'In contrast to *all* other interpretations known to me . . . there is no EPR type non-locality.'

It is certainly true that proofs of Bell's theorem assume that a single result is obtained in each wing of the apparatus; they would thus seem to be inapplicable at least directly to treatments using an MWI. Yet there must be a time, although it may be delayed somewhat following the actual measurement, when observers in the two wings, but in the same world or in the same branch of the wave-function, may compare their results. In each world, these results must be identical to one of the possibilities for a von Neumann interpretation following collapse. The room for manoeuvre is thus extremely limited, and discussion of the EPR problem and locality in MWI-type interpretations [242, 243, 254] may be subtle and rely on detailed nuances of the particular interpretation.

I now mention an approach that, in a sense, uses the Everett–MWI type of mathematics, but assumes none of the metaphysical baggage. It's described, for example, by Peierls [53]. Essentially you begin with the wave-function at $t = 0$, and, still at $t = 0$ if you like, calculate its development for t greater than zero through whatever measurements are scheduled. Measurements do not require special treatment – the Schrödinger equation is obeyed at all times, there is no collapse, and all branches of the wave-function develop. The wave-function in no sense describes 'what happens'; rather it gives an account at $t = 0$ of all future possibilities.

When does anything actually happen? One may perhaps say that, when the Universe ends God may decide which possibility was realised, the possibility referring to a complete history of the system, and anything else with which it interacts, from $t = 0$. Since we don't want to wait till then, we may note that, at a perfect measurement, although all branches of the wave-function continue to evolve, only one can actually be activated. It appears that we can gain privileged early information on which branch this is.

It might be said that this way of looking at things genuinely accepts what other interpretations shy away from in one way or another – that quantum theory is probabilistic; the wave-function clearly does express probabilities, not any 'real' state of affairs. It is close to, maybe equivalent to, Stapp's account of the Copenhagen interpretation [72]. Tony Sudbery [255] has given a good account of this approach, without particularly endorsing it. Whether the idea appeals to one is mainly a matter of taste, and whether it is in any sense 'the answer' depends on what one thinks 'the question' is!

Many-worlds interpretations: recent developments

Over the last few years, there has been a sustained attempt to put the MWI (or simply the Everett interpretation as it is often called, though always involving actual worlds) on a more rigorous footing. This may be said to answer the complaints of Bell, Kent [256] and others that the argumentation for the interpretation had been shoddy. The scientists mainly involved have been

Vaidman [257, 258] and a group centred on Oxford including Deutsch [259], Simon Saunders [260–264], David Wallace [265–267] and Hilary Greaves [268].

I will sketch the various arguments. Bell inquired why, following a measurement of the z-component of spin of a spin-$\frac{1}{2}$ particle, there are branches for the eigenstates with s_z equal to $+\hbar/2$ and $-\hbar/2$, but no branches for arbitrary linear combinations of these eigenstates, which are equally acceptable in quantum theory.

A partial answer is provided by the theory of decoherence. This is discussed in detail a little later in this chapter. Here I will just say that interaction of the system with its environment causes decoherence, so that the appropriate states, such as the states with definite values of s_z following a measurement of s_z, are stable, whereas linear combinations of these decay in an extremely short time.

Yet decoherence, though an essential component of the argument, cannot be the whole answer. Wallace [265] argues that it is vague, picking out preferred states only approximately, and it is ambiguous in that no precise set of rules has been produced to pick out the states. More fundamentally, though, he believes it rather misses the point; the Everett interpretation is designed to interpret the Schrödinger equation, not to be the gateway to new and difficult physics, such as the detailed mechanism of decoherence.

For Wallace, with the *ontology* of a theory defined as what the theory says actually exists, it is essential to distinguish between the fundamental ontology, which is written into the theory from the start, and the higher level or emergent ontology. The former is easy to find and discuss; finding and analysing the latter is a very much more subtle process.

In the Everett interpretation, the fundamental ontology is the wave-function. The 'many worlds' that we discuss are not part of the fundamental ontology, but of an emergent ontology. Thus much of the questioning of the interpretation, which we have summarised as asking where in the formalism one finds the worlds, is deemed irrelevant. They are just not in the formalism.

To find them, we must search for patterns in our mathematical description enabling us to build up an interpretation that includes macroscopic objects like cats, people and so on. We are not actually obliged to introduce these concepts, but without them it will scarcely be possible to appreciate the formalism at all, and we will almost certainly miss much of what it is telling us.

We may accept that classical physics is not correct, but unless we can interpret the wave-function in ways understandable in general terms from a classical viewpoint, the wave-function will just remain uninterpreted. This approach, of searching for macroscopic objects as patterns in the underlying microphysics, is due to Daniel Dennett [269].

Such a search for pattern, such a process of interpretation of the full wave-function, leads us inevitably to a description in terms of many worlds. This description is complete, since we can trivially recover the entire wave-function from it. The worst that could conceivably be said of it is that it is somewhat arbitrary.

Wallace [267] sums up the present status of many-worlds interpretations as follows. There have been three main types. He says that the first, due principally to de Witt, in which worlds are explicitly added to the formalism, has no remaining defenders. Adding to the formalism, he says, is precisely what the Everett approach was designed to avoid. The two surviving types are those based on some variety of many minds mentioned briefly earlier, and the decoherence-based approaches defended by Wallace himself and just discussed.

Let us now turn to the question of probabilities. The apparent difficulty in even discussing probability in an MWI is as follows. There are two ways in which probability normally comes into physics. The first is via lack of determinism as in orthodox interpretations of quantum theory. The second is through what will be called (deliberate) ignorance; in Boltzmann statistical mechanics or in de Broglie–Bohm versions of quantum theory, even though there is underlying determinism, we do not attempt to specify values of all physical quantities, but work in terms of probability distributions. The difficulty is that the MWI is specifically set up to have neither indeterminism nor ignorance.

Clearly we require a different apporoach to probability. There are many such [97], but the one used in this context particularly by Deutsch [259] and Wallace [266, 267] is called *decision theory*. They argue that we may use probability in two ways. First, we may make inferences; if we repeat an experiment a large number of times, and in 60% of the times a particular event X occurs, we infer that the probability of obtaining X is 60%. Secondly, we may take decisions; if we think a particular team will win a football match, we may bet on it to do so at certain odds.

This may be called the subjective approach to probability [270]. It argues that probability is merely a way of encoding what individuals believe and how they act. To understand this, suppose that a particular individual has a rational set of beliefs. The rationality is ensured by a number of conditions; for instance if one would bet on X in preference to Y, and Y in preference to Z, then one should bet on X in preference to Z. Mathematically it can be shown that this rational individual will act as if ascribing a probability to any event.

This is a subjective and individual probability. But there are also objective probabilities. Physics talks of the probability that a particular radioactive nucleus will decay in a given period of time, and this is not a matter of opinion. Clearly one would like subjective probability to be equivalent to objective probability; one might even be inclined to demand it. Nevertheless it is extremely difficult to justify this belief [264].

The great feat of the rather elaborate analysis of Deutsch and Wallace is to show that in an MWI, if the Born rule leads to a probability of P for a given event, then one should also assign a subjective probability of P to that event. Thus the Born rule gives both subjective and objective probabilities.

Saunders [264] stresses that the proof works only for an MWI; it breaks down if one restricts oneself to a single world and tries to solve the usual problems of quantum theory by adding either hidden variables or a spontaneous-collapse mechanism (the latter is discussed later in this chapter). Under the MWI, subjective and objective probabilities are, he says, seamlessly joined, an achievement unprecedented in classical physics or other interpretations of quantum theory. Far from being a weak point of the MWI approach, he suggests, the question of probability is one of the strongest points in its favour.

More about ensembles

Earlier in this book I have said a good deal about ensembles, particularly in connection with the work of Einstein and Bell. These have been almost entirely *Gibbs ensembles*, in which all physical quantities have precise values at all times. Thus, if the wave-function of a free particle is an eigenfunction of momentum, *all* members of the ensemble will have the corresponding value of momentum, but in addition each has a precise value of position, though these values will all be different. The values of position must be called hidden variables, because they are not related to the wave-function. There is also a more general type of hidden-variable ensemble, where the hidden variables do not relate directly to physical quantities, but determine the values that will be obtained *if* any physical quantity is measured.

In this section, though, we examine a different type of ensemble interpretation, which does *not* entail hidden variables, but may be introduced by a statement such as that of John Taylor [53] – '[W]hen we're making a measurement of any observable [quantity] in a system what we're actually doing . . . is that we're making a measurement on an aggregate or *ensemble* of identically prepared systems . . . Hence our results take the form of a probability distribution of particular values for that measurement.' So it is the ensemble, not the individual system, that is the basic unit we must work with in quantum theory. Such an ensemble interpretation without any hidden variables may be called a *minimal ensemble* [97]. In his major review, Ballentine [88] particularly considers Gibbs ensembles, but also allows minimal ensembles. A detailed analysis of Ballentine's position is given in [97].

As such, he gets fairly short shrift from Shimony [271], who says 'There is, for example, Ballentine . . . He says "I am not a hidden variable theorist, I am only saying that quantum mechanics applies not to individual systems but to ensembles . . ." I simply do not understand that position. Once you say that the quantum state applies to ensembles and the ensembles are not necessarily homogeneous you cannot help asking what differentiates the members of the ensembles from each other. And whatever are the differentiating characteristics these are the hidden variables. So I fail to see how one can have Ballentine's interpretation consistently. That is, one can always stop talking, and not answer questions, but that is not the way to have a coherent formulation of a point of view. But to carry out the coherent formulation of a point of view, as I think Einstein had in mind, you certainly have to supplement the quantum description with some hypothetical variables.'

Similarly d'Espagnat [272] makes the title of his section on these interpretations 'The (so-called) ensemble interpretations', and suggests that Ballentine's argumentation is 'unclear and, on the whole, inconsistent'.

Indeed one may take the example of the most important physical quantity which definitely *does* relate to an ensemble, not to an individual system – the temperature. Temperature may be defined only for a large number of entities, for example molecules, in thermal equilibrium. (See Chapter 2.) But though the temperature is certainly a property of the ensemble, the molecules taken as a whole, it is related to a distribution of the energies of the individual molecules. It is difficult to see how such a property could relate to an ensemble if the various members of the ensemble themselves do not differ in any way. (And if they *do* differ, we have a hidden-variable ensemble, *not* a minimal ensemble.)

In fact I have to say I find Taylor's defence of the minimal ensemble in [53] rather unconvincing. I start from the point that the results of a quantum-mechanical measurement are, in general, statistical; there is certainly need to consider a *post-measurement ensemble*. That is not a matter of interpretation; it is just quantum theory.

Interpretation must tell us how or why this range of measurement results appears. Given that measurements on individual systems can certainly be done, it seems no explanation or interpretation at all to say that we should use a *pre-measurement ensemble* as well. It just adds nothing to the statement of quantum theory in the previous paragraph. To this extent, we might feel inclined to say that we are studying the minimal-ensemble *non-interpretation*. In [53], Taylor consistently declines to consider what happens at the actual measurement – which is surely what quantum interpretation should be all about. (At one point, Paul Davies, the question-master, is provoked to ask 'Isn't that a bit of a cop-out?')

Taylor's explanation of EPR (–Bohm), incidentally, is that 'we're looking at a whole ensemble of such systems. Some 50% of them may have (when we're measuring them) nearby particles with spin up and far away ones with spin down, while the other 50% have the opposite spins. But we can't say in any particular case what that spin of the far away particle is from the measurement nearby, because we don't know about it; we only know about ensembles of such situations.'

But surely, even from the ensemble point of view, what we have is an ensemble of systems, in *each* of which one spin is up, one down. This common feature, and how it comes about, requires an explanation, but Taylor and the minimal-ensemble interpretation provide none.

Despite, though, these rather negative features, it must be admitted that such ensembles are frequently used – for the reason that they provide an approach to quantum measurement that seems, at least superficially, quite attractive.

Ensembles and measurement

To explain this, I need to introduce (I am afraid) a new mathematical idea, one that will be used quite frequently in the rest of this chapter, the *density-matrix*. A density-matrix may refer to a single system, in which case it provides exactly the same information as the wave-function of the system. The same density-matrix may also refer to an ensemble of different systems, each with the same wave-function. (Technically we may call this a *pure ensemble*.) But in addition, the density-matrix approach can do something that the wave-function approach cannot, at least not in a convenient way – it can describe an ensemble of different systems that do *not* all have the same wave-function. (Technically we may call this a *mixture* or a *mixed ensemble*.)

I shall explain this in terms of a spin-$\frac{1}{2}$ system (or an ensemble of such systems), for which the density-matrix is always a 2×2 array of numbers. The so-called diagonal elements are top-left and bottom-right. (I'll write them as $(1, 1)$ and $(2, 2)$, respectively.) The off-diagonal elements are bottom-left and top-right, $(2, 1)$ and $(1, 2)$, respectively.

Let us suppose that we wish to measure the z-component of spin, s_z. *Before* any measurement, the usual case would be that all four elements of the density-matrix are non-zero. Now let us consider the situation following a perfect measurement of s_z, assuming a von Neumann collapse. In all cases, we should find that the density-matrix is diagonal. If we first consider a *single particle*, we can say more. If the result of the measurement is that s_z is equal to $+\hbar/2$, then element $(1, 1)$ must be equal to 1, and all the other elements zero; if, on the other hand, the result is $-\hbar/2$, element $(2, 2)$ must be 1, and again all the others zero.

But as I have said, the density-matrix can also handle the situation where we have an ensemble of *many particles*. In this case, following a measurement both diagonal elements, (1, 1) and (2, 2), will be non-zero (though the off-diagonal elements will again be zero). The two diagonal elements must actually add up to unity, and their relative sizes tell us the relative numbers of particles for which S_z has been found to be $+\hbar/2$ and $-\hbar/2$, respectively.

While this has been admittedly (and unfortunately) rather technical, it does enable us to say fairly simply what constitutes a measurement (with collapse) in the density-matrix approach. We start with a density-matrix with diagonal *and* off-diagonal elements non-zero, and the interesting case is where we have a *pure* state, as defined above, so that each system is quantum-mechanically identical and *no system* has a precise value of s_z.

The *first stage* in the measurement procedure must be a stripping-off of the off-diagonal terms, to give us a diagonal density-matrix. If we are thinking of an ensemble, this, in fact, does allow us to speak of a measurement having been made, the sizes of the diagonal elements giving the probabilities of the various results. Each member of the ensemble now has a precise value of s_z, some $+\hbar/2$ and some $-\hbar/2$. The members of the ensemble are no longer to be considered as identical; the ensemble is now mixed.

If, though, we are considering a single system, there must be a *second stage*, in which *one* of the diagonal elements becomes unity and the other zero. Only when *both* stages have taken place do we have the von Neumann collapse for a single system as expressed by the density-matrix.

One can now see opportunities for using the density-matrix and ensemble interpretation to discuss and even 'explain' measurement. For the first stage, 'all' one has to do is to strip off the off-diagonal elements, and there are mathematical constructions that can do this readily. To be explicit, one can carry out the mathematics of interacting the system (or ensemble of systems) under consideration with a second system (perhaps a measuring device). One can then do the best possible job of separating out the density-matrices for the two systems, and (hey presto) what one obtains for the system under consideration, which is technically known as its *reduced density-matrix*, is indeed diagonal! This looks promising, and of course, if we restrict ourselves to ensembles, we may avoid the second stage altogether. A *pure* ensemble would have become a *mixed* ensemble.

Many take this line; Tony Leggett [127] describes it as 'probably most favoured by the majority of working physicists', and refers to it as the 'orthodox' resolution of the problem. Yet it faces considerable difficulties. Once the two systems mentioned above have interacted we cannot strictly use individual density-matrices at all – just a combined density-matrix for the joint system, and this will *not* be diagonal. (To explain this a little more, following the measurement the wave-function of the combined system is entangled, as discussed in detail in the previous chapter; in this situation, the reduced density-matrix for *either* individual system may be useful computationally, but has no direct physical significance. D'Espagnat [273] suggests that the appropriate state for either system should be called not a (proper) mixture, but an improper mixture.)

After all, this is what we should really expect. As has been stressed in the previous chapters, the Schrödinger equation cannot provide collapse in wave-function terms, and changing the mathematics to density-matrices should not really be expected to alter this. And, as stated in this chapter, it is in any case not clear what the real justification for use of ensembles is – apart from appearing to legitimise the mathematical manipulations above [97, 274].

One physicist who definitely does not accept the argument is Leggett himself. He also attacks the common assumption that the off-diagonal elements in the density-matrix for the combined system somehow don't matter, because their practical effect is to cause interference between macroscopically different states, which it is assumed does not occur.

What he calls '[t]his "solution" to the quantum measurement paradox (which seems to be re-discovered and re-published, in some variant or other, about once a year on average', he considers to be 'no solution at all . . . [A]s a solution of the *conceptual* problem it is a non-starter.' The correct quantum description may give the same experimental predictions as the desired result after measurement (for the ensemble case), but this does *not* mean that the two descriptions are identical, which, Leggett says, would be 'in transparent violation of the interpretation of the quantum formalism as applied at the macroscopic level'.

Leggett agrees, of course, that, if the ensemble interpretation is reduced to being no more than a method of calculating probabilities, it will be self-consistent, but since it cannot even provide a quantum description of a single macroscopic object, he feels that not many would find it satisfactory.

He concludes that '[the technique] only looks plausible at all because of a *fundamental ambiguity* in the interpretation of the density-matrix formalism [my italics] . . . To repeat: the conceptual (as distinct from the practical) problem is not whether at the macroscopic level the Universe (or the macroscopic apparatus) behaves *as if* it were in a definite macroscopic state, but whether it *is* in such a state.'

Bell put forward very similar arguments in rather a polemical article titled 'Against measurement' [76] which was published in 1990 shortly before his death. Bell admits that quantum theory is fine FAPP – *for all practical purposes*, but wants something more – an exact formulation, leaving no discretion to the theoretical physicist, and, in particular, no *special* techniques to handle the nebulous idea of 'measurement'.

Bell particularly criticises the discussion of measurement in the classic text-book of Landau and Lifshitz [275], the influential book of Kurt Gottfried [276] and a much-referenced paper of Nico van Kampen [277]. In each case he argues that the authors are illegitimately smuggling in the idea of collapse of wave function, though he suggests that [276] and [277] achieve this in a more sophisticated way than does [275].

I will concentrate on the argument against Gottfried, who uses the density-matrix and ensemble argument much as described above. Gottfried suggests that he is free to drop the off-diagonal terms, 'safe in the knowledge that the error will never be found'. Bell is suspicious of this 'butchering' of the density-matrix, as he calls it, but prepared to see where the argument leads. However, he is totally unprepared to follow Gottfried into believing that the diagonal density-matrix may then be read off directly as giving the probabilities of measurement.

'I am quite puzzled by this', he says. 'If one were not actually on the look-out for probabilities, I think the obvious interpretation of even [the butchered density-matrix] would be that the system is in a state in which the various [wave-functions] somehow co-exist . . . This is not at all a probability interpretation, in which the different terms are seen not as co-existing, but as alternatives.'

The removal of the off-diagonal elements of the density-matrix may be called *elimination of coherence*, or *decoherence* (words we shall meet a lot more in the next few sections). Bell comments that 'The idea that elimination of coherence, in one way or another, implies the replacement of "and" by "or" is a very common one among solvers of the "measurement problem". It has always puzzled me.'

Indeed, for those readers who may have struggled with the idea of the density-matrix, this puts the point less mathematically. The central strange idea of quantum theory is that different states of a system may be somehow combined in a wave-function in a totally non-classical way. This feature is required in order to explain the typical quantum experiment – interference of electrons or photons. (This is Bell's 'and', or what we defined as a pure state.) Measurement, though, requires that we obtain one result or the other – Bell's 'or', and our mixed state. Bell, like Leggett, rejects the mathematical manipulations of Gottfried and many others for getting from 'and' to 'or'. He suggests that the original theory 'is not just being approximated – but discarded and replaced'.

Several replies to Bell were published in the months after Bell's death; they included criticism of Bell's position by Gottfried [278], van Kampen [279, 280] and Peierls [281], and support for him from Squires [282].

Gottfried was an old friend of Bell from CERN. In June 1990 they had both taken part in a workshop on the foundations of quantum theory at Amherst College, and there they had argued long and hard, though always with respect and good humour. Gottfried remarked [278] that Bell's devoting great efforts to discussing his ideas, albeit highly critically, was the highest compliment he had ever been paid.

His initial reply [278] was initially presented at a Symposium on Quantum Physics in memory of Bell held at CERN in 1991, the lectures at which were later published [283]. In this reply he analysed the approach in his book and Bell's criticisms. He admitted that some of Bell's charges were fair, though he pointed out that he was not squeamish about QED FAPP. He doubted whether a crisis in physics was imminent, as Bell had suspected. If in fact there was, he had told Bell in their last conversation, he would wager that any new theory would be further from our philosophical and conceptual prejudices than quantum theory, and Bell would be left pining for the time when he had 'good old' quantum mechanics to kick around.

Gottfried remained extremely interested in these matters, and acted as joint editor for a new collection [284] of many of Bell's papers in 1995. Gottfried was responsible for the area of quantum foundations, Mary Bell, John's wife, for papers on accelerators, and Marcus Veltmann for papers on particle physics. In this collection there is a slightly different selection of papers on foundations from that in [76]. Also, while in [76] the papers are conveniently re-formatted in common style, in [284] they are directly copied from the original; readers can choose which they prefer.

Gottfried gradually became dissatisfied with his initial response [278, 283] to Bell's paper, and in 2000 he published what he considered a much more satisfactory argument [285]. The same analysis was included in the second edition [286, 287] of the book initially criticised by Bell. (Tung-Mow Yan is the co-author of this second edition.)

In this paper, Gottfried says that Newton could have understood the meaning of the electromagnetic field, had he been given Maxwell's equations and the equation for electromagnetic force. He then asks whether, in similar fashion, had Maxwell been given the Schrödinger equation, he could have derived the statistical interpretation of quantum theory, that is to say the fundamental role of probability and the Born rule.

Gottfried argues that, in the classical limit, the Schrödinger equation describes not a single system but a population of such systems. He says that, had Maxwell been told about the result of the Stern–Gerlach experiment, he could have come very close to the uncertainty principle, and finally he would have been able to deduce the statistical interpretation of quantum theory for discrete variables such as spin, though not (at least in the present argument) for continuous variables such as position or momentum. Gottfried makes the surprising claim that the decoherence process (discussed in the following section) occurs without the need for an environment.

Van Kampen has also continued his debate with Bell in further publications [288, 289], which refine his views stated in [277]. Van Kampen's view is that quantum mechanics, including quantum measurement, is a perfectly logical coherent theory that may be understood entirely rationally. He is dismissive of 'philosophising' and of such notions as many worlds, hidden variables, the mind of the observer and manoeuvres of GRW type (discussed later in this chapter).

Van Kampen uses the idea of collapse, but, in contrast to von Neumann, who considers it a process outside and in conflict with the Schrödinger equation, van Kampen believes that it is fully described by the Schrödinger equation for system and apparatus together. He provides what he claims are explicit models for collapse, and a potentially nice feature of his analysis is that, rather than using the rather cumbersome density-matrices, as discussed earlier in this section and as used by most of those working in this area, he restricts himself to wave-functions.

Nevertheless, I have argued [290] that his actual analysis, rather than demonstrating collapse, only goes as far as producing the superposition of terms involving measuring and measured systems from which collapse should occur. The actual collapse process is not discussed in a rigorous manner.

To finish the discussion of ensembles on an entirely different note, I mention that, in his original article on ensembles [88], Ballentine was unenthusiastic about the Bohm theory. By 1988, though, he [291] regarded it as '[a] good example of a structure added within the existing framework of the statistical quantum theory' and suggested that generalisations should be sought. This may be another example of how, as an interpretation becomes tautened by conceptual and empirical pressures, it looks more and more like that of Bohm.

Zurek and decoherence

In the previous section, I mentioned *decoherence*. We have *coherence* when the wave-function corresponds to a sum of contributions from classically rather distinct situations; this is a specifically quantum concept, leading to typical quantum behaviour such as interference. In these circumstances we say that the state of the system is *pure*. *Decoherence* implies that the various contributions are effectively decoupled so that interference can no longer take place. Zeh and Erich Joos [165, 292] are mainly responsible for initially drawing attention to the topic of decoherence, and understanding much about the way it occurs.

Wojciech Zurek [166] is also prominent among those who seek to explain decoherence in terms of an interaction of the system *with its environment*. A macroscopic quantum system (such as a measuring device) can never be isolated from its environment. It is not a *closed* system consisting of a fixed complement of atoms unaffected by anything external to itself, but an *open* system, freely interacting with its environment. (It should be mentioned that, while the term 'environment' naturally includes *external* influences, it may also include *internal* factors, the behaviour of *internal* variables of the system that do not appear in the quantum description. Gottfried, we just noted, rather unusually believed that decoherence could take place *without* any environment.)

So macroscopic objects (except rather special ones to be discussed in the final section of this chapter) should *not* obey Schrödinger's equation, which should actually apply exactly only to closed systems. Rather, additional mathematical terms should be included, relating to coupling between system and environment, and Zurek says that 'As a result systems usually regarded as classical suffer (or benefit) from the natural loss of quantum coherence, which "leaks out" into the environment.'

If this is taken for granted, it is relatively easy mathematically to describe a movement towards a diagonal density-matrix as discussed in the previous section, and this would constitute the *first stage* in the measurement process. Zurek performs the calculations for a particle moving in a given direction and subject to a simple form of interaction with the environment. Initially the system is in a highly non-classical state, its wave-function being the sum of two fairly localised wave-functions with a comparatively large distance between them.

Two times appear in the solution – a *relaxation time* in which the energy is dissipated, and a *decoherence time*, in which it becomes true that the particle is in *one or other* location, not some spooky combination of both. For a fairly typical macroscopic situation, Zurek obtains the extremely satisfying result that, while the relaxation time may be as long as 10^{17} s, the decoherence time may be as short as 10^{-23} s. Decoherence seems to work!

An interesting way Zurek uses to describe the decoherence process is in terms of *information*. The coherent pure state has *excess information*, which must be disposed of in the decoherence process. Subsequently information *gain* takes place when the observer intervenes, observer states becoming correlated with those of the detector and the system.

Zurek [166] is very upbeat about the success of the method. He says that 'recent years have seen a growing consensus' that progress is being made with the measurement problem in this way, and that 'there is little doubt that the process of decoherence is an important fragment central to the understanding of the big picture' – the transition from quantum to classical.

Yet it is clear that not everyone would agree – particularly if Zurek's argument is taken as a claim to resolve entirely the measurement problem. After all, the argument appears precisely of the type described in the previous section, and criticised so severely by Bell and Leggett. Zurek's 'environment' plays exactly the role of what I called there the second system, with which the system under investigation is caused to interact.

So it is not surprising that eighteen months after the publication of Zurek's article [166], *Physics Today* published an extensive selection of comments, together with Zurek's reply [293].

Several of Zurek's critics did adopt the stance I just outlined. Ghirardi, Grassi and Pearle, for instance, said that 'Zurek's consensus certainly didn't include John Bell.' Gisin likens Zurek's suggestion that the discrepancy between pure and mixed states can never be detected *in practice* after the coherence time to his own children's pragmatic reasoning – it is not forbidden to do silly things, just to get caught. Gisin asks 'Can our basic understanding of our fundamental theory rely on such pragmatic pseudophilosophy?'

Actually Zurek's claims [166, 293] may be somewhat different, one might say somewhat less, from what is often thought. He acknowledges that his system is left in what Bell called the 'and' stage rather than the 'or' one. His approach 'does not force all of the wavefunction of the Universe into a unique state corresponding directly to our experience'. It uses, in fact, what he calls an 'Everett-like framework', which 'assumes that the observers are an integral part of the Universe and analyzes the measurement-like processes through which perception of the familiar classical reality comes about, thus showing why one *can be aware of* only one alternative [my italics]'.

So we do not have – do not require – just one distinct state of the system in the final wave-function after decoherence, since we have only set up the mathematics ready for some variety of many-worlds approach. The important point is that *each* component of the wave-function is (or rapidly decoheres to become) classical in nature. In part Zurek is aiming to dispel the worry of Bell (and others) mentioned earlier in the chapter about any MWI approach – how is it that the various branches of the wave-function following a measurement correspond to distinct results of the measurement? In such situations, it is decoherence that assures this, and corresponds, in fact, to a practically instantaneous collapse of the wave-function being perceived *in each branch of the Universe*.

Perhaps more fundamentally, though, Zurek aims to create 'a fixed domain of states in which classical systems can safely exist, but superpositions [combinations] of which are extremely unstable'. Schrödinger's cat, for instance, will, practically instantaneously, become dead or alive for any given observer. In general, 'states that are localised will be favored . . . Events happen because the environment helps define a set of stable options that is rather small compared with the set of possibilities available in principle [in quantum theory]'. Thus Zurek hopes to explain why, for instance, planetary orbits remain totally classical with no quantum smudging, in a more genuine way than the rather sweeping discussion of the *classical limit* in a Copenhagen approach.

Indeed, though Zurek adopts the Everett-type approach, he says that '[T]he picture that emerges in the end – when described from the point of view of an observer – is very much in accord with the views of Bohr. A macroscopic observer will have recording and measurement devices that will behave classically. Any quantum measurement will lead to an almost instantaneous [collapse] of the [wave-function], so that the resulting mixture can safely be regarded as corresponding to just one unknown measurement outcome.'

The clash between Zurek and his opponents (and for another example of the latter see comments of Shimony and Bernard d'Espagnat in [271]) may perhaps be viewed, rather loosely, as a clash between Zurek, as a follower of Bohr, and followers of Bell and hence of Einstein. I shall expand on this remark in the last chapter of the book.

I will now sum up the implications of this section. We have already seen earlier in this chapter how the Everett or MWI scheme of, for example, Wallace [265] relies on decoherence. Here we are seeing the reverse of this: decoherence can be a part of a solution to the measurement problem only if buttressed with many worlds.

Decoherence performs the supremely important task of preventing a classical object from losing its classical form as a result of constant quantum perturbations. It means that a dead cat remains an (unsmudged) dead cat; a live cat remains an (unsmudged) live cat. But it allows a superposition of dead cat and live cat to remain as a superposition of dead cat and live cat; it certainly has no power to collapse the superposition down to one or the other. (As we have said many times, such collapse would involve violating the Schrödinger equation, and decoherence does not allow us to do that.)

Despite this, and highly unfortunately, it has become extremely common in popular and semi-popular books on quantum theory [294] to assert, without any idea of assuming many worlds, that decoherence provides collapse, and thus has solved the measurement problem. This idea is quite false.

Over the last few years there have been many technical advances in the theory of decoherence, and Zurek has prepared an updated version of his 1991 paper [166]. This is on the Los Alamos quantum-physics archive http://arxiv.org/find/quant-ph at quant-ph/0306072.

Consistent or decoherent histories

The considerations of the previous section cannot apply directly to the whole Universe, for which there can be no environment. Hence, when one wishes to discuss the behaviour of the Universe, Zurek [166] refers to use of an approach that makes use of decoherence in rather a wider context, the method of *consistent histories* or *decoherent histories*, a way of thinking about quantum theory that has come to prominence in the last two decades, with strong supporters and probably equally strong detractors.

The initial idea of the *history* (to be explained shortly) came from Robert Griffiths [295] in 1984. The concepts were developed by Griffiths himself [296, 297], and also by Roland Omnès, who introduced decoherence, which was not used by Griffiths, and who wrote a particularly thorough, though very technical, review [298] of the method in 1992. Omnès has also published three books [299–301], which, though wide-ranging in content, have a common perspective of the decoherent-histories approach. The work of both these authors was directed towards achieving a consistent interpretation of quantum theory.

A similar research programme has been developed by Murray Gell-Mann and James Hartle [302, 303], but for rather more ambitious ends – 'The theory of Gell-Mann has to be understood, from the very outset, as dealing with nothing less than the entire history of the Universe from the beginning to the end.' [125]. It is attempting to give a quantum description of cosmology.

I shall start to explain these theories by describing the fundamental concept – the *history*. The idea is actually quite straightforward – a sequence of properties holding at a sequence of times. For example it might give us the position of a particle at one time, its momentum at a later time and its position again at a later time still, each value being within a given range.

As Jonathon Halliwell says, in a brief accout [304] of the approach, 'Histories are the most general class of situation one might be interested in. For example, a typical experiment might be of the form, a particle is emitted from a decaying nucleus at time t_1, then it passes through a magnetic field at time t_2, then it is absorbed by a detector at time t_3.' As he goes on to say, the idea of histories is an attractive one. Rather than restricting ourselves to studying the past through present records, it allows us to discuss past events directly. In cosmological terms it may help us understand how an apparently classical Universe has emerged out of quantum uncertainties. And it may perhaps give us a better understanding of what the world is actually like. Halliwell admits that 'the minimal pragmatic aim of theoretical physics is to explain the data' but continues as follows: 'Yet many feel that our theories should explain more than just the numbers; they should supply us with a picture of the world the way it really is. Histories arguably supply us with that picture.'

Attractive as the idea of histories may be, it is immediately clear that its application must be restricted – by the laws of quantum theory themselves. For, in the Bohr–Einstein debate on two-slit interference, it was essentially a *history* that Einstein demanded – which slit did the photon or electron go through? Bohr convinced him he couldn't have it!

Griffiths [295] showed that the kind of histories which *could* be considered were what he called *consistent histories*. To explain this term, I shall introduce the idea of the *weight* of a particular history. This is an easy mathematical construction, one corresponding to what we would be very inclined to call the probability of the particular history being followed. I shall refrain for the moment from using the term 'probability' for the following reason. It turns out, according to the rules of quantum theory, that, if A and B are two particular histories, then the weight corresponding to (A and B) is not usually equal to the sum of the weights of A and B individually. (It will be realised that the two-slit problem is just an example of this fact; the probability distribution of the interference pattern on the screen when *both* slits are open is *not* the sum of those when *either* slit is open.)

To discuss the term 'consistent history', one must define a *family* of individual histories. The *family* has a consistent history *if* (but only if) the *weight*, as defined above, of the family *is* equal to the sum of *weights* of the individual histories. In these circumstances it does make good sense to re-christen the *weight* as a *probability* for that family of histories. Omnès [298] expresses what has been achieved as follows: 'Among all the conceivable histories, only a few make sense, insofar as one can assign a probability to them . . . Accordingly, histories give us a grasp of what is meaningful or meaningless in quantum mechanics.'

The term *decoherent history* expresses much the same idea as *consistent history* (though mathematically the qualifying conditions are somewhat different). The new term conveys the idea that such histories do not *interfere*, and may evolve separately, and hence, one might hope, classically.

The approaches described here require no arbitrary distinction between microscopic and macroscopic. They do not assume a classical region, but such a region may emerge from the initial assumptions under certain conditions. Similarly wave-function collapse, and the idea of measurement in general, are not fundamental features of the method, though they may be described by it under certain circumstances. (A measurement, or measurement-like interaction, always causes decoherence.)

Omnès in particular adopted a strictly logical approach to his analysis. 'The logical structure of quantum mechanics,' he says [298], ' . . . led to a theory of classical phenomenology containing the proof that classical determinism is a consequence of quantum mechanics. Together with decoherence, it generated an apparently complete and consistent interpretation [of quantum theory].'

In discussing measurement, Gell-Mann and Hartle go considerably further. They discuss how the observer, whom they refer to as an *IGUS*

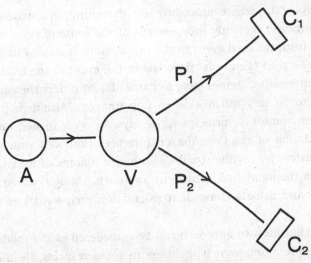

Fig. 8.1. A particle in region A travels towards scattering centre V, from which it may travel along path P_1 or P_2 to counter C_1 or C_2.

(*information-gathering and -utilising system*) has evolved through history. The IGUS, they say, must evolve so as to focus on situations of decoherence, because it is only then that useful predictions, which are essential for survival, may be made.

Let us now see, in the case of a very simple example considered by Griffiths [295] and shown in Fig. 8.1, how a consistent-histories approach may yield different information about the world, perhaps more sensible than that available from an orthodox approach. A particle in region A at time t_0 is travelling towards a scattering centre V, from which it emerges at time t_1. Accessible to it now are two routes, one travelling along path P_1 towards counter C_1, the other along path P_2 towards counter C_2.

According to an orthodox interpretation, after time t_1, the wave-function must grow more and more *grotesque*, as Griffiths calls it, being straddled between paths P_1 and P_2, that is, between two regions of space each of which is itself fairly localised, but which are separated by a macroscopic distance. Only when either counter C_1 or C_2 clicks does the wave-function collapse to the region of the particular counter.

With the histories approach, one must first check that the two paths do indeed constitute consistent histories. Once that is done, one may make sure that, if counter C_1 clicks, immediately afterwards the particle is in the region of C_1; this is just as on the orthodox view. In addition, though, and totally at variance with the orthodox view, once one knows that C_1 has clicked, consistent histories will tell us that, immediately *before* that, the particle was on path P_1; in fact the particle never has Griffiths' 'grotesque' wave-function.

As Griffiths says, 'The whole procedure for determining consistency and assigning probabilities is explicitly independent of the sense of time: a history read "backwards" from the final event to the initial event is treated in precisely the same way as one read "forwards" from the initial event to the final.'

Despite these differences, Omnès [298] in particular considers the consistent-histories method to be very much akin to Copenhagen. 'Although [the new theory] relied upon some new principles,' he says, 'it soon turned out to be in fact a reformulation of the Copenhagen interpretation well suited to the treatment of consistency . . . [T]he usual Copenhagen axioms of measurement theory reappear as theorems in a consistent approach, though [wave-function collapse] is now found to be a convenient calculation recipe with no specific physical content.'

Omnès implies that his own approach can be considered as the modern version of Bohr's interpretation, even if it differs in some respects. He mentions that 'nothing can inspire a greater respect for Bohr's insights than rediscovering them with the help of a deductive method. One gets the feeling he was able to keep in parallel all the necessary tenets by the sheer power of thought.'

Griffiths [295] too says that, 'despite very important differences in approach and outlook, the orthodox and consistent history approaches seem to be basically compatible, with the latter providing a number of insights into the successes and conceptual difficulties of the former, while at the same time getting rid of some of the less satisfactory philosophical conclusions which orthodoxy seems to imply (temporal asymmetry, need for conscious observers), but with which the orthodox themselves have never been very content. But, equally importantly, the consistent history approach goes well beyond the orthodox interpretation by providing a much more general regulatory principle which makes possible a greatly increased number of precise statements about events in the microscopic domain, in a controlled and sensible way.'

Now to the limitations and problems of the approaches, first the Gell-Mann and Hartle theory. A difficulty, or at least a hard fact, is pointed out by Fay Dowker and Adrian Kent [305]. '[A]lthough they naturally use everyday language in many of their arguments and illustrations, they do not always intend the words to be read in their everyday sense. The statement that "dinosaurs *did* roam the earth many millions of years ago" is *not* to be read as a statement about a definite, observer-independent physical event; it is meant as a shorthand for the claim that there is a particular consistent set which includes the actual facts presented to us and in which quasiclassical dinosaur density projection operators followed largely deterministic equations of roaming through the Cretaceous era.' But there are other arrangements of the facts, which are incompatible with this picture – '[A]ll these pictures are just theoretical constructs – a vast library of consistent historical fables, of which one is free to choose any or none.'

Similar remarks may be made about the Griffiths or Omnès approach to quantum theory, which Squires [304], in particular, has discussed fairly critically. There may again be a question of what the consistent histories actually are. One might be inclined to think that they represent possibilities, *one* of which is actually realised in the world we live in, and the allocation of probabilities would seem to confirm that view. The 'greatly increased number of precise statements' must then correspond, in normal terminology, to *hidden variables*. Yet Griffiths [295] specifically says that 'there are no "hidden variables"'.

Indeed, he [297] says that 'Quantum mechanics allows us to tell a variety of different stories about a system.' It seems that even to say that a chosen history 'is realised', as Dowker and Kent [305] say of a slant of their own, may mean only that it 'correspond[s] to nature' as a solution of an equation. Thus any direct connection between consistent history and 'actual life' seems perhaps a little obscure, and a better title may even have been *consistent story*.

Because hidden variables may be a useful way of discussing measurement, this suggests the following important question: does the consistent-histories approach aim to solve the measurement problem, and, if so, does it succeed? Omnès' claim of achieving a consistent and complete interpretation of quantum theory, complete in that it 'offer[s] a precise prediction in every experimental situation,' would lead one to expect it at least to attempt to solve the problem. Yet Squires [304] points out that Omnès says that '[T]he theory can only give some *a priori* probabilities. Now, all of a sudden, one of them becomes real and the others fade into oblivion. How is this?' Having thus posed the measurement problem, he does not answer it, but decides that it would be unreasonable to expect an answer. Griffiths [295] explicitly declines to discuss whether his proposal solves the problem, whereas Halliwell states categorically that such approaches do not do so.

Since solving the measurement problem might normally be regarded as the central difficulty of quantum interpretation, it seems clear that the status and degree of success of the consistent-histories method will continue to be debated for some time.

I would mention Squires' view that, considered as a hidden-variable theory, 'the consistent histories approach is a poor relation to the Bohm model. By explicitly requiring *classical* probabilities, the theory loses all the subtle contextuality of quantum measurements [for which see the previous chapter] which the Bohm model treats so beautifully.' This is the third suggestion in this chapter that an interpretation of quantum theory regarded by some as naive would, if sharpened, look more like the Bohm theory. This seems to suggest that, while open to criticism in many ways, the Bohm theory may go some way towards maximising the type of approach in which one works with an unadulterated Schrödinger equation.

The connection between Bohm and consistent histories has been examined by Griffiths and by Hartle. Griffiths [306] says that the two interpretations contradict each other, both in the absence and in the presence of a measuring device. Consistent histories, he says, are closer to standard quantum theory, and avoid the problem of non-locality. If this is the case, it is quite clear that they indeed cannot be regarded as hidden-variable theories, since hidden-variable theories must be non-local.

However, Hartle [307] argues that, although the two interpretations represent a given history by different operators, and their predictions for closed systems generally differ, there is an idealised model of measurement for which the predictions for measurement outcomes are identical. Thus the two may be difficult to distinguish experimentally, though their accounts of the past history of the Universe may differ.

I said earlier that the Bohm theory had been tested to see whether it might provide information on problems involving time, an area where orthodox interpretations can say very little. The consistent-histories approach has been used in the same way. Yamada [308] claimed that the method gives a simple solution for the time taken by a particle to tunnel through a barrier. He says that a probability distribution of times cannot be defined, but the range of times can be, although it is very broad.

Sokolovski [309], though, contradicts this view. He says that the consistent-histories approach cannot itself provide meaningful probabilities, and he suggests that instead the system's quantum state should be decomposed in certain appropriate ways. Egusquiza and Muga [310] have also examined problems involving time by the method of consistent histories, including, in particular, the quantum Zeno effect, which is discussed later in this chapter.

Lastly, Griffiths has looked at the problem of the quantum delayed-choice experiment. In the previous chapter we saw how this experiment, which was so difficult to understand using an orthodox interpretation, was very easy to understand in a Bohmian approach. Griffiths [311] showed that consistent histories could explain this just as well. Any idea that the future might influence the past disappears when one takes proper account of all the possible quantum descriptions of the world, and their mutual compatibility or incompatibility. Griffiths [312] has also collected much of the work in the area in a recent book.

The quantum state diffusion model

Another technique for studying quantum processes and quantum measurement has been developed in recent years, principally by Nicholas Gisin and Ian Percival [313–315], though following earlier work of Gisin himself [316] and Lajos Diósi [317].

As I have said in earlier sections, the most common description of quantum systems uses the density-matrix, and relates, not to a single system, but to an ensemble of systems. But, Gisin and Percival [314] say, 'What the density-[matrix] gains in mathematical elegance it loses in physical directness.' Using the density-matrix, one may certainly discuss the evolution of quantum systems, and quantum measurement, with great effect, but the physical meaning of the various mathematical steps is not always clear, and, as suggested in earlier sections, their justification could be questioned.

Gisin and Percival [313] contrast this rather abstract approach with the fact that 'many physicists, particularly experimenters, have insisted on treating quantum jumps of individual systems as real, and the [wave-function] as representing the behaviour of an individual system, as exemplified by a single run of a laboratory experiment . . . The experimenters' picture has given them valuable physical insights, which have sometimes escaped the theoreticians with their relatively elaborate mathematical tools based on density-[matrix] evolution.'

Like the methods of the section before last, the quantum state diffusion method considers a quantum system *in interaction with an environment*, but works with an *individual quantum system*, not with an ensemble. The Schrödinger equation for this system is modified by adding extra terms corresponding to the interaction with the environment, these extra terms having a random element because the interaction itself is random (or 'stochastic').

Typically one then finds that the system exhibits sudden transitions between discrete quantum states; these transitions are not instantaneous, but take a short 'jumping time'. Measurement becomes a simple example of the interaction of the measured system with its environment, and may be treated in a straightforward way, by solving the Schrödinger equation modified with appropriate terms. What emerges is an irreversible movement of the measured quantity to one of its allowed values in a measurement. We are seeing a process analogous to collapse of wave-function, not, though, imposed arbitrarily in the manner of von Neumann, but appearing directly from a suitable equation.

It is dramatic and exciting to see what are (or what, to put it at its most cynical, seem to be) pictures of the behaviour of individual systems. (They look just like experimental observations!) Figure 8.2 (from Gisin and Percival's paper [313]) shows a system that has two states, with no photon and one photon, respectively,

Fig. 8.2. The behaviour of a system with two states according to the quantum state diffusion model. Photon number makes transitions between 0 and 1, but the jumps take a non-zero time. From [313] by kind permission.

and absorption and (stimulated) emission processes are inserted into the appropriate equation. The figure shows fairly sudden transitions between the two states; the jumps take a time that is short but greater than zero.

Figure 8.3, from the same paper, shows a measurement process. The initial state is pure; mathematically it consists of a sum of states with one, three, five, seven and nine photons. There is a clear progression to systems with one *or* three *or* five *or* seven *or* nine photons, and this looks just like Bell's 'and' to 'or' process, the key part of what must be achieved in a measurement.

It is interesting to compare the clarity of these pictures with what would be obtained from the more conventional density-matrix approach, which averages over many systems. The equivalent to Fig. 8.2 would be a fairly dull smooth curve heading to a steady *average* photon number of $\frac{1}{2}$. In Fig. 8.3, the density-matrix would lose its off-diagonal terms, but one would still be left at Bell's 'and' stage, not his 'or' stage.

In the discussion of Zurek's work, I introduced the idea of decoherence producing localisation, for instance localising a planet in its orbit, and it is interesting to see whether the method of this section can do anything analogous. Gisin and Percival demonstrate mathematically that this will happen very often, but

Fig. 8.3. A measurement process according to the quantum state diffusion model. A pure state at $t = 0$ becomes effectively mixed. From [313] by kind permission.

actually localisation should be regarded in more general terms. They [314] say that 'Depending on the nature of the system and its interaction with the environment, localisation may take place with respect to many different [physical quantities], but localisation in position is particularly important because *interactions* [with the environment] are localised in position space to a greater extent than for other [quantities].'

An example of 'localisation' in its more general sense is demonstrated by molecules with *chirality* or *handedness*. These are molecules, such as sugars, that can exist in two *distinct* forms, one the mirror image of the other. Conventional quantum theory would expect individual molecules to have their wave-functions the sum of those for the left-handed and right-handed states; the method of this section suggests that 'localisation' leads to molecules being of one form *or* the other, and this is what is found in practice.

Given that the method seems to give vivid insights into what is 'actually happening', what are its potential difficulties and defects? One might mention that the position of the boundary or *cut* between 'system' and 'environment' is arbitrary. (This is reminiscent, of course, of the arbitrarily positioned cut between 'measured system' and 'measuring system' for Bohr.) In the present case, it would not seem that important difficulties of principle result.

On a more questioning note, it *might* be suggested that the whole method is empirical rather than rigorous – even that the results, exciting as they are, may reflect little more than the particular interactions inserted into the initial equation. However, Gisin and Percival [315] consider that 'There is no need to carry out a detailed physical analysis of these interactions' for any particular example, any more than one would refrain from using the idea of *electrical resistance* in the theory of elementary electric circuits, although it may be very difficult to calculate the resistance of any particular material.

I will now mention some studies by Diósi, Gisin and Percival with Halliwell [318], to establish an overlap between the methods of *quantum state diffusion* and *decoherent histories*. (It could be suggested that this might link the empirical strength of the first with the formal rigour of the second.) I also mention an interesting exchange of views between Hans Dieter Zeh [319], who is broadly in agreement with Zurek and an Everett-type interpretation, and Gisin and Percival [320], who wish to modify the Schrödinger equation explicitly. From the latter paper I take the words 'The Schrödinger equation is no longer the best for all practical purposes.'

Over the last few years, much work has been put into the theory of quantum state diffusion, mostly technical improvements. One interesting extension has been made by Diósi, Gisin and others, from Markovian to non-Markovian quantum state diffusion [321–323]; in the former the behaviour of the system at a particular time depends on its state at that time only, whereas in the latter it may depend on its state at all previous times as well. Some improvements over the Markovian case are found.

There has also been an extension to the relativistic case [324], though there has been some controversy as to whether this attempt is correct [325, 326]. Lastly I would mention two general accounts of the method written by Percival [327, 328] and an interesting review of [328] by Tim Spiller [329].

Peierls and the knowledge interpretation

Having followed a path from ensemble interpretations to some quite recent developments in the previous part of the book, I now return to much older ideas in this and the next section.

The first is that the problems of quantum theory may be explained by the idea of *knowledge*. I have touched on this point of view in Chapter 5 in particular. It is a very natural idea for someone getting to grips with some of the difficulties of quantum theory to decide that the solution is as follows. I may not *know* the value of, say, s_z, the z-component of spin for an electron, just as I may toss a coin and not *know* whether it has landed heads or tails, just because I have not looked! I may then look at the coin, and discover that it is, in fact, heads, and was so, of course, before I had looked. Similarly I may measure s_z for the electron, and obtain the value of $+\hbar/2$ (rather than $-\hbar/2$), and I may again deduce that it must have been so before I measured it! That is all, from this point of view, that collapse of the wave-function is about. I gain *knowledge* of something that existed all the time, and I do not need to disturb the system to do so. Thus the wave-function denotes *knowledge* of the system.

A rather more sophisticated variant might be that I *do* disturb the system. In measuring s_z, I may disturb the value of s_y, for example, so I could never *know* the values of both simultaneously. But this does not rule out the comfortable idea that *both* s_z and s_y may simultaneously *have* precise values.

This view may be encouraged by conventional statements of orthodox quantum theory. Much is made of 'measurement', whereas, for von Neumann, and more explicitly for Wigner, it is 'consciousness' that collapses the wave-function. These ideas might not seem to be very far divorced from the view that quantum theory discusses our *knowledge* of something, and only indirectly the thing itself.

And indeed we find Rudolf Peierls, a theoretical physicist of massive achievements, seeming to advocate just such a position in a short article titled 'In defence of "measurement"' [281], published in reply to Bell's 'Against "measurement"' [76]. (Peierls, I am sure, would describe his views as being orthodox, just those of Bohr; see his contribution to [53].)

He says 'In my view the most fundamental statement of quantum mechanics is that the wavefunction, or more generally the density matrix, represents our knowledge of the system we are trying to describe.' If our knowledge is complete – in the sense of being the maximum permitted by the uncertainty principle and the general laws of quantum theory, we use a wave-function; for less knowledge we use a density-matrix. Uncontrolled disturbances may reduce our knowledge; measurement may increase it, but if we start off with the wave-function case, any new information gained in a measurement must be counterbalanced by losing some of the information we already have.

He says that 'it is clear that upon a change in our knowledge the density matrix must change. This is not a physical process, and we certainly cannot expect it to follow from the Schrödinger equation. It is just the fact that our knowledge has changed, and this must be represented by a new density matrix.'

Peierls remarks that, at an observation, 'for all practical purposes' the apparatus makes the off-diagonal matrix elements of the density-matrix disappear, but, in principle, they disappear, and we may describe the measurement as being completed, only when we *know* the result of the measurement. He says, incidentally, that 'the system to which we apply our description . . . cannot include the mind of the observer and his knowledge, because present physics is not able to describe mind and knowledge (and it is not obvious that this is a proper subject for physics)'.

He answers possible objections as follows. 'How can one apply quantum mechanics to the early Universe, where there were no observers around?' The observer does not need to be contemporary with the events; when we draw conclusions about the early Universe, we are, in this sense, observers.

'[W]hose knowledge should be represented in the density-matrix?' Many people may have some information about the state of a system, and they may all use their own density-matrices. However, all descriptions must be consistent; we cannot have one density-matrix giving a value of s_z, and another a value of s_y, since simultaneous knowledge of both is forbidden, he says, by the laws of quantum theory.

Until comparatively recently, Peierls' position would not have been a popular one. It could not be said that it was in any way actually wrong, but it seemed to focus on the wrong issues. It seems to lump together fundamental restrictions in a quantum universe, such as the fact that we cannot know both the momentum and position of a particle, with things that we *could* know but have not made sufficient effort to find out.

Mermin [330] was present with Gottfried and Bell at the 1990 Amherst meeting. Here, and in the paper produced shortly after, 'Against "measurement",' Bell made it quite clear that, as well as the term 'measurement', there were several others that he considered should not be used as primary terms when discussing quantum foundations. These included 'system', 'apparatus', 'environment', 'microscopic', 'macroscopic', 'reversible', 'irreversible' and 'observable'. They did not include 'knowledge' but did include 'information'. Bell's comments on the use of this term were '*Whose* information? Information about *what*?'

Mermin suggests that the words 'knowledge' and 'information' should be regarded as synonymous in this context. In any case he was for some years totally on Bell's side in being suspicious about the use of either word as a primary term. In the late 1990s, however, there came the rise of quantum information theory, which is discussed in detail in the following chapter, and with it a cadre of people who considered it obvious that quantum theory was all about information and nothing more. Mermin felt obliged to consider Peierls' approach more favourably

He has come to consider one of Bell's questions, 'Information about *what*?', as metaphysical. We cannot know, he suggests, whether information is information *about* something objective, or just plain information. However, the other question, '*Whose* knowledge?', Mermin regards as very important. It is essential to clarify this point, but it is also very tricky. He has actually shown that Peierls' answer cannot be correct in detail, and has gone some way towards finding a better answer [331], but there is still much that needs to be thought out.

Zeilinger and Časlav Brukner [332–334] have particularly argued for the view that Mermin mentioned, though they say they are following the lead of Carl von Weizsäcker [335] and Wheeler [336]. They claim that the basic concept of quantum physics is indeed information, that quantum physics is only indirectly a science of reality, but fundamentally a science of knowledge. To avoid Bell's problems, they refer to a total amount of information, rather than the information that might be carried by any particular subject.

They are able to show that their ideas lead to a natural explanation of complementarity, a full understanding of entanglement and a derivation of the quantum evolution of a two-state system. They also suggest that their approach could lead to answers for the most fundamental questions concerning quantum physics, in particular Wheeler's famous question 'Why the quantum?' Information is necessarily quantised in integral numbers of propositions; hence our description of information, which is just quantum theory, must also exhibit quantisation.

Stochastic interpretations

The mathematical form of the Schrödinger equation exhibits considerable similarily to that of the classical equation for diffusion, the kind of process by which gases mix. The latter equation may be studied via a theory of *probabilistic* (or *stochastic*) processes; Einstein's theory of *Brownian motion* mentioned in Chapter 1 was one of the first examples of such an approach.

It appeared very natural, then, to attempt to show that the probabilistic nature of quantum theory was analogous to that of Brownian motion, and that quantum theory itself might perhaps be, at a fundamental level, a *classical* theory of stochastic nature. The conceptual struggles which have dominated so much of this book might, then, perhaps be avoided.

Schrödinger himself, in 1932, was the first to study the analogy, and he has been followed by a very large number of physicists and mathematicians over the succeeding 70 years or more [67, 97], Edward Nelson [337, 338] making particularly important contributions.

In 1975, Luis de la Peña and Ana Cetto [339] produced and analysed a set of equations general enough to give Brownian-motion behaviour for one set of parameters, quantum behaviour for another. In both cases, the particles behave stochastically as a result of interaction with a *stochastic* surrounding medium acting in addition to the applied forces. So they conclude [340] that '[Q]uantum mechanics may be interpreted as a Markov process [the kind of mathematical structure used in analysis of Brownian motion] but *irreducible* to a Brownian type stochastic motion', and [341] '[T]he electron follows a stochastic trajectory, broadly reminiscent of Brownian motion, but at the same time differing esssentially from it, due to the entirely different nature of the background in which electrons and Brownian particles move'.

The theory of *stochastic electrodynamics* investigates the physical cause of the stochasticity – the interaction with the surrounding medium. After early work, the major development of the theory was undertaken by Marshall [342] in the 1960s. Many important contributions have been made since then, including those of Braffort, Boyer and Santos.

Santos [343] says that a stochastic theory is just classical physics *without* the idea of isolated systems, in the sense that we must *always* include the interaction of a given system with the rest of the Universe. Thus Santos discusses the stochasticity of the Rutherford atom as follows. Usually we say that the atom should radiate energy and thus collapse. However, if we remember to include all the other hydrogen atoms in the Universe, we will recognise that the energy radiated by one will be absorbed by another. Hence we should have *overall* a situation of equilibrium, and hence stability. The hydrogen atom should thus be regarded as consisting of a proton, an electron and a background of electromagnetic radiation filling all space. Since the radiation comes from a large number of sources, it must be thought of *statistically* (or *stochastically*). Thus stochastic electrodynamics studies systems of charged particles in a random electromagnetic field.

The central role in the stochasticity is played by the zero-point field. In Chapter 4, we met the idea that the lowest energy of the simple harmonic oscillator is not zero but $hf/2$ – the zero-point energy (where h is Planck's constant and f is the frequency of the oscillator). Exactly the same idea applies to the electromagnetic field, but it is usual to regard this zero-point energy as more formal and mathematical than of direct physical interest. In stochastic electrodynamics, though, it takes centre-stage, and relates to a *real physical field*. An atomic system may exchange energy with this field, and thus its own energy will appear to fluctuate randomly. Santos [344] writes of '[t]he hypothesis that the zero-point field causes the quantum fluctuations and it is, on the other hand, created by these fluctuations in a self-consistent manner'.

Much detailed work has been done using stochastic methods; in many cases good agreement with orthodox quantum methods is obtained, though there are, in some calculations, relatively small discrepancies [340, 343]. Fundamental problems also continue to be investigated. For example, Piotr Garbaczewski and Vigier [345] have explained the dependence of the stochastic interaction on the motion of the particle itself, by using the principle of conservation of momentum between the particle and the stochastic medium.

Thus stochastic interpretations have achieved much. There are, though, still fundamental questions. Schrödinger pointed out some of these in 1932 [67] – the wave-function is *mathematically complex*; in classical physics the probability density is subject to a differential equation, whereas in quantum theory it is the amplitude, *not* the probability density. And Timothy Wallstrom [346] has shown that, in all such methods, a *quantisation condition* must be applied in order to reach the Schrödinger equation. He claims that there is nothing in stochastic mechanics that would seem to justify such a condition.

Indeed, there is a more fundamental difficulty for stochastic interpretations that are aimed, as the ones we have discussed so far have been, at providing a realistic (albeit stochastic) model of physics. It is interesting how many of those who have produced the theories mentioned in this section have been among those we called the realists in the previous chapter – Marshall and Santos, but Vigier would also be included in this category.

They have sought to avoid Bell's theorem, and the stochastic interpretations may be seen as one aspect of that aim, but Bell's theorem must apply to the stochastic interpretations themselves. Since they are realistic (and again, even though stochastic) they cannot be local if they are to be faithful to quantum theory [225].

Indeed, by 1985 Nelson [338] had realised that his theory could not reproduce the results of EPR-type experiments in a local way. He wrote 'I have loved and nurtured Markovian stochastic mechanics for 17 years, and it is painful to abandon it. But its whole point was to construct a physically realistic picture of microprocesses, and a theory that violates locality is untenable.'

There is no reason why others with different aims should not maintain an active interest in such theories, and in 1999 Guido Baciagaluppi [347] produced an interesting account of their properties. However, this paper makes it clear that they are not to be considered as attempts to ascribe a broadly classical (though stochastic) picture to quantum theory. Rather, being non-local, they are on the same footing as Bohm theory, and Baciagaluppi directly compares their properties with those of the Bohmian model.

Models of spontaneous collapse

Until now, the various interpretations of quantum theory considered in this chapter and the last have not fundamentally contradicted the Schrödinger equation. Bohm *added* the idea of particle trajectories, but did not change the equation itself or any consequences. Stochastic interpretations endeavour to reach the Schrödinger equation from a particular starting-point. Although the quantum state diffusion method adds extra terms to the Schrödinger equation *of the system under investigation*, the idea is that they relate to the *total* Schrödinger equation of system plus environment. And the other interpretations – many worlds, ensembles, consistent histories and so on – stick to the basic Schrödinger equation, claiming that the correct conceptual approach can avoid the usual difficulties.

In this section, and for the first time, we meet theories that explicitly alter the equation; on a fundamental basis, unexplained by any discussion of interaction with an environment, an extra term is added.

We may motivate this extra term by discussing the last of John Bell's possible worlds of quantum mechanics [76] – the last of the unromantic ones he approved of because they required work by physicists. Bell suggests 'we accept Bohr's insistence that the very small and the very big must be described in very different ways, in quantum and classical terms respectively. But,' he goes on to say (as reported in Chapter 5), 'suppose we are sceptical about the possibility of such a division being sharp, and above all about the possibility of such a division being shifty [its position being arbitrary]. Surely the big and small should merge smoothly with one another? And surely in fundamental physical theory this merging should be described not just by vague words but by precise mathematics? This mathematics would allow electrons to enjoy the cloudiness of waves, while allowing tables and chairs, and ourselves, and black marks on photographs, to be rather definitely in one place rather than another, and to be described in "classical terms".'

Bell thought it might be necessary to add statistical (or *stochastic*) features to the Schrödinger equation. Certainly it would be necessary to introduce *non-linearity*. The Schrödinger equation is what is known as *linear* in mathematical terms – each term contains the wave-function to the first power only. The result is that rather different solutions may be added together to give new solutions, and, outside the measurement situation, this is one of the main mathematical glories of quantum theory, giving rise to the typical quantum behaviour of interference. But it cannot provide a unique measuremental result, it cannot give the classical side of things, so, Bell said, something different would have to be added to it.

In this 1985 paper, Bell reported that there had been 'interesting pioneer efforts in this direction, but not as yet a breakthrough'. The 'pioneer efforts' were presumably those of Pearle [168, 348], who spent many years developing theories that describe the collapse of the wave-function as a physical process. It seems, though, that Bell did regard as a breakthrough a paper published in the following year by Ghirardi, Rimini and Weber (GRW) [169]. Bell [76] publicised this theory at the Schrödinger Centenary Conference held in London in 1987, and indeed characteristically he was able to re-express it in what GRW [76] themselves (in a remark attached to his own 'Against "measurement"' paper) have described as a 'brilliant and simple presentation'. Bell [76] continued to advance the GRW approach, and it has been discussed quite widely [71, 125, 349].

Any proposal of GRW type clearly has to achieve two things – it has to create change for the measurement process where change is required, but it must avoid noticeable change nearly everywhere else, where the Schrödinger equation manages very nicely.

The GRW theory achieves this by requiring that the wave-function is always governed by the Schrödinger equation, but, in addition, is subject, at random intervals, to spontaneous *localisation* processes. In such processes, if we are dealing with a composite system of N particles, the constituent particles are localised; the spread of their wave-functions is reduced to a small range that I shall call d. (One cannot make d actually zero, which would be *complete* localisation, because that would imply no uncertainty at all in position, and hence complete uncertainty in momentum, which would be a totally unstable situation! But we can virtually eliminate such problems, and still have d quite small – GRW estimate about 10^{-7} m.)

The clever point in GRW is that the rate of these localisation processes depends on how big the composite system is, or, in other words, on the value of N. In fact the average time between localisations is equal to a new constant, τ, *divided by N*.

A suggested value for τ is about 10^{15} s, which is about 10^8 years. This implies that there will be no noticeable effect for microsystems, for which N is very small; the time taken for *spontaneous collapse*, as we may call it, will be of order τ, 10^8 years. But macroscopic objects, for which N may be something like 10^{20}, will suffer collapse in about 10^{-5} s. For the purpose of measurement, then, a pointer will effectively point in one direction only, and, as Bell says, '[Schrödinger's] cat is not both dead and alive for more than a split second.'

In 1987, Bell had found another good sign in the GRW paper. Though it was formulated non-relativistically, he was able to detect a 'residue, or at least an analogue, of Lorentz [relativistic] invariance'. He concluded his own paper by commenting that 'I am particularly struck by the fact that the model is as Lorentz invariant as it could be in the non-relativistic version. It takes away the ground of my fear that any exact formulation of quantum mechanics must conflict with fundamental Lorentz invariance.'

Since the publication of GRW, many improvements have been made to the method – by G, R and W themselves, and by Pearle, Renata Grassi, Angelo Bassi, Diósi and Gisin. A first attempt at introducing a relativistic approach [350] was made in 1990. This succeeded in localisation, but unfortunately resulted in an infinite rate of increase of energy, though it now seems that a new approach by Pearle [351] may solve this problem, and another recent suggestion is that of Nicrosini and Rimini [352]. It was also shown how the random spontaneous jumping could be replaced by a continuous process known as *continuous state diffusion* (CSL) [353]. The CSL process also solved another difficulty of the original GRW; it was able to handle indistinguishable particles correctly.

In mathematical content, CSL may end up being very similar to the *quantum state diffusion* model discussed earlier in the chapter; in both cases one adds continuous processes to the basic Schrödinger equation. However, there are fundamental conceptual differences, since the localisation process for CSL is introduced a priori, and does not simulate an interaction with any environment.

Two much discussed problems of GRW and related techniques have been the 'tail problem' and the related 'counting problem' (or 'enumeration anomaly'). The first was pointed out by Shimony, and by Albert and Loewer at the same meeting [354] in 1990. They showed that the wave-functions produced by the dynamics of reduction cannot be perfectly localised, but rather they have tails extending to infinity. Thus they do not give an objective classical world at the macroscopic level.

The second problem is closely related to this one. The details of the localisation process imply that we may have situations in which n marbles are all in a particular box, and yet the probability that all n of them may actually be found in the box is very small. This argument has been put forward especially by Lewis [355, 356], and by Clifton and Monton [357, 358]. Ghirardi and Bassi [359–362] have continually challenged both of these arguments, and Cordero [363] and Parker [364] have broadly supported their position.

Another criticism of GRW-type theories has come from David Albert and Lev Vaidman [365, 366], who argue, in terms of detailed models of measurement, that perfectly proper measurement processes that do not require a macroscopic change in the *position* of anything may take place. Thus, they say, GRW *cannot* provide a unique result. Albert and Vaidman have pursued the argument as far as the nervous system of the observer; at this point, Ghirardi and colleagues [367] argue that the states of the brain *do* differ in such a way as to provide a GRW collapse. We might feel that such a basic activity as perception should not depend on the precise details of brains that *happen* to exist, but that might be missing Gell-Mann and Hartle's point mentioned earlier, that human beings may have evolved in such a way as to be able to process useful, that is classical, information.

More recently, in a major review of the whole topic of dynamical reduction models, Bassi and Ghirardi [362] have analysed the situation in greater detail. They estimate that the number of photons required for an act of vision to be accomplished by the brain, or, as they call it, to instigate psycho-physical parallelism, may be as few as six. They then study (a) the multiplicative chain in the photoceptor cell; (b) transmission of the electrical signal along the fibres of the optic nerve; and lastly (c) excitation of neurons in the cortical visual area.

They argue that, in stage (a), the numbers of particles involved increases by a factor of 10^9. This, they say, makes correct the conviction of everybody analysing the nervous system that the classical description will be adequate by the end of this stage. In stages (b) and (c), the number increases by a further factor of 10^4. Overall they claim that the number of particles involved by the end of the whole process must certainly be large enough to cause a spontaneous collapse of the system following the GRW prescription.

The GRW type of theory is designed to reproduce the results of quantum theory to the very good approximation to which they have been experimentally verified, while also providing a collapse mechanism. Since the theory must therefore differ from standard quantum theory, it must make some predictions that differ from the standard ones, even apart from the collapse.

Far from being a weak point of the theory, it is a strong one. We will remember that Bohm theory was generally disregarded, just because its predictions were identical to the standard ones. With GRW, there is a hope of searching for these non-standard predictions, and then submitting them to experimental testing. If deviations from the standard results are obtained, this could serve to encourage support for GRW theories, and also possibly to help to refine the theories. However, this process has not been easy to implement.

In 1994, Pearle and Squires [368] reviewed the GRW and the CSL theories; in particular they considered the values of the parameters used in the theories, like the d and the τ above, in the light of the experimental evidence of the time. They concluded that CSL would have a much greater possibility of both providing collapse and remaining within the bounds set by experiment than would GRW. This possibility would be much more likely if the rate of collapse were made proportional to the mass of the object. This might support the idea of a gravitational mechanism for collapse, as suggested by several authors, including Roger Penrose. (See the next chapter.)

Since that time, Pearle has been involved in two other studies [369, 370], examining the generation of 'spontaneous' X-rays from the ground state of the germanium atom under the CSL scheme. These X-rays are caused by a ground-state electron occasionally being knocked out of the atom as a part of the CSL process. From known experimental results, it was possible to put limits on the CSL parameters. It was also shown that the collapse mechanism for electrons should be of strength 0.35% or less of that for neutrons and protons, again suggesting a collapse mechanism proportional to the mass of the particle involved.

Another possible area of experimental testing is that of superconductivity and SQUID rings [371, 372]. (Superconductivity and the SQUID ring are discussed briefly in the last section of this chapter.) According to the famous Bardeen–Cooper–Schrieffer (BCS) theory of superconductivity [373], the cause of super-conductivity is Cooper pairs of electrons that are strongly coupled together. The GRW collapse process will occasionally break Cooper pairs, and in principle there should be experimental effects. However, it is shown in these papers that these are far too small to be observed, and testing any effects of collapse is well beyond present technology.

After 20 years, GRW-type theories have reached quite a high level of rigour and sophistication, and this is good for those who are keen to show that well-argued alternatives to the Copenhagen interpretation are available. However, unless and until they are able to demonstrate some experimental consequences, they will probably fail to attract very great levels of interest and support.

The quantum Zeno effect

Having spent most of this chapter looking at different *interpretations* of quantum theory, I now turn to another puzzling *prediction* of the theory, often known as the *quantum Zeno effect*, which until recently was often called the *quantum Zeno paradox*.

In the case that is usually discussed, the effect is a result of the nature of the rate of decay of radioactive nuclei over very short times. The usual discussion of radioactive decay is in terms of *half-lives* – of a system of unstable nuclei, half will decay in one half-life, a quarter will survive for two half-lives, an eighth for three, and so on. This corresponds to a constant probability of decay, or we may say a constant *rate* of decay for any nucleus, for as long as it survives.

Then the number of surviving nuclei decreases *exponentially*, that is ever more slowly and tending to zero only after an infinite time. From the present point of view, the most important deduction is that, for a *short* decay-time t, the prob-ability of any nucleus decaying in that time t is proportional to t itself. The story told in this paragraph and the previous one is the unambiguous deduc-tion from all the experimental evidence. If it were the *whole* truth, there could be no quantum Zeno effect!

Stubbornly, quantum theory insists that it is *not* the whole truth. For the shortest of time-periods, in fact, the *rate* of decay cannot be constant, but must start at zero, and at time t (within this range) it is *proportional to* t. This means that, again within this range, the probability of decay within a period up to time t is proportional not to t, but to t^2. (I would mention that, for *very long* times too, the quantum prediction is different from that given by the simple argument above, but that does not concern us here.)

While the quantum prediction has been known in general terms since the fairly early days of quantum theory, it is only comparatively recently that it has been made more specific. This is a result of beautiful computations performed by Serot and co-workers [374] for alpha-decay from a particular isotope. For times less than about 10^{-21} s, the decay-rate is indeed found to be proportional to time; after a short period of adjustment around that time, it then quickly reaches a constant value.

Are there any systems for which the appropriate time might be much greater than 10^{-21} s, large enough even to be directly experimentally accessible? Greenland [375] showed how difficult finding such a system would be, but he did have hope for such processes as the detachment of electrons from negative ions using highly stabilised lasers.

Progress has been slow. While much theoretical work has been carried out, a recent example being [376], finding experimental evidence has been much more difficult. In two detailed experiments, Eric Norman [377, 378] was unable to find any deviations from exponential decay, but in 1997, an investigation of quantum tunnelling (not nuclear decay) [379] did reveal non-exponential behaviour. Recently there has been a claim of non-exponential behaviour [380] in a nuclear decay, though this was achieved only by combining two genuinely quantum-mechanical results, and, in fact, the claimed departure from exponential behaviour actually occurs for long times, rather than the short times relevant for the quantum Zeno effect. Still, it is a start!

Before I actually explain the quantum Zeno effect (or paradox; I don't distinguish the terms at the moment), I should mention the original *Zeno paradox*. The ancient Greek philosopher worried about how one could *ever* complete a walk of a given distance; first one must walk half the distance, then half the remainder, then half of that . . . It must, he felt, take for ever! The orthodox reply is that there may be an infinite number of time sequences in Zeno's calculation, but they may still sum to a finite total time. The reader will recognise broad similarities with Zeno's argument when we reach the *quantum* Zeno effect, though I would make it clear that the latter is in no sense a quantum version of the original paradox.

Now I come to the quantum effect itself. I shall at all times be discussing behaviour for very short decay-times – times for which the probability of decay is proportional to t^2, and *provisionally* I shall use the idea of collapse of wavefunction at an observation. First let us imagine a system of nuclei decaying in the usual way. After a time t, the probability of a given nucleus having decayed

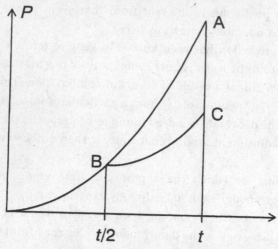

Fig. 8.4. P denotes the probability of decay of an unstable system in (short) time t. If no observation takes place, P will be proportional to t^2 at all times, and the system will reach point A at time t. However, according to the quantum Zeno effect, an observation at time $t/2$ will cause the system to move subsequently along path BC, and the probability of decay by time t is halved.

will be proportional to t^2 – call it bt^2, where b is a constant. This probability will be very small, since the t^2-region is very short.

Now let us imagine a slightly different sequence of events – the period of decay is split into two sub-periods, and in the middle there is an observation of whether the nucleus has decayed or not. (I shall not, for the moment, discuss exactly how this might be done, but I shall return to this point.) After the first sub-period of $t/2$, the probability that the nucleus has decayed will be $b(t/2)^2$. At time $t/2$, following this observation, and assuming it has survived, we have observed that it has done so, so we must collapse its wave-function back to its original form, and we must effectively start the clock at $t = 0$ again. So in the second sub-period of $t/2$, *again* the probability of decay must be $b(t/2)^2$. (See Fig. 8.4.) So when there is an observation at $t/2$, it seems that the *total* probability of decay is the sum of two terms, each equal to $b(t/2)^2$, or just $bt^2/2$. The observation at $t/2$ appears to have reduced the probability of decay between zero time and time t by a factor of two!

(The alert reader may have noticed that the argument is not *exactly* true. Since the probability of the nucleus surviving the first sub-period, and so starting the second sub-period at all, is *slightly* less than 1 (by $b(t/2)^2$, in fact), its probability of decay during the second sub-period must be *slightly* less than I have said – but by a very small amount I can ignore; mathematically it is proportional to $(t/2)^4$, which is exceptionally small because t is so small.)

To complete the argument, if the period is divided into n equal sub-periods separated by observations, the probability of decay is predicted to be reduced to bt^2/n. Now imagine the decaying system being observed *continuously* (though putting off for the moment discussion of what this might actually *mean*); it would seem that n should be infinity, and the probability of decay should be reduced to zero. This is a *most* peculiar result, certainly worthy of the name of quantum Zeno *paradox*; an alternative name is the *watched-pot paradox* since, as is well known, a watched pot never boils!

It may be worth spelling out precisely *why* the prediction is so strange. It is *non-local*, in the sense that the presence of a detector at one point is predicted to affect the decay of nuclei elsewhere. More than that, though, it just seems difficult to believe that the mere presence of a detector should affect decay-rates at all. (Of course we may feel that the decaying system should be *totally surrounded* by detectors, so that any decay would *definitely* be observed, but that does not in any way solve the basic conceptual problem.) Lastly, at least in the case of *continuous observation*, when no decays at all are predicted, one has a *negative-result experiment*.

Negative-result experiments are well known in quantum theory [67, 381, 382]. Imagine, for example, an electron known to be *either* in region 1 *or* in region 2. Quantum-mechanically its wave-function will be a sum of wave-functions specific to the two regions. We now use a photon to observe whether it is in region 1, and, let us say, find that it is not. According to collapse ideas, the wave-function must now be localised in region 2. Except in simple hidden-variable theories, where all that this shows is that the electron was in region 2 all along, the photon seems to have had a considerable effect on the electron *without interacting with it!* Of course, in the quantum formalism everything works out all right [381], but from most points of view the effect seems very strange. Similarly, in the quantum Zeno prediction for continuous observation, the detector appears to have a strong effect on the decaying nuclei without detecting anything at all!

The quantum Zeno paradox was apparently discussed informally for many years. (It is sometimes called *Turing's paradox* after the pioneer in computers.) But it was only in the 1970s that it received serious attention and its current name [383–386]. Perhaps not unexpectedly, some have thought the prediction somewhat naive, easily removable by a little more mathematical rigour. Many have seized on the use of the collapse postulate in the original treatments. (Following these, I used it myself above, of course.) Ballentine [387], who, as we have already seen, is a strong opponent of the idea of collapse, regards the paradox as 'amusing', but a 'fallacy', resulting from the fact that it uses collapse.

Others, though, have pointed out that the quantum Zeno prediction survives even in interpretations of quantum theory that retain the entire wave-function, and so do not use collapse [388, 389]. Rather it is suggested that the states of the observing apparatus must be taken fully into account. Then the quantum Zeno effect is the result of the process of measurement itself, which couples states of observed and observing systems irreversibly (in an entangled wave-function) and thus effectively separates the wave-functions of decayed and surviving nuclei [390–391].

Another conceptual problem in the discussion as presented so far is the use of *continuous measurement*. It has been pointed out [386] that any detector takes a finite time to respond to an incoming particle, and this time could be designated a *time between measurements*. More boldly the same authors suggested that time itself could possibly be thought of as being atomic in nature. Either of these possibilities would abolish any idea of continuous measurement, and hence the quantum Zeno prediction in its most extreme form.

Home and myself [391] suggested a thought-experiment that renounces any attempt at continuous measurement, but allows well-characterised discrete interactions between the decay-products of decaying nuclei and the detector. The scheme is based on the idea that, while total freezing of the decay is obviously the most dramatic feature of the quantum Zeno predictions, the most conceptually significant is that the presence of the detector has *any* effect on the decaying nuclei; this prediction may be called the *generalised quantum Zeno paradox*.

The thought-experiment to examine this uses, as before, an array of detectors covering the inside surface of a sphere surrounding the system of decaying nuclei, but now the surface is made of a balloon-type material, and its radius may be changed extremely rapidly, while maintaining its shape and the position of its centre.

If we want to test decay or survival at t/n, $2t/n$, . . ., t, the radius of the detector is held initially at a high enough value that no decay-products of decaying nuclei can reach it by time t/n. At time t/n, the radius is rapidly reduced to a very small value, and immediately returned to the high value it had before. This process corresponds to a measurement of whether any nucleus has decayed in a time up to t/n, since in such cases there will be a detection during the sweep-in. Similar contractions and immediate expansions will be made at time $2t/n$ and so on. Another experimental run could use a different value of n, and *any* dependence of the decay statistics on the value of n would be taken as evidence for the *generalised* quantum Zeno paradox.

Clearly any experiment of this nature would be exceptionally difficult to perform, but the scheme does seem to have the advantage of being clear-cut, and getting away from the difficult concept of continuous measurement.

Another suggestion for a relevant experiment was made by Seizo Inagaki, Mikio Namiki and Tomohiro Tajiri [392]. This suggestion is particularly interesting because it moves right away from the area of radioactive decay, to the behaviour of neutron spins; this is perhaps the simplest context in which quantum Zeno processes may be discussed.

Like an electron, a neutron has a spin of $\hbar/2$, so it has two spin-states, with z-component of spin $\hbar/2$ and $-\hbar/2$. (Let us call these 'spin up' and 'spin down', respectively.) If a neutron starts with spin up, and there is a magnetic field along the x-axis, the wave-function of the neutron's spin will become a sum of both spin-states, the amount of spin up initially decreasing, and that of spin down initially increasing. And if, *after a fairly short time, t,* a measurement of the neutron's spin-state is made, the probability of obtaining the spin-down state will be proportional to t^2. This is the connection with the quantum Zeno effect for decaying nuclei.

The proposed experiment uses *spin-flippers*, which transform a give spin-state (up or down) to a sum of both (as described in the previous paragraph), and magnetic mirrors, which split a beam containing a sum of spin-states into two components travelling in different directions. The crucial component would be a non-destructive detector for neutrons, which unfortunately does not exist at present, but the authors suggest that a time-dependent chopper would probably suffice. The basic idea of the experiment is to see whether detection of the neutron's spin inhibits change from spin up to spin down. Further theoretical and practical comments concerning this suggestion have been made [393, 394]. It will be noted that the two experiments just mentioned attempted to meet the original terms of the quantum Zeno effect; the changes in decay-rate are a result of a genuine measurement.

Having mentioned proposed experiments, it is surely time that I gave an account of a 1990 experiment, which was described by its architects as demonstrating the quantum Zeno effect. I should say at the outset that this experiment which was carried out by Wayne Itano and colleagues (IHBW) [395], and based on a proposal of Richard Cook [396], was brilliantly conceived and undertaken.

In the experiment, *trapped ions* were used. IHBW comment that 'Trapped ions provide very clean systems for testing calculations of the dynamics of quantum transitions. They can be observed for long periods, free from perturbations and relaxations. Their levels can be manipulated easily with [radio-frequency] and optical pulses.' In the experiments of IHBW, the temperature of the ions was maintained at about a quarter of a degree above absolute zero by *laser cooling*, a process that uses the mechanical forces exerted by light to restrict the thermal motion of the atoms [397, 398]. (We will return to some discussion of the ion trap and laser cooling in the following chapter.)

The main transition observed was between level 1, the ground-state of the ions, and an excited state, level 2. This transition was driven by a field of the appropriate frequency; if an ion was in level 1 at $t = 0$, a *short* time t later a measurement would have a probability proportional to t^2 (again!) of finding it in level 2.

In the IHBW experiment, though, this process is disturbed at times t/n, $2t/n$, ... by what IHBW call *measurements*; these are intense pulses of radiation causing transitions between level 1 and a third level, level 3, which is considerably higher in energy. IHBW's justification for calling these measurements is that, during a pulse, there *may* be photons emitted with frequencies corresponding to the energy difference between levels 1 and 3, or there *may* not. We may say that the pulse acts as a measurement by telling us whether the ion is found in level 1, and during the pulse cycles between levels 1 and 3, emitting appropriate photons, *or* in level 2, where it will remain. IHBW speak of the fact that it *is* found to be in one level or the other, rather than in a combination of both, as a *collapse of wave-function*.

The principal result of the experiment is that, in typical quantum Zeno fashion, the 'measurement' inhibits the transition from level 1 to level 2, and the title of IHBW's paper [395] was just 'Quantum Zeno effect'.

While there was general agreement on the importance and interest of the experiment, a number of groups of theoreticians [399, 400] soon questioned the use of the terms 'measurement', 'collapse' and 'quantum Zeno effect', on the grounds that the results of the experiment could be explained precisely by merely following the Schrödinger equation for microscopic systems, but including explicitly the photons emitted in the various transitions. There is no need to involve any macroscopic measuring apparatus.

In 1992, it was suggested [391] that the term 'quantum Zeno *paradox*' might be reserved for the kinds of experiments discussed earlier, which involve negative-result measurements performed by a macroscopic measuring device, and the term 'quantum Zeno *effect*' would be appropriate for experiments like those of IHBW, where the 'measurement' is actually an interaction between microscopic systems. (The latter type of experiment is extremely interesting, but not at all paradoxical, and in fact may be understood in a straightforward way.)

Thirty years after the (modern) discovery of the effect, and fifteen years after the IHBW experiment, it may be said that this is actually what has happened. In fact the former type of experiment has practically and rather unfortunately been forgotten, but on the positive side a very great amount of excellent and important work, theoretical and experimental, has been performed in the latter area. And indeed today it is very unusual to hear the word 'paradox' used in the quantum Zeno context.

While very many workers have been involved, Saverio Pascazio and Paolo Facchi of the University of Bari have been extremely prominent in the theoretical development. Discussion of the quantum Zeno effect has become important in all the many areas where the evolution of quantum systems is considered, and here only a few examples can be given. Various reviews [148, 401–403] written over the last few years will be useful for further reference.

Among general work, analysis of 'quantum Zeno subspaces' has been important [404]: for complicated systems, the quantum Zeno effect might not forbid evolution from an initial state altogether, but may allow it to occur only to a certain set of final states. Another recent paper [405] has demonstrated that Zeno-like measurements on one system may drive another system, coupled to the first, into a pure rather than a mixed state. It has also been shown [406] that, if one takes into account the fact that a detector has a finite energy band of detection, a quantum Zeno effect may occur even if decay is precisely exponential.

The quantum Zeno effect may be spoken of as suppressing decoherence in quantum systems [407]. This is exactly what is required in quantum computation to prevent error detection, and the use of the quantum Zeno effect in this context is discussed in the next chapter.

An interesting fairly recent discovery [408–412] has been that, by variation of the time between measurements, one may achieve an acceleration of decay, rather than its slowing down. This may be called an inverse quantum Zeno effect, or a quantum anti-Zeno effect.

The last application of the quantum Zeno effect is far more speculative. Stapp [413] has put forward a comprehensive account of the role of mind in nature and the part quantum theory plays in explaining this. He pays special attention to the way in which the state of the mind may affect the activities of the brain, and argues that a crucial role is played by the quantum Zeno effect.

The neutron interferometer

The neutron interferometer has been the device, more than any other, that has succeeded in turning thought-experiment into actual experiment over the last 30 years or so. From the point of view of orthodox interpretations of quantum theory, this may actually be a double-edged sword. On the one hand, it certainly brings to reality in the most dramatic way some of the more abstract of Bohr's theoretical arguments. On the other, there is always the possibility that it may be used for experiments that some may claim demonstrate the limitations and contradictions of Bohr's approach.

The first neutron interferometer was constructed by Helmut Rauch, Wolfgang Treimer and Ulrich Bonse [414] in 1974. The neutrons used were so-called thermal neutrons from a nuclear reactor (which produces copious streams of such neutrons). Now neutrons, like electrons, and indeed any other type of 'particle',

Fig. 8.5. A neutron interferometer. (*a*) is a view from the top, showing splitting at A, reflection at B and C, interference at D. Neutrons may be detected at E and F. (*b*) is a view from the side, showing the three 'ears' of the interferometer.

also display a wavelike nature, as explained in Chapter 4, and these neutrons had a wavelength of about 10^{-10} m, roughly the distance between any two atoms in a typical solid. This is also the range of wavelengths of X-rays, and the neutron interferometer was developed from a similar instrument for X-rays, which had been constructed by Bonse and Michael Hart in 1965 [415]. There have been several readable accounts of the neutron interferometer and its use in studying fundamental aspects of quantum theory [140, 416–419].

As shown in Fig. 8.5, there are three active regions in the neutron (or X-ray) interferometer. In the first, the incoming neutron beam is split into two divergent beams at A; in the second, each of these beams is reflected at B and C, so that they subsequently converge, and interfere in the third region at D. Neutrons in the two beams may be detected at E and F.

In principle, one could use three separate crystals for these three regions, but in practice, it is essential that there is practically perfect lining-up of the atoms in each region. The solution of Bonse and Hart was to make all three regions from a large single crystal of silicon, cut so as to yield three slabs or 'ears' [417], with enough of the rest of the crystal left to provide structural support.

I would remind the reader that, in a single crystal, the periodicity of the atomic spacing must remain perfect throughout; there must be no faults at all. The production of such crystals, which were of length 10 cm or so, was possible only because of the great advances in the technology of crystal growth achieved by the solid-state semi-conductor industry. Thus fundamental science has gained from technology, but on the other hand, were it not for developments in quantum theory in the 1930s to 1950s, there would *be* no semi-conductor industry, so we have a pleasant example of mutual co-operation and benefit.

Technically, construction of the neutron interferometer was a great achievement. The thicknesses of the ears, and the distances between them, had to be accurate to about 10^{-4} cm, or about 10 000 atomic layers. The crystal had to be supported carefully to avoid distortion under its own weight, and it also had to be shielded from temperature gradients and vibration. Vibration is actually a much more important problem for neutrons than for X-rays, because, while X-rays move at the speed of light, and so travel through the interferometer in about 10^{-9} s, neutrons are much slower and take about 10^{-4} s. The slightest vibration on this timescale will wipe out any interference pattern.

Daniel Greenberger [417] points out that the cross-section of the beam is about 1 cm^2, and we may think of each neutron as being represented by a wave-pulse about 10^{-3} cm long. As Greenberger says, the wave-pulse is thus about the size and shape of a small postage stamp. We further note that the typical reactor produces about one neutron per second, so neutrons usually pass singly through the interferometer, rarely interacting with each other. In addition the two beams may be several centimetres apart. In contrast, Greenberger points out that it would be immensely difficult 'to confine a photon within the interferometer', while in such experiments electron beams can be separated only by fractions of a millimetre, not by centimetres. Hence the great advance conceptually in working with neutrons.

So we see that, from the classical point of view, the neutrons seem to have been prepared very much as particles. And yet the fact is that one *does* achieve interference at the far end, suggesting that the neutron has sampled (or travelled along) *both* of two well-separated paths – very much wavelike behaviour. Of course I have been discussing this kind of idea right through the book, so I shouldn't expect you to be too amazed at this point. But it *is* important to recognise that the experimenters had achieved a totally unambiguous and direct demonstration of what had, till then, been assumed on the basis of less direct evidence.

A particularly nice additional feature is that a wedge of suitable material may be inserted into *one* of the beams. As the thickness of this wedge is increased, the shift in phase it produces also increases, and the interference pattern at D changes in a way that can be calculated precisely.

It will be remembered that a very important feature of the Bohr–Einstein debate was the argument that, if one attempted to determine which path the electron/photon/neutron took, the interference patern would be destroyed. At the time this argument was convincing enough, but not very specific. For the neutron case, Greenberger has been able to produce an exact theoretical model of the interferometer, and says that '[T]he detailed process by which this destruction [of the pattern] takes place is itself fascinating, and provides insight into how the laws of quantum theory manifest themselves in actual physical situations. In our interferometer model, we can follow every step of the calculation and keep track of the coherence of the neutron beam.'

A range of other experiments can be performed with the neutron interferometer to demonstrate other features of quantum theory and other areas of physics [416–418]. The negative-result idea which was discussed in the previous section is frequently involved in neutron-interferometry experiments. For example, particles may pass through a partial absorber. Classically these particles have not interacted with the absorber, so their properties should be entirely unaffected by the absorber. Quantum mechanically, though, we must think in terms of the amplitude of the wave; this is reduced, so the behaviour in interference experiments of the emerging particles will be very much changed by the absorber, with which they have not interacted! Greenberger has also shown how it is possible to set up a delayed-choice experiment, as discussed in the previous chapter.

Another interesting experiment demonstrated explicitly the totally non-classical fact that, if a neutron is rotated by 360°, its spin wave-function changes sign; it takes a *double* rotation to return the wave-function to its original state. (This is a result of the spin of the neutron, like that of the electron and the proton, having a quantum number of $\frac{1}{2}$; particles whose spins have *integral* quantum numbers *do* return to their original state in a 360° rotation.)

The neutron spin may be involved more explicitly in the interferometry experiments. One may use a *polarised* incident beam, for example, in which all the spins are in the same direction. It is then possible to use a *spin-flipper* in one of the beams to rotate the directions of the spins in that beam. When the two beams subsequently interfere, the differing directions of the spins of the beams must be taken explicitly into account. Theory agrees precisely with experiment *provided* that the wave-functions for the spins are added together *coherently*, *not* as a mixture. This is a very important point because it shows that separation of beams with different spin-states does not *of itself* constitute a measurement of the spin-state. Such experiments had been suggested (as thought-experiments) by Wigner as early as 1963, but it was not till the development of the neutron interferometer that they could be performed [420].

Further important experiments bring gravitation into the discussion. Here the beams are oriented with one vertically above the other. Thus there is a very small difference in gravitational potential between the two beams, which in turn has a small but measurable effect on the interference pattern. These experiments actually confirmed experimentally for the first time that quantum arguments could be applied in discussions of gravity [417].

So far all has been good news for Copenhagen – confirmation, and explicit demonstration of tricky points. Dewdney [421] was able to show that Bohm theory could also explain the results; the well-defined trajectories were, as usual, extremely interesting, but it was, of course, no surprise that the Bohm theory reproduced the orthodox results.

Rather more challenging to Copenhagen was some work of Vigier, together with Rauch [422]. Neutron interferometry is the ideal setting for detailed analysis of wave–particle complementarity. A particle picture allows one to discover which path (or *Welcherweg*, which way) a neutron took through the interferometer. A wave picture allows one to have interference at the end. Wave–particle complementarity says that you cannot have both.

In the experiments discussed, single polarised photons are sent into the interferometer. In *each* beam there is a spin-flipper, arranged to invert the spin direction of the neutron. Now when a neutron flips its spin, it exchanges one photon with the flipper, the energy of the photon being hf_1 or hf_2, where f_1 and f_2 are the frequencies at which the flippers operate. Experimentally it turns out that, if f_1 and f_2 are the same, the interference pattern is unchanged by the presence of the flippers, but if they are different, the pattern oscillates with a frequency given by the difference of f_1 and f_2.

Rauch and Vigier argued that, if one assumes absolute conservation of energy and momentum, the photon must have been exchanged *either* in beam 1 *or* in beam 2, since there are no half photons. They hence argued that 'neutrons inside a perfect neutron interferometer choose a certain beam path although the decision which beam path has been chosen remains unknown'. They then referred to the Bohr–Einstein debate, and said that their argument was in direct opposition to Bohr, but supported Einstein's '*Einweg*' (one-way) assumption.

There was controversy [423–425] on the exact interpretation of the experiments, *and* on how Bohr would, in fact, have interpreted them. On the basis of further experiments of Rauch and co-workers [426], Vigier [427] was prepared to go even further, and claimed that not just '*Einweg*', but also '*Welcherweg*' (which-way) knowledge is at least potentially available from neutron-interferometry experiments, and hence that Bohr had been shown to be unambiguously wrong!

However, on the basis of more recent experiments, and to an extent re-examination of older ones [428, 429], the situation has been clarified. The dogmatic either/or of Copenhagen was an over-simplification. In fact there is a duality relation between predictability (which-way information) and visibility (intensity of interference fringes). The higher the predictability, the lower the visibility, and vice versa.

In recent years there has been constant technical improvement of the neutron interferometer, and many interesting experiments have been performed. A good account is provided in the book of Rauch and Samuel Werner [430]. We will just mention two applications, which relate to ideas discussed earlier. The first [431] relates to carrying out the type of interaction-free measurement discussed in connection with the Hardy experiment in the previous chapter.

The second [432] is a violation of a Bell inequality. In this experiment, which is based on a previous theoretical argument [433], the necessary entanglement is not between different particles, but between the spin and spatial parts of the wave-function of any given neutron. Since the detection of neutrons is very much easier than that of photons, it might seem that this experiment would not be vulnerable to the detector loophole. Unfortunately, though, only half the neutrons produced in the experiment actually travel the path that allows them to be used in the Bell-type analysis. Obviously one cannot satisfy fair sampling; the unused neutrons could conceivably have entirely different properties from those of the neutrons that are used.

Leggett and macroscopic quantum mechanics

In the last section of this chapter, I deal with another body of work, experimental as well as theoretical, which aims to challenge orthodox views on quantum theory. This body of work has been inspired, and, at least on the theoretical side, largely accomplished, by one scientist – Tony Leggett, winner of the 2003 Nobel Prize in Physics for his work on the properties of liquid helium. (At low temperatures, helium loses all viscosity and becomes *superfluid*.)

Aside from this important work, for 25 years Leggett has produced a stream of papers on the problems of quantum theory when it considers macroscopic bodies. He set out his general ideas in the 1980s [434–439, 77], and has continued to develop them and report on experimental progress in their evaluation [440–444].

Leggett is quite prepared to concede a substantial level of success to Bohr *in the microscopic region*. Since Bohr's death, he says [77], '[T]here have been dramatic verifications of some of the more counter-intuitive predictions of quantum mechanics *at the atomic level* – and hence, one might perhaps reasonably conclude, of the correctness of his interpretation *at this level at least* [both italicisations mine].' (Leggett mentions some of the neutron-interferometry experiments described in the previous section, and also Aspect's work.)

Yet, he suggests [436], 'gnawing doubts persist ... about the ultimate meaning and indeed the self-consistency of the quantum-mechanical formalism – doubts so severe that some physicists believe that in the end a much more intuitively acceptable world-picture will supersede the quantum theory, which will then be seen, with hindsight, as merely a collection of recipes which happened to give the right answer under the experimental conditions available in the twentieth century'.

As for himself, Leggett [77] is prepared to make 'an awful confession: If you were to watch me by day, you would see me sitting at my desk solving Schrödinger's equation ... exactly like my colleagues. But occasionally at night, when the full moon is bright, I do what in the physics community is the intellectual equivalent of turning into a werewolf: I question whether quantum mechanics is the complete and ultimate truth about the physical Universe. In particular I question whether the superposition principle really can be extrapolated to the macroscopic level in the way required to generate the quantum measurement paradox. Worse, I am inclined to believe that at *some* point between the atom and the human brain it not only may but *must* break down.'

It may be suggested that working physicists' approach to the quantum theory of macroscopic bodies is ambivalent. On the one hand [436], they would believe that, *as a matter of principle*, understanding of the fundamental entities of matter – electrons, quarks and so on – automatically generates an understanding of the assemblies of these entities that constitute macroscopic objects; that *complexity* of itself does not yield fundamental novelty; and hence that macroscopic objects must, in principle, obey the laws of quantum theory. In particular, one should be able to give a meaning to a wave-function for a macroscopic object that is a sum of wave-functions for states that are macroscopically distinct (or to describe the state of a macroscopic system as being 'a superposition of macroscopically distinct states').

On the other hand [437] – 'Despite years of schooling in quantum mechanics, most physicists have a very non-quantum-mechanical notion of reality at the macroscopic level, which implicitly makes two assumptions. *Macroscopic realism* [or *macro-realism*, which will be called MR]: a macroscopic system with two or more macroscopically distinct states available to it will at all times *be* in one or the other of these states. *Non-invasive measurability at the macroscopic level*: It is possible, in principle, to determine the state of the system with arbitrarily small perturbation on its subsequent dynamics.'

Your average physicists could attempt to justify the apparent conflict by remarking that, *in practice*, it would be *extremely* hard to distinguish a genuine *superposition*, in which the wave-function of *each* of an ensemble of N particles is a sum of two macroscopically distinct wave-functions, ψ_A and ψ_B, from a *mixture*, where $N/2$ have wave-function ψ_A, and $N/2$ wave-function ψ_B. To demonstrate the superposition, it is necessary to establish that there may be *interference* between ψ_A and ψ_B.

Their argument is spelt out for them by Leggett [436]. It relies on the complexity of the macroscopic system, and hence the difficulty of performing measurements that show interference between ψ_A and ψ_B; on the complicating effect of the environment (which actually means that the measurements have to couple the system with the environment); and on the inevitable lack of detailed knowledge of the initial state of a macroscopic system (which serves to obscure the effects being sought).

Thus it is quite usual to state that (1) it is impossible to distinguish superposition from mixture (for macroscopic systems), and hence that (2) the first may be replaced by the second in a measurement situation, thus 'solving' the measurement problem. (And cats will always be *either* dead *or* alive.) We have already discussed this decoherence-type 'solution', and Leggett's criticism of it (really of (2)), earlier in the chapter. Here, following Leggett, we concentrate on (1), which is usually, as he says, not regarded as a matter of much controversy.

It would be natural for us to lump Bohr in with the 'average physicists' just discussed – after all, for him a measuring device had to be macroscopic and behave classically. Leggett [77], though, considers Bohr as certainly believing that *in a measurement* there must be irreversibility 'simply to stabilise the result of the measurement', but leaving open the possibility that a measuring device *when not being used for measurement*, or a macroscopic object *not capable of being used as a measuring apparatus*, might be able to demonstrate interference, and thus a macroscopic superposition of states. (Complementarity would demand that the system could display *either* interference *or* macrorealism, but not both.)

To study the MR hypothesis, we must look for *quantum interference between macroscopically distinct states*, which Leggett [442, 443] calls QIMDS. At least a rough definition of 'macroscopically distinct' is required, and Leggett couples together two criteria. The first he calls 'extensive difference' – how different the two states are in terms of some chosen variable, such as the position of the centre of mass.

On its own, this would allow the (spatially well-separated) states of a particle in the two paths of an interferometer to be considered as macroscopic, which is not what is required. Leggett therefore supplements this criterion with a second, which he calls 'disconnectivity', D, which measures how intrinsically different the states are in terms of the microstates of the elementary particles which constitute the macroscopic systems. Disconnectivity is related closely to the degree of entanglement, which is discussed particularly in Chapters 7 and 9.

It should be noted that the ordinary states of lasers, superconductors and superfluids do not have high values of D, so their undoubtedly exciting quantum properties do not provide any evidence as to whether quantum theory is true at the macroscopic level.

Leggett distinguishes [443] among three stages of any experimental programme. At stage I, the experimentalist produces what theory would call a quantum superposition of macroscopically distinct states, but the actual data can equally be explained in terms of a mixture of states. At stage II, one performs experiments that exclude the possibility of a mixture at the macroscopic level, but only if one uses the standard prescriptions of quantum theory.

One may say that experiments at this stage uphold quantum theory in circumstances under which it may have been thought vulnerable, but do not rule out MR. Only at stage III are there experiments that rule out the MR hypothesis, irrespective of whether one assumes these prescriptions, that is to say independent of interpretation.

Many interesting and important experiments involving macrosystems do not, Leggett says, move beyond stage I Only recently, in fact, have experiments that genuinely reach stage II been performed, and by far the most important type of experiments has been that performed using SQUIDs.

We have seen that all experiments in this area must be extremely difficult because of decoherence. When he initially proposed such experimental work, Leggett recognised that the key to any possible success would lie in thinking of a type of system that had a *macroscopic variable* effectively decoupled from the microscopic variables of the system. An obvious example would be the centre of mass of a large body. A much more useful one from the present point of view is the magnetic flux (essentially the amount of magnetic field) trapped in a *SQUID ring*.

A SQUID is a superconducting quantum interference device consisting of a superconducting ring interrupted by a *Josephson junction*. A superconductor is a material whose electrical resistance drops to zero at very low temperatures. A Josephson junction uses the *Josephson effect*, which was predicted theoretically by Brian Josephson in 1962. He showed that, at a junction of two superconductors, a current should flow even when there is no voltage drop; when there *is* a voltage drop, the current should oscillate at a frequency related to this voltage, and there should also be a dependence on any magnetic field. The Josephson effect has led to a great amount of very important physics. It has become the best way of determining the constant e/h, and has led to an international standard of voltage. SQUIDs themselves have been used in low-temperature and ultrasensitive geophysical measurements, and in computers.

In the present context, the important point about the SQUID is that it allows control of a *macroscopic* variable by a *microscopic* energy. This is required to avoid the trap of the *correspondence-principle* limit. If we do become trapped in the *classical region* (Chapter 4), the correspondence principle tells us that classical and quantum results must be identical, and so we cannot hope to distinguish between them, as Leggett's proposed experiments must do. Another important point is that we must not be swamped by thermal energy; the temperature must be no higher than about 0.5 K (half a degree above absolute zero). When he initially advocated such experiments, Leggett [436] noted that the possibility of this type of experiment would depend on 'rapid advances in cryogenics [low-temperature techniques], noise control [electronic signal detection] and microfabrication technologies'.

With all this achieved, the flux should, to a good approximation, obey a Schrödinger equation for an individual isolated particle; we must examine how the potential energy varies as the flux is altered, and the interesting point is that it may display minima or *wells*. (In the equation, incidentally, the role usually played by the mass of the subject of the equation is played by the capacitance [charge-storing ability] of the Josephson junction.) There are now two typical quantum effects that may be sought. The first is called *macroscopic quantum tunnelling* (or *decay*) from a single well, MQT, and is analogous to radioactive decay. The second is QIMDS which has already been defined, and which Leggett considers the important sign of quantum theory surviving at the macroscopic level.

While much good work at the stage-I level has been performed on SQUIDs, only comparatively recently have experiments that may justifiably be claimed to have reached stage II been performed [445, 446]. These experiments deal with superpositions of states in which the electron current is circulating clockwise (L) and anti-clockwise (R), respectively. If this superposition were in fact a mixture, there would be a symmetric state for which the two lowest energies should be equal, so the frequency of radiation absorbed for transitions between these two energy-levels should go to zero.

For the case of a genuine superposition, the two energies for this state are not identical, so the frequency remains non-zero, and this is what is found in the actual experiments. Leggett points out that the value of D for this situation is about 10^9, so this represents considerable progress (taken logarithmically, or according to power of 10) from 10^0 (or just 1) for a single atom, to 10^{24} for a real-life cat.

There are three other areas where work has reached stage II in recent years. First, Arndt and collaborators [447, 448] in Zeilinger's group have succeeded in demonstrating two-slit interference with molecules as large as C_{60}. This corresponds to a value of D of order 10^3. (Until recently the most complex objects on which interference experiments had been performed were small clusters of helium or hydrogen atoms, with a value of D of about 30.)

The second set of experiments [449] actually dates from the 1990s, and they were on the magnetic material ferritin. In this material there are two equilibrium states. In each the magnetism is small compared with that in most similar materials, since the magnetic moments of most of the atoms cancel out, but there is a small resultant magnetisation, which may point in either of two opposing directions. Though small, this magnetisation is still macroscopic.

If quantum theory is obeyed for this macroscopic magnetisation, there will be two different superpositions of these opposing magnetisations, and the experiments have been claimed to reveal absorption of energy due to transitions between these different superpositions. Provided this interpretation of the data is upheld, the value of D in these experiments would be of the order of 10^3.

The last type of experiment which succeeds in reaching stage II is in the area of quantum optics, another area where many stage-I experiments have been performed. In the particular experiment [450], a very large degree of entanglement has been obtained for two macroscopic samples of caesium atoms. The samples contain about 10^{12} atoms, and the entanglement is achieved by interaction of the samples with a pulse of light performing a non-local Bell measurement (for more information on which, see the following chapter). The entanglement can be maintained for about 0.5 ms (5×10^{-4} s). The value of D will be of the order of the square root of the number of atoms, so it is about 10^6.

Let us assume, then, that QIMDS has been demonstrated at least reasonably convincingly at the stage-II level. According to quantum theory, one definitely has a superposition, not a mixture. An important question was asked by Leggett and Anupam Garg [437, 439]: does that automatically rule out MR, or, to put it in the terms defined above, have we also reached stage III? The immediate answer is certainly not, because MR may be able to mimic these results if one does *not* assume that quantum theory is necessarily true.

This argument is very much analogous to the similar one in the Bell case, which says that, while admittedly quantum theory is incompatible with local realism, quantum theory itself might not be correct in Bell-type experiments. What was required was an experimental criterion to rule out local realism, which would be completely independent of the truth of quantum theory. This was provided by Bell and by CHSH.

Similarly Leggett and Garg were able to perform much the same analysis as Bell, but with the magnetic field or polariser direction replaced in the analysis by time. As well as MR, they made another assumption, that of NIM, or non-invasive measurability at the macroscopic level. NIM is a very common-sense type of assumption; it says that, if you search for something at the macroscopic level, but fail to find it, you have not influenced its subsequent behaviour.

Much as CHSH did, they were able to obtain a quantity that, for simple two-state systems such as the SQUID, must take values less than 2, if MR and NIM are assumed. Again in analogy to CHSH, quantum theory makes this quantity as high as $2\sqrt{2}$ in some cases. Experiments testing the value of this quantity might then be able unambiguously to decide that the conjunction of the MR and NIM hypotheses has to be abandoned. Leggett suggests that the step to stage-III measurements of this type need not be dauntingly large.

Incidentally, there have been arguments [451–453] that, in any clash between quantum theory and the combination of MR and NIM, it is NIM that causes the problems. In other words, quantum theory can co-exist with MR, but not with NIM.

In 1988, Leggett [439] argued that, if it were possible to set up experiments in which QIMDS was unambiguously predicted, the results of the experiments must be interesting irrespective of whether they are positive or negative. He says that 'A direct confirmation that such superpositions exist would knock out at least some possible solutions of the quantum measurement paradox and present that paradox in considerably sharper form: for if it is not the transition from microscopic to macroscopic which converts "potentiality" to "actuality", as Bohr and Heisenberg in their writings so often seem to imply, then what is it?'

But 'an outcome in clear violation of the quantum-mechanical predictions could . . . be even more exciting'. It would show that one cannot argue that a complete understanding of the 'building blocks' yields a complete understanding of the whole; the whole *is* more than the sum of the parts.

It seems that we are coming close to the first of Leggett's possibilities. Quantum theory will have survived another test, but only at the expense of removing as possibilities various proposed solutions of the measurement problem, such as GRW. Survival of many tests does not mean, of course, that quantum theory will not be brought down eventually. In 1986 Leggett [77] opted, particularly 'on nights when the moon is *very* bright', for a completely new theory, 'taking us even further from classical notions in a way in which at present we can hardly imagine'.

The achievement of QIMDS need not tell against this in the long run. In 1986 Leggett [77] admitted that 'If you asked me today I would certainly bet myself that the solution will not be at the level of "inert" physical devices such as SQUIDs, but would more likely be found at the level of complexity and organisation which must be necessary for biology, let alone psychology. But one has to start somewhere!'

9

Quantum information theory – an introduction

The rapid rise of quantum information

For the last four chapters I have been describing what may be called the 'fundamental' or 'foundational' aspects of quantum theory. I have been explaining what happens when you penetrate deeper into the ideas of quantum theory than when you merely make use of it, when you try to understand its basic meaning and what, if true, it tells us about the physical nature of the Universe.

As I said in Chapter 5, once Bohr had given his approach to these matters in his Como paper of 1927, general opinion was that his views were sacrosanct, and that no self-respecting physicist needed to return to discuss them further. It was Bell's great achievement to dent this complacency at least a little, and from the very end of the 1960s for the next quarter of a century or so, there were a few experimentalists keen to test the Bell inequalities with increasing stringency, and a few theoreticians with fresh ideas about how quantum theory should be interpreted.

While it was generally admitted that Bell's work was a genuine and important contribution to our understanding of quantum theory, and eventually the ideas actually reached some of the textbooks, in the 1970s and 1980s the field was still very much a minority interest, and even those who found it interesting would scarcely have felt that it had, or was likely to have in the foreseeable future, practical application.

To fairly general surprise, and in a very short period of time around 1994, things changed dramatically. Till then, the general attitude to Bell's theorem and entanglement had been that they were big worries. One may picture physicists asking each other whether things were really like that, and, if so, how could they cope with the situation? Then suddenly entanglement was not a problem, but 'an information resource', in the words of Andrew Steane [454] in his beautiful survey of quantum information theory published in 1998. (Other good introductory accounts of quantum information theory are in the book of Julian Brown [455], and that of Colin Williams and Scott Clearwater [456].)

Whether by conscious 'spin' (of a non-quantum-mechanical kind) or merely by a shared desire to take the subject forwards, such phrases as 'non-determinism', 'lack of realism' and 'non-locality', which could be perceived, at least by outsiders, as having negative connotations, were sidelined, and replaced by the kinds of exciting schemes that could be put into place using these facts of quantum life – 'quantum computing', 'quantum cryptography', 'quantum teleportation'.

Within a matter of little more than months, and with the crucial aid of a small number of dramatic discoveries, quantum information theory, and in particular (for reasons to be explained shortly) quantum computation, had become the hottest of hot topics. Money was flowing in comparatively rapidly, and the most prestigious physics journals contained numerous articles both on quantum information theory itself and on the quantum foundations that had played and were still playing a major part in its stimulation.

In this chapter, I shall give an introductory account of what is already a very large topic. In the next section, though, I shall sketch a rather different approach to explaining why we must grapple with quantum computation.

'Moore's Law'

Perhaps the most striking and continuous technological development over the last 45 years or so has been the remarkable increase in computing power [455]. A schoolchild today has a computing power available that would outclass a mainframe computer of a few decades ago.

This idea was formulated as long ago as 1964 by Gordon Moore, co-founder of Intel. He suggested that computing power, to be specific the number of transistors on a silicon chip, was doubling roughly every 18 months. This trend, celebrated as Moore's Law, has continued ever since, and the number of transistors per chip has increased from something like 1000 in 1964 to something like 100 million by the year 2000.

As a corollary it is clear that the size of each transistor must have fallen steadily through the period, and in fact the number of electrons in a transistor has dropped from millions in the 1960s to less than 100 by 2004. At this rate of progress, by some time after 2010 the number will be perhaps three or four, and we will be well inside the quantum regime. Like it or not, if we want to keep up with Moore's Law, we will have to learn to handle the quantum computer. Shortly I shall show that we may indeed like it!

A sketch of the classical theories of information and computation

When anyone is told of the subject of quantum information and quantum computation, the first thought – a very natural one – is that information and computation are surely nothing to do with quantum physics *or* classical physics, in fact nothing to do with physics at all. Rather one feels that the information that the computer computes is purely abstract, just like the subject of mathematical theorems and problems. It seems very natural to us that, in a university setting for example, the subjects of Mathematics and Computer Science may be close together, a little removed from Physics, Chemistry and the other sciences. If this were the case, there could be no quantum information or computing, no quantum or classical computing, just information and computing.

Indeed the major analysis of information and computation that took place in the 1930s and 1940s was presented in exactly this way, in the form of a series of developments in pure mathematics. The central ideas of classical information theory [457] are due to Claude Shannon [458] in the 1940s. He produced a robust expression for the amount of information contained in a sequence of numbers. The central result of classical information theory, known as Shannon's Theorem, deals with error correction.

This addresses the fact that a channel of information will always produce some errors when a message is passed along it; in other words it is noisy. Error codes enabling the receiver of the message to detect, locate and, finally, correct errors may be constructed. Shannon's Theorem tells us that, however noisy the channel, it is possible to produce a satisfactory error code, though the more noisy the channel, naturally the more elaborate the code must be. We shall see that error correction will be perhaps even more important in quantum computation than in classical computation.

The central element of the classical theory of computation is the universal Turing machine, named after Alan Turing [459], who was particularly famous for his work at Bletchley Park [460] during the Second World War, specifically the breaking of the code used by the German Enigma cipher machine, but more generally hugely important throughout the early history of computation.

A Turing machine is a conceptual mechanical device that performs computations. It uses a tape on which there is a sequence of binary symbols (0s and 1s). The machine has a finite number of internal states. When the machine reads a character on the tape, any of three actions may follow according to rules for a particular computation: the character may be changed, the state of the system may be changed, and the tape will move one step left or right. The computation consists of an input sequence on the tape being transformed to an output result.

Turing was then able to describe what is called the universal Turing machine, which can simulate the action of any other Turing machine. Almost simultaneously in 1936 Turing [461] and two other mathematicians, Alonzo Church [462, 463] and Emil Post [464], produced versions of what is usually called the Church–Turing hypothesis. These three versions emerged from very different conceptual frameworks, but the fundamental meanings were all the same. In Turing's terms, the hypothesis says that the universal Turing machine could compute 'any function naturally regarded as computable'. The last phrase may seem rather vague, and we will indeed shortly see how it may be sharpened, but it implies a very great range of tasks.

In addition any other means of computation, including those actually studied by Church and Post, could be shown to be computationally equivalent to the universal Turing machine. Universality is a central feature of computation; it implies that, given sufficient memory, any computer, no matter what its form of design and construction, can perform the same range of tasks, though some computers will obviously be much faster than others.

Thus was produced a general, powerful and seemingly final abstract theory of computation. Turing turned some parts of this theory into practice in his work at Bletchley Park during the war, though the machines he constructed there were restricted to performing one particular task.

If one defines the computer in today's terms as a machine that is programmable and has data and programs stored in exactly the same way, then that did not appear until 1945 when Turing and John von Neumann both came up with the idea, which seems so obvious in retrospect. From then on right up to the present day, progress, of course, has been dramatic.

It is worth mentioning that, though the path towards the digital computer is certainly the most important facet of the Church–Turing–Post work of 1936, at least from the present point of view, the questions being asked and answered were deep problems in mathematical logic [465]. The *Entscheidungsproblem*, included in the titles of the papers of both Church and Turing, was posed by the great mathematician David Hilbert, who called it the fundamental theorem of mathematical logic. If answered positively, it would imply that any mathematical problem could be solved by means of an algorithm, a clearly defined sequence of mathematical operations.

The work of 1936 showed that this was not the case. Turing, for example, showed that the so-called halting problem was insoluble; in general it is not possible to construct a computable function telling us whether a Turing machine with a particular input will eventually halt or continue operating indefinitely.

The fact that some functions are non-computable only added to the discomfort for mathematicians, and particularly for Hilbert, already caused by the well-known Gödel theorem of 1932. In reply to a call by Hilbert in 1900, as one of his famous list of 23 challenges to mathematicians, asking them to produce a complete and consistent set of rules of inference for mathematical proofs, Gödel was able to show that this could not be done. In fact he showed that mathematics is essentially incomplete; for any formal system containing arithmetic, there must be true statements of the system that cannot be recognised within it.

Here, though, we concentrate on the positive side of Turing's work, which made a central contribution to the seemingly highly satisfactory state of information theory and computation theory attained by the 1950s. In retrospect, though, we can now see that unnecessary and limiting assumptions were still being made. It was taken for granted tacitly that information and computation were abstract in nature. Inevitably, of course, they were manifested physically. Information would be conveyed by physical marks on paper, fluctuations in air pressure due to spoken words, or electronic signals. Computation would be performed physically by means of an abacus, a digital computer, or just by the firing of neurons in our brains.

However, since the physical representation was assumed not to be fundamental, its general form did not need to be discussed, and it was again tacitly taken for granted that classical ideas would suffice, as in the highly classical description of the Turing machine given above. As Deutsch [252] says, 'computable' was taken to mean computable by mathematicians, again limiting consideration to classical methods.

The process of questioning these assumptions was a slow one. Rolf Landauer [466, 468] of IBM, who was to be one of the pioneers of quantum computation, stressed from the 1960s on that 'Information is physical.' (Landauer actually gives the credit for first recognising this to Leo Szilard.) In [468] Landauer wrote that 'Information is inevitably tied to a physical representation. It can be engraved on stone tablets, denoted by a spin up or down, a charge present or absent, a hole punched in a card, or many other physical phenomena. It is not just an abstract entity; it does not exist except through physical embodiment. It is, therefore, tied to the laws of physics and the parts available to us in our real physical universe.'

Clearly, if information must necessarily be considered explicitly in terms of its physical representation, we must pay much more attention to the actual physics. Rather than a tacit assumption of classicality, clearly we must recognise that Nature is in fact quantum-mechanical.

It was David Deutsch [469] in 1985 who applied the same argument to computation. His book *The Fabric of Reality* [252] is a fascinating account of his views of the Universe, how what he calls the four principal strands of explanation – quantum physics, epistemology, evolution and computation – constitute the first genuine Theory of Everything. In this book he writes that 'The theory of computation has traditionally been studied almost entirely in the abstract, as a topic in pure mathematics. This is to miss the point of it. Computers are physical objects, and computations are physical processes. What computers can or cannot compute is determined by the laws of physics alone, and not by pure mathematics.'

Assuming this to be the case, clearly one must consider the possibility of quantum computation. In fact, one may go further. Just as classical mechanics is always just an approximation to quantum mechanics, often admittedly a very good and satisfactory approximation, so one may deduce that any computation is always actually quantum in nature, and classical computation is always just an approximation, though again often an extremely good one.

For his recognition of this argument, and his analysis of 1985, which effectively re-worked the arguments of Turing and Church, but using the laws of quantum physics rather than classical physics, Deutsch should be considered the founder of quantum computation as we would discuss it today. However, there was an extensive period prior to 1985 during which important preliminary results were obtained, and we will discuss the developments of this period before studying Deutsch's work in detail.

To finish this section, though, we should mention a very important dissenter from the viewpoint of Landauer and Deutsch. This is Roger Penrose, certainly among the most celebrated of today's applied mathematicians, among whose many achievements are fundamental work on black holes and the Big Bang, some carried out with Stephen Hawking, and the discovery of aperiodic tiling, which may have started as practically recreational mathematics, but has turned out to have considerable importance in solid-state physics.

Over the last 15 years or so, principally through three extremely well-known popular books, *The Emperor's New Mind* [470], *Shadows of the Mind* [471] and *The Road to Reality* [472], Penrose has carried on a sustained argument with what may be taken as the accepted view of many scientists and probably nearly all computer scientists, that the mind may be thought of as a computer, and mental activity consists of merely carrying out algorithmic processes. This view encourages the search for artificial intelligence, which we may think of as a computer program that has the attributes of a human mind such as intelligence, understanding, consciousness, free will, creativity and emotions.

Penrose believes that there are facets of the human mind that cannot be duplicated by a computer, so there can be no artificial intelligence. In particular, he argues from Gödel's theorem that mathematicians may somehow transcend the Church–Turing hypothesis. He believes that some of their work is not computational in nature at all – there are things in mathematics that cannot be calculated algorithmically, but come from direct access to some form of mathematical knowledge, or what we may wish to call insight. In contrast to Landauer's 'Information is physical', Penrose [470] believes that 'Devices can yield only approximations to a structure that has a deep and "computer-independent" existence of its own.'

Penrose believes that a completely new physics is required, a radical upheaval in the very basis of physical theory. This would be of the same magnitude as the change from Newtonian gravitation to general relativity, though, just as with that change, observable consequences would be extremely limited except in a few areas where they would be extremely important. The new physics would explain the collapse process of quantum measurement, and also contribute to an understanding of the features of human minds that he believes are not algorithmic.

In the absence of such a theory, in the absence even of a theory of quantum gravity, as mentioned towards the end of Chapter 4, Penrose's ideas of how the new physics may work may be outlined as follows. A quantum superposition is effectively a separation of mass from itself, so in terms of general relativity it is equivalent to simultaneous but very adjacent spacetime curvatures in opposite directions. This, according to Penrose, must be unstable, leading to the collapse of wave-function discussed so much earlier in this book. The timescale for this collapse is related to the gravitational energy E of the effectively separated masses; by the time–energy uncertainty principle of Chapter 4, it is given roughly by Planck's constant h divided by E.

Penrose suggests that this collapse process is non-computable, the influences from spacetime being neither random nor algorithmic. He also believes that this process may relate precisely to non-algorithmic aspects of the behaviour of the mind – consciousness and so on. The seat for this reduction he sees as microtubules, protein structures found in the cytoskeleton which organises activity in the brain, including that of the neurons.

The objective collapse process, he believes, may serve to produce subjective instants of consciousness, consciousness itself being discrete in nature. Again individual collapses may provide the opportunity for understanding and the making of conscious choices. Another possibility is that the subjective time flow may derive from the sequence of irreversible collapses.

Penrose's views have fairly naturally stirred up great hostility on the part of those computer scientists who have spent years of their lives working towards artificial intelligence, and perhaps see Penrose as an arrogant mathematician (or theoretical physicist) who believes he can use his discipline to criticise theirs. It is important to realise that he is as critical of the majority in his own discipline who do not share his view, and do not recognise that quantum theory as it stands is as much in need of correction as aspects of computer science.

In this context it is interesting to study a debate between Penrose and Hawking in a joint publication [473]. In the course of this debate, Hawking describes himself as a positivist, unconcerned with whether a physical theory corresponds to reality, and satisfied provided that it agrees with observation. He calls Penrose a Platonist, but Penrose replies that he would prefer to be described as a realist.

Penrose likens the Hawking–Penrose debate to the Bohr–Einstein debate. Though Hawking supports some version of many-worlds theory rather than Bohr's complementarity, Penrose considers that they both in reality require a theory only to make predictions. He likens his own views to those of Einstein in that both believe in the existence of a real world, possibly beyond what one may obtain from any wave-function.

Feynman and the prehistory of quantum computing

In this section, I outline work done and suggestions made concerning quantum computation principally in the 1970s and the first half of the 1980s before Deutsch's seminal work of 1985 [469], though with a brief excursion right back to 1959. I call it, a little arbitrarily perhaps, prehistory rather than early history to emphasise that, although many central ideas and important results were obtained earlier, quantum computing as we think of it today does date back to Deutsch.

In accounts of this period, the name of Richard Feynman is central. We met Feynman towards the end of Chapter 4 as one of those who, around 1947, purged quantum electrodynamics of its unwanted infinities by the technique of renormalisation.

For his version of the solution, Feynman invented the technique of Feynman diagrams, a visual means of representing a possible way in which a particular physical outcome of an interaction involving sub-atomic particles may be achieved. For any outcome there may be several ways of achieving it, and hence several different diagrams. The technique was attractively simple in principle, but at the same time profound in that quantitative results for such quantities as interaction rates could be obtained, in some cases with much less work than that needed using other methods. Ever since, Feynman diagrams have been central in all theoretical work on elementary particles.

The ability to make incisive and crucial contributions to a wide range of areas of physics, coupled with the vision to come up with and promote entirely new fields, made Feynman a colossus of twentieth-century physics [474–476]. He demonstrated his deep and wide knowledge of physics, and his ability to explain well-known topics in novel and thought-provoking ways, in his course on fundamental physics, published in 1965 as *The Feynman Lectures on Physics* [477], which has inspired generations of students.

With a charismatic personality, and a gift for self-publicity [478, 479], Feynman became a figure of considerable renown and great influence, both inside and outside the world of physics, right up to his death in 1988. In the early 1980s he was an able publicist for the view that the quantum nature of computers should be studied.

In a general way, Feynman had made an entry into the area long before. In an after-dinner speech at a meeting of the American Physical Society in December 1959, he talked on 'There's plenty of room at the bottom' [480]. The talk was a visionary and wide-ranging plea to miniaturise. The first section was titled 'An invitation to enter a new field of physics', and Feynman hoped that the field of study he was laying out could be as interesting and as technically useful as the fields of low-temperature physics, the physics of high pressures and the study of high vacuum.

Among the possibilities he suggested were writing and reading the *Encyclopaedia Britannica* on the head of a pin (with little more than 1959 technology), putting all the content of all the books in the world in a cube something like 10^{-2} cm across, and, by improvement in the electron microscope, enabling biologists to solve many of their problems by direct inspection of the molecule or structure of interest.

Feynman offered two prizes of $1000 each – the first for production of an operating electric motor in a cube of side 1/64 inch (very roughly 0.005 cm), the second for reducing the area taken by the information on the page of a book by a factor of 25 000 in linear scale in such a way that it could be read by an electron microscope. He paid out for both prizes, for the first as early as 1960, but for the second not till 1986.

This paper may be regarded as the founding of the field of nanotechnology [481], one of today's hottest topics in terms of scientific interest and technological application. Feynman discussed the construction of layered materials atom by atom, choosing the nature of each individual atom; during the subsequent half century considerable progress has been made in this direction in the manufacture of chips for computers.

Perhaps more speculatively, he discussed the making of miniature machines by working downwards in scale by successive factors of perhaps four – as Feynman puts it: 'training an ant to train a mite to do [your requirement]'. Here he hit on precisely the aspect of nanotechnology that a few claim to find disturbing – the possibility that the ants or possibly the mites might get out of hand and decide to take over the world!

Feynman's comments on computing in this paper are preliminary, but very interesting. First he noted that, if the density of information storage in a computer were sufficiently high, we would not need to erase it; this could have implications for achieving reversibility, as we shall see shortly. Secondly he called for computers, which in 1959 still occupied very large rooms, to be made smaller and smaller. In particular, he said, they might be able to make judgements on how to do their tasks, rather than having to be told every step, and they could perform such tasks as voice recognition.

In part this is merely a call for just the decrease in size that has occurred since 1959 and is given by Moore's Law. However, Feynman's demand is for more than this – it is for components of little more than atomic size, with the obvious but tacit corollary that quantum methods of analysis would be called for.

We now move on to the 1980s. From 1981 to 1986, Feynman, who, according to the great computer scientist and close friend of Feynman, Marvin Minsky [482], loved every aspect of computation, gave an interdisciplinary course at Caltech called 'The Physics of Computation', in some of those years in collaboration with John Hopfield, Carver Mead and Gerald Sussman. These lectures covered a wide range of topics, including computer organisation, the theory of computation, coding and information theory, and physical aspects of computation, but also some discussion of quantum computers and also of reversible computation, which is important in the study of quantum computation.

Since Feynman's death, tapes of these lectures, together with Feynman's notebooks, have been used by Tony Hey, together with Robin Allen, to produce an edited version of the lectures [483]. Feynman's contribution to the course was supported by invited lectures by several of his friends, experts in various aspects of computing. Hey has encouraged many of these to provide contributions for a companion volume [467]; these articles provide some accounts of collaboration with Feynman during the 1980s, but also bring the discussions up to the present day. Hey [484] has also written a short account of his whole project.

Contributors include Hopfield, Mead, Sussman, Landauer, Minsky, Charles Bennett, Paul Benioff, Norman Margolus, Ed Fredkin and Tom Toffoli. During the 1980s and in some cases much earlier, Feynman interacted with all these people, learning from them, inspiring them, building on their foundations.

In 1981 Fredkin, Landauer and Toffoli organised a conference at MIT on the Physics of Computation, which was, according to Landauer [468], an almost invisible field at that time. Feynman was persuaded to give the keynote lecture; Landauer comments that 'Feynman's mere participation, together with his willingness to accept an occasional lecture invitation in the area have helped to emphasize that this is an interesting subject'. His lecture [485] was a great step forwards and was probably the first to discuss using quantum ideas in computation and in fact to use the term 'quantum computer'.

One problem in retrospect with using Feynman's renown was that unfortunately the contributions of other workers were either not noticed or soon forgotten. For example, Benioff [486] had stimulated this work of Feynman by his own work of 1982 in which he studied whether quantum systems could be used for efficient simulation of classical computers. He was able to show that, contrary to what was often supposed at the time, the uncertainty principle would not be a problem.

Feynman in a sense asked the opposite question – can a classical computer simulate quantum-mechanical systems exactly? He did not mean merely replacing physics by a set of numerical algorithms and solving these approximately, which of course is always possible. He meant *exact* simulation. In addition he required that the number of computer elements required should increase only in proportion to the size of the physical system to be simulated; it should not explode and increase exponentially.

Feynman first showed that there is no problem in simulating classical systems, but, if we have a quantum-mechanical system and try to compute the probability of each of R particles being at each of N points, the amount of memory required is proportional to N^R – it increases exponentially.

Feynman then asked whether a classical computer could simulate a quantum system *probabilistically*, but here again the answer is negative. Interestingly the reason is precisely Bell's inequalities – if one could represent the quantum-mechanical system by hidden variables, one would effectively be making it classical and could use the classical result above, but of course this is not possible.

Feynman suggested, however, that quantum systems *could* be simulated by other quantum systems. He talked of a quantum computer, by which he meant a quantum system behaving as a computer, specifically different from a Turing machine. He asked whether there is a new type of universality, whereby different classes of quantum systems are intersimulatable, and *some* quantum systems may simulate anything, including the physical world, and thus are universal quantum simulators.

This paper was clearly an important step on the road to quantum computation. It recognised the inadequacies of classical computers for certain types of problems. Specifically it showed that the problem was that the magnitude of the computational task increased exponentially with the size of the problem, and this is extremely important in modern developments. (To be precise, for

Feynman it was the amount of memory that had to increase exponentially, whereas in modern quantum computation it would be the amount of time or the actual number of computer operations, but clearly the concepts are highly likely to be intimately connected.)

Where Feynman fell short of the final achievement, though, was that his ideas were restricted to simulation; though he used the word 'universal' in terms of his simulator, the work of Deutsch would suggest that, to be truly universal, a quantum computer must be able to perform a much greater range of tasks.

Feynman and reversible computation

Feynman was to publish another extremely important and influential paper [487] on quantum theory and computing in 1985. While the 1981 paper was somewhat vague about the details of the proposed quantum computer, the 1985 paper was much more detailed. It relied quite heavily on work on the reversibility of computers that had been carried out over a period of nearly 25 years by Landauer, Bennett, Margolus, Toffoli, Fredkin and Feynman himself, and I will first sketch the ideas involved in this work.

In computation we may look at the problem of reversibility and irreversibility from two points of view that seem very different but are actually closely connected – physical and logical reversibility or irreversibility.

The first physical point of view is the well-established one of thermodynamics, which we touched on in Chapter 3. Natural processes are almost inevitably irreversible; they proceed in one direction but will not naturally return to their starting-point. For example, an ice-cube placed in a glass of water will melt; the resulting water is at a common temperature, lower than that of the water present in the glass at the beginning but obviously higher than that of the ice (except perhaps if the ice was exactly at the freezing point of water). If the system is left on its own, the process will never run in the opposite direction; we will never find that a lump of ice has formed at the top of the water.

We may characterise the type of process of which this melting is an example in various ways, all known and much discussed for well over 100 years, and mentioned in Chapter 3. We may say that the one-way nature of the process is just a statement of the Second Law of Thermodynamics. We may say that it demonstrates the arrow of time [24].

We may alternatively say that it is a process in which entropy increases (which is actually just another way of talking about the Second Law). Entropy is an elusive concept, but it increases in many types of natural process – for example heat flow or dissipation of energy by friction or viscosity. Entropy is a measure of disorder; in the melting, the relatively ordered state at the start, where some molecules are in the solid ice, some in the liquid water, is clearly lost by the conclusion of the process.

Another important example of this type of process is that of a volume of gas divided into two sub-volumes by a removable partition. Suppose that the system is set up with the partition in place and with one sub-volume containing gas at a certain pressure, the other containing a vacuum. When the partition is raised, there takes place a process, which is complicated when viewed in terms of the actual molecules of gas, at the end of which the gas is distributed evenly through the total volume, the pressure, density and temperature being, of course, the same throughout.

This is again an irreversible process; we will never see the reverse process in which all the molecules of a gas spontaneously find themselves in one half of the chamber they are occupying. For the naturally occurring process, all the remarks about the previous example apply, including the increase of disorder.

Reversible processes are much more difficult to achieve. In the first place they must be carried out extremely slowly. Secondly, we must avoid the processes that increase entropy – heat flow, dissipation of energy. A typical textbook example is a process in which a piston is moved to increase the volume of a gas. It must be carried out extremely slowly, and the system must be thermally isolated so that there is no heat flow to or from the surroundings. (The process may be called *isentropic* or *adiabatic*.)

This process may then be reversed, returning the system to its initial state. In each stage of the combined process there is no increase in entropy. (In full the Second Law says that the change in entropy of an isolated system must be positive for irreversible changes, or zero for reversible ones.) But we see that reversible changes always seem to be exceptional and, in some ways, an idealisation.

We will now see how these ideas apply to computation, but it is helpful also to study their application to the famous or infamous Maxwell's demon, since the analyses of the two problems complemented each other and were carried out mostly by the same people [488–490].

Maxwell's demon was a troubling figment of Maxwell's imagination from 1871. In our divided chamber of gas above, the partition was to be complete, except for a small gap where the demon stood ready. In Maxwell's original version, the demon would allow fast molecules to move from right to left, slow ones from left to right, thus causing the temperature to increase on the left and decrease on the right. More simply the demon may just allow any molecule to travel from right to left but prohibit travel in the other direction. The Second Law of Thermodynamics is violated in either case; entropy decreases, order increases.

The ramifications of this little problem have rumbled down through the years, culminating in the appearance of a volume [491] collecting together some of the most important papers on the subject. (The total number is over 550.)

Before describing how these issues have been resolved, I will just remind you from Chapter 3 that the scale in which energies of atoms, molecules and elementary processes must be discussed at absolute temperature T is as multiples of kT, where k is Boltzmann's constant. This doesn't mean that the energies should necessarily be integer multiples of $kT/2$, as in the cases we studied in Chapter 3, but they should be of this order of magnitude.

The first analysis of Maxwell's demon we wish to discuss is the important study of Szilard [492] in 1929. The important insight was that the demon had to perform a measurement on each molecule in order to decide whether or not to let it through. In other words, the demon had to gain *information*. This was perhaps one of the first serious discussions of information in physics; this is the paper which Landauer credits with the first recognition that information is physical.

Szilard in fact stated that making a measurement requires the expenditure of a certain amount of energy, $kT \log_e 2$. (The logarithm is to the base e, the exponential number; $\log_e 2$ is about 0.693.) It might not have been entirely coincidental that in 1949 von Neumann came up with the same value for what seemed to be a rather different problem, that of energy expenditure in computing. In a lecture at the University of Illinois [493], which was not published till 1966, after von Neumann's death, von Neumann suggested that $kT \log_e 2$ was the minimum amount of energy required to manipulate one bit of information. A manipulation is a decision between two paths, or a transmission of energy. This seemed reasonable enough, because the information had to stand out over the background noise which increased with temperature, and the result became conventional wisdom. In fact, though, it was flawed.

The first step forwards was taken by Landauer [494] in 1961. Leaving aside the question of measurement, he analysed the way in which computer operations were actually performed. He came to the conclusion that, rather like the adiabatic expansion mentioned above, if everything was carried out slowly and steadily, there was no requirement for any energy expenditure during the actual computing process. Everything may be reversible. However, energy is dissipated whenever information is erased or discarded; this would include such actions as re-setting a register to an initial value or over-writing the value in a register. The amount of energy dissipated is $kT \log_e 2$; this is often called Landauer's constant, and the general idea is known as Landauer's principle.

The reason why routine computer operations may be reversible is broadly the same as for the adiabatic expansion of a gas mentioned above; provided that one is cautious and slow enough, one may avoid dissipation. Erasure is entirely different; in erasure the two states of the system for a particular bit, 0 and 1, must both end up as 0, but the machinery to do this must be the same for each starting-point. It is not possible to come up with physical models to do this that do not involve dissipation of energy.

However, it becomes easier to see this on moving from physical to logical irreversibility. In general terms, if two different initial states of a system are transformed to the same final state by a process, then the process must be logically irreversible; in the case under consideration any hypothesised reverse process would not know which of the initial states should be recovered [495, 496].

We leave reversible computation briefly to look at the analogous resolution of the Maxwell's demon problem due to Bennett [490]. The key feature is that energy is dissipated and there is an increase in entropy, not when the demon performs a measurement, as was implied by von Neumann, but when it discards the information it obtains in the measurement. Without this contribution, the overall entropy balance is negative, violating the Second Law; it is this extra positive contribution which makes the overall change in entropy zero, so the Second Law is upheld, and ideally the whole process may be reversible.

This resolution of the problem of Maxwell's demon was also obtained by Oliver Penrose [497], brother of Roger, and Leff and Rex [491] call it the Landauer–Penrose–Bennett solution.

One might ask the following question – what if the demon does not discard the information but just keeps hold of it? Here we must insist on the point that, according to the basic principles of thermodynamics, the Second Law applies to cyclic processes; in each cycle the system must finish in the same state as that from which it started, so if we are to consider the Second Law at all, we must insist that the demon does discard the information it obtains earlier in the cycle.

We now return to computation. While an individual element of the computation process might not require energy expenditure, it still seemed inevitable that anyone attempting to design a reversible computer would run into the same problem as that just explained for the demon. Whenever the computer memory was re-set to undertake a new computation, information would be discarded, energy dissipated, and any hope of reversibility lost. If one opted not to discard information, one must eventually be 'choked by one's own garbage'.

It was Bennett again [495] who showed how to overcome this apparently insurmountable problem. He found that it was possible to perform any computation in three parts. The first generates the required answer and also other information that was not required. In the second part, the desired answer is copied onto a blank tape; this stage is logically reversible. The third part of the process essentially reverses the first; all information generated during the first part is automatically removed by the program itself, not by an erasure. At the end of the computation, all aspects of the computer have been returned to their initial values and no energy has been dissipated, but we have the answer we require on the spare tape.

At the time these results seemed of conceptual rather than of practical interest since contemporary computers used between $10^6 kT$ and $10^{12} kT$ of energy per elementary manipulation of information, vastly more than the Landauer limit, much of it expended in keeping different parts of circuits at fixed voltages. Production of heat was a problem for the earliest computers, which used valves, but this problem was removed when transistors were introduced, at least while computers remained large. However, as Moore's Law has come into play and components have been packed closer and closer together, the heating problem has become more and more significant, and development of reversible computation has been a central topic even for classical computers.

But it is in the field of quantum computation that reversible computation has always been an essential feature. In part this is for the same reason as in the previous paragraph – in a quantum computer the components must inevitably be packed close together. More fundamentally, though, since the laws of quantum theory are themselves reversible, quantum computing must use reversible methods.

Back in the 1970s it actually seemed far from obvious that the ideas of Landauer and Bennett were correct; the belief that computing did require dissipation of energy was very strong. When Minsky [482] was asked to referee one of the key papers by the journal to which it had been submitted, he says that he read it every day for a month, then reported that the result seemed impossible, but since he could not find the mistake the journal had better publish it.

Feynman was also highly dubious, Minsky says, but reacted in a typical way. He first came up with a number of independent proofs of the same result from very different points of view. Once he was convinced reversible computation was possible, he worked with others to find out what components would be required to construct a classical reversible computer. Fredkin was already extremely interested in the same question, though his motivation was somewhat different from Feynman's. Fredkin wished to develop the idea that the Universe was just the workings of a big computer program, but, if this were to be the case, it must be possible for computers, like the laws of physics, to be reversible.

The difficulty was that it seemed almost part of the nature of a calculation, whether carried out by brain, calculator or computer, to be irreversible. One wishes to calculate $17 + 6$, and having obtained the answer as 23, it seems

A	A	NOT A
0	0	1
1	1	0

Fig. 9.1. Truth tables for single-bit operations. The first column gives the initial bit, 0 or 1. The second and third columns show the results of applying the identity operator to retain A, and operating with the NOT gate to obtain NOT A.

A	B	A AND B	A OR B	A NAND B	A NOR B	A XOR B
0	0	0	0	1	1	0
0	1	0	1	1	0	1
1	0	0	1	1	0	1
1	1	1	1	0	0	0

Fig. 9.2. Truth tables for double-bit gates. The first two columns show the input bits, A and B. The other five columns show the output obtained by use of the various gates.

almost inevitable to drop the initial 17 and 6. Once this is done, it is clear that the calculation has been performed irreversibly because one cannot recover the input from the output.

This feature of calculation was mirrored in the behaviour of all but the simplest individual functional components of a classical computer, the *gates* which perform elementary operations on bits.

This is not the case for the simplest possible gates, those that act on a single bit of information. There are two possible operations here, the identity and the NOT. In the truth table shown in Fig. 9.1, the input is written as A, the identity also as A, and the effect of the NOT gate is given by NOT A.

Clearly both operations are easily reversible, the first trivially since there is no change to undo, the second simply by applying a second NOT operation. This reversibility is achievable because there is one input and one output.

However, when one moves to the standard two-bit gates of classical computation, one does not have reversibility. This feature is obvious and unavoidable with these gates because there are two inputs but only one output. The truth table for these gates is shown in Fig. 9.2.

The gate A AND B produces the output 1 only if both A and B are 1 in the input. A OR B outputs 1 if either A or B is 1 (or both). NAND and NOR are just

NOT AND and NOT OR, so for cases where AND or OR give 1, NAND or NOR give 0 and vice versa. XOR is the exclusive OR, and gives 1 if either A or B is 1, but *not* both.

The gates are obviously capable of performing the necessary manipulations for computation provided that only two bits are involved. In fact one may go further than this; the NAND and NOR gates are each individually universal, in the sense that either may be used to construct any required computer logic.

Successful as these gates had been up to this point, for the present purpose they suffer from the fact that they are clearly irreversible. If, for example, one obtains the output 1 for A AND B, one does indeed know that, prior to the gate being applied, both A and B had the value 1, and one could attempt to reverse the operation; however, if A AND B has the value 0, it is clearly impossible to find out the values of A and B before the operation, which must therefore be irreversible. All the other gates behave in much the same way.

Clearly we are bound to find this result for gates that have two inputs but only one output. Information must be discarded. If there is to be any hope of reversibility, a gate must have at least as many outputs as inputs. (Reversible gates we consider will all have the same number of inputs and outputs.) This is the kind of gate Fredkin and Feynman looked for. The most important they found, from the point of view of reversible classical computation, the Fredkin gate, has three inputs and three outputs; it is universal in the sense that any classical reversible logic can be constructed using Fredkin gates. However, we will first look at a simpler gate, which is not so important for classical computing because it is not universal there, but extremely important for quantum computing.

This is the controlled-NOT or CNOT gate, which has two inputs and two outputs. Like the Fredkin gate, it uses a control input C; a control input does not change its value during the operation, but does determine the behaviour of another input. For the controlled-NOT gate, the other input A keeps the same value if C is 0, but changes value if C is 1. The truth table shown in Fig. 9.3 gives the initial values of the two quantities, C and A, and their final values, C' and A'.

It is easy to see that the controlled-NOT gate is reversible. Just as knowledge of the values of both C and A determines the values of C' and A' uniquely for the forward process, so knowledge of both C' and A' determines the values of C and A uniquely for the reverse process. Alternatively, it is easy to check that applying the controlled-NOT to C' and A', treated as inputs, returns the system to its original state given by the values of C and A. In other words we may say that the controlled-NOT gate is its own inverse.

C	A	C'	A'
0	0	0	0
0	1	0	1
1	0	1	1
1	1	1	0

Fig. 9.3. The truth table for the controlled-NOT gate. C is a control input and its value is not changed by the gate, but its value determines how A is changed by the gate; it changes value if C is equal to 1, but not if C is equal to zero.

C	A	B	C'	A'	B'
0	0	0	0	0	0
0	0	1	0	0	1
0	1	0	0	1	0
0	1	1	0	1	1
1	0	0	1	0	0
1	0	1	1	1	0
1	1	0	1	0	1
1	1	1	1	1	1

Fig. 9.4. The truth table for the Fredkin gate.

The Fredkin gate has three inputs and three outputs, again including one control C. The other two inputs swap values if and only if C is equal to 1. The truth table is shown in Fig. 9.4.

Like the controlled-NOT gate, the Fredkin gate is clearly reversible because, just as any particular input leads to a unique output, so any particular output is a result of a unique input. Another way of seeing the same point is that, if the output is passed through another Fredkin gate, we regain the input state; in other words, the Fredkin gate, like the controlled-NOT gate, is its own inverse.

Both the Landauer–Bennett argument and the truth tables for reversible gates were still a little abstract. Could a physical model be imagined? This was Fredkin's next self-imposed task, and, with some help from Toffoli, he produced a blueprint for the so-called billiard-ball computer [455, 498]. In this scheme, billiard balls are rolled along a smooth table along directions making angles of

45° either side of a principal axis. Timings are arranged so that the billiard balls collide appropriately, and mirrors parallel to the principal axis cause reflections.

By suitably arranging the precise lines along which the input balls travel and the positions of the reflecting mirrors, it can be arranged that the pattern of balls emerging from the machine gives precisely the information corresponding to a particular gate for the particular input, working up in complication as far as the Fredkin gate.

This worked in theory, though, of course, it was not to be expected that a billiard-ball computer could be made in practice. With real systems, energy would be lost through friction as the balls rolled, and because the collisions would not be perfectly elastic; timings of collisions and the precise directions of travel of the billiard balls would inevitably depart from the ideal, and the functioning of the computer would gradually deteriorate through any computation.

Nevertheless the work showed quite directly that the laws of physics themselves did allow reversible computation. Indeed, Fredkin invented independently of Bennett the way of removing garbage by copying the answer and then running the program in reverse. This was especially important for the billiard-ball computer, for which leaving garbage is effectively losing billiard balls!

In the second of his important papers [487], Feynman expounded much of the contents of this section, and sketched his ideas of how the various gates might be put together to implement a quantum computer – albeit a computer using quantum mechanics but solving tasks of classical computation.

How near did Feynman come to genuinely quantum computation as we shall describe it in the following section? Landauer [468] suggests that, if he had only put together the contents of his two papers [485, 487], he would have anticipated Deutsch's invention of what Landauer calls 'quantum parallelism', and that it is remarkable he didn't do so.

However, Hopfield [499] believes that Feynman had no concept that quantum computers could be better in principle than classical computers. His main vision was that the scale and speed of computation need not be limited by the classical world; he realised that reversible computation and the use of atomic systems were possible, but he had no understanding that quantum computers were different from classical computers; that insight, Hopfield says, came later and to others, not to Feynman.

Julian Brown [455] has suggested that the period of the 1980s, what we may call the Feynman period, was rather analogous to the period of the old quantum theory before Heisenberg and Schrödinger, as described in Chapter 4. Many important and indeed lasting results had been obtained but there was no deep understanding. That had to wait for David Deutsch.

Deutsch and quantum computing

Deutsch's great achievement [469] was to recognise that quantum computation requires a theoretical base of its own. He rewrote the argument for the Church–Turing thesis, mimicking, as he puts it [252], Turing's constructions, but replacing the classical physics that Turing had implicitly assumed by quantum physics. He was thus able to deduce the possibility of a universal quantum computer, which could compute anything that any other quantum device could compute. This was analogous to, but distinct from, the universal Turing machine, which could compute anything any other Turing machine, essentially anything any *classical* computer, could compute.

This is certainly important from a conceptual point of view. As Deutsch says [252], the classical theory of computation, unchallenged as the foundation of computing for more than 50 years, is now obsolete, except, like all classical physics, as an approximation scheme. The theory of computation *is* the quantum theory of computation. Quantum computers may be designed and, at least in principle, constructed.

An obvious question though is whether quantum computation may be especially useful, whether, in fact, it may be able to perform tasks impossible or effectively impossible with classical computation. This will be discussed in detail in the following section, but I will make a few introductory remarks here, for which I will first need to describe some basic features of the quantum computer.

For classical computation we have, as basic unit of information, the bit, which may take either of the values 0 and 1. The analogous quantity in quantum computation has been christened the qubit by Schumacher [500]. Like the bit it has two states, which we will write as $|0\rangle$ and $|1\rangle$. In quantum computing, the two states may be the states for a spin-$\frac{1}{2}$ system with m_s equal to $+\frac{1}{2}$ and $-\frac{1}{2}$, respectively (usually called spin up and spin down); or the two directions of the polarisation of a photon (horizontal or vertical); or the two lowest energy states of an atom (ground-state and first excited state); or two states for any other system that may be convenient.

The qubit is much richer in possibilities for use than the bit, and it could well be said that the whole practical importance of quantum computation emerges from that simple remark. While the bit can take *only* the values 0 and 1, the qubit may be in the state $|0\rangle$ or the state $|1\rangle$, but it may alternatively be in any state $c_0|0\rangle + c_1|1\rangle$, where $c_0{}^2 + c_1{}^2 = 1$. Examples would be $(1/\sqrt{2})(|0\rangle + |1\rangle)$, $(1/\sqrt{2})(|0\rangle - |1\rangle)$ and $(1/\sqrt{3})|0\rangle + \sqrt{2}/\sqrt{3}|1\rangle$. (Just for those readers who are familiar with mathematically complex numbers, I would remind them that c_0 and c_1 may be complex, and $c_0{}^2$ and $c_1{}^2$ should actually be $|c_0|^2$ and $|c_1|^2$, where $|c_0|$ is the modulus of c_0.)

To see the full significance of this, let us move on from a single qubit to a *register*, which contains a number of qubits. For example, since the number 11 (eleven) is 1011 in binary form, it may be represented in a quantum register by a register containing four qubits, which we may conveniently represent by $|1011\rangle$. A four-state quantum register may represent 16 or 2^4 different numbers, 0 to 15 in decimal notation.

In general an n-state quantum register may represent 2^n different numbers and we will represent a general one as $|x\rangle$. If we were to perform a computation with our starting point each one of these numbers, then classically we would need to start with $|00\ldots00\rangle$ and run our program, then repeat the process with the starting point of $|00\ldots01\rangle$, and so on. In all we would have to do 2^n computations.

With a quantum computer, on the other hand, our input could be $\sum_x c_x |x\rangle$, and we imagine that all the c_xs are non-zero, so all 2^n states of our register are included. We may then perform a single computation, and examine the behaviour of all possible inputs in one run of the computer. If the output for input $|x\rangle$ is $|f(x)\rangle$, then for reversibility we may keep both input and output using two quantum registers, and obtain the final state of the system as $\sum_x c_x |x\rangle |f(x)\rangle$. It seems as though we are doing 2^n calculations all at the same time, and this is the meaning of Landauer's term 'quantum parallelism'.

However, things are not as easy as we have so far suggested. At the end of the computation we must obtain a result. In quantum-mechanical terms, we must perform a measurement on the output from the computer. But when we do this, we know that we obtain just one of the answers that the computation has provided; all the others are lost. (Of course the probabilities of obtaining various results need not be equal; the probability of obtaining the result $|x\rangle |f(x)\rangle$ is just $c_x{}^2$, or technically $|c_x|^2$. But, provided that all the c_x's are non-zero, any result may be obtained.)

This highly frustrating conclusion tells us that it is far from clear that quantum computing will be useful at all. Deutsch was able to make a first tentative step towards demonstrating that quantum computation *does* have a use, but, as we shall see in the following sections, it was to be practically a further ten years before it became understood just how powerful a quantum computer could be.

Having studied qubits and quantum registers, we now move on to the other main components of the quantum computer. These are the quantum logic gates, which perform the operations on the qubits, just as the gates in a classical computer operate on the bits. The quantum logic gates are then assembled into quantum networks to perform the actual computation.

The quantum logic gates must be reversible, and may be based on particular classical gates, provided that the latter are reversible. However, just as the qubit has richer properties than the bit, so there are many more possibilities for quantum logic gates than for classical reversible gates.

To start with gates acting on single bits or qubits, for the classical case the only possibilities for output are 0 and 1. The only two interesting possibilities are the identity and the NOT gate. We showed truth tables for the classical case in the previous section. The same operations may be applied to qubits, and here we summarise the results for the identity as $|0\rangle \to |0\rangle$ and $|1\rangle \to |1\rangle$, and those for the NOT gate as $|0\rangle \to |1\rangle$ and $|1\rangle \to |0\rangle$.

However, in the qubit case, the output is not restricted to $|0\rangle$ or $|1\rangle$; it may be $c_0|0\rangle + c_1|1\rangle$ with a considerable range of possibilities for c_0 and c_1. Three single-qubit gates are particularly important in quantum computing. The first two do not produce superpositions of $|0\rangle$ and $|1\rangle$; they just create a relative phase difference between the two. The $\pi/8$ gate achieves the following:

$$|0\rangle \to |0\rangle, \qquad |1\rangle \to \frac{1+i}{\sqrt{2}}|1\rangle,$$

whereas the phase gate achieves

$$|0\rangle \to |0\rangle, \qquad |1\rangle \to i|1\rangle.$$

(For those unfamiliar with complex numbers, the factors involving i do just what was stated – change the phase of the appropriate term.)

The third single-qubit gate does produce superpositions of $|0\rangle$ and $|1\rangle$. It is called the Hadamard gate and performs the following operation:

$$|0\rangle \to \frac{1}{\sqrt{2}}(|0\rangle + |1\rangle), \qquad |1\rangle \to \frac{1}{\sqrt{2}}(|0\rangle - |1\rangle).$$

The only double-qubit gate we need to refer to is the controlled-NOT or CNOT gate we met in the previous section. We may re-state how it operates as follows:

$$|00\rangle \to |00\rangle, \qquad |01\rangle \to |01\rangle, \qquad |10\rangle \to |11\rangle, \qquad |11\rangle \to |10\rangle.$$

In other words, the second qubit swaps between $|0\rangle$ and $|1\rangle$ if but only if the first control qubit is in state $|1\rangle$. The control qubit itself remains unchanged.

This set of three single-qubit gates and one double-qubit gate is very important [501] because it is a set of universal quantum gates, in the sense that any reversible operation may be approximated to arbitrary accuracy by a combination of these gates.

In fact, if one were to include *all* single-qubit gates, then *exact* universality would be achieved with these and the CNOT gate – in other words, single-qubit and CNOT gates are (exactly) universal. To this we add the remark that any single-qubit operation may be approximated as accurately as one wishes using $\pi/8$, phase and Hadamard gates, and the statement of the previous paragraph follows.

C_1	C_2	A	C_1'	C_2'	A'
0	0	0	0	0	0
0	0	1	0	0	1
0	1	0	0	1	0
0	1	1	0	1	1
1	0	0	1	0	0
1	0	1	1	0	1
1	1	0	1	1	1
1	1	1	1	1	0

Fig. 9.5. The truth table for the Toffoli gate. C_1 and C_2 are control inputs; A swaps between 0 and 1 only if both C_1 and C_2 are equal to 1.

The small number of different types of gate required for quantum computation, and in particular the restriction to single-qubit gates and one double-qubit gate, is crucial in making it possible even to aim at making useful quantum computers. As we shall see in some detail, quantum computers are exceptionally vulnerable to many types of error, but especially those from interaction of the computer with its environment. Of the required gates, the CNOT gate is the most difficult to construct because it is difficult to achieve the required interaction between two qubits while totally avoiding further interaction outside these qubits. If a gate involving three or even more qubits were required, clearly the difficulty would be increased very much further.

We will actually say a little more about one particular three-qubit gate: the Toffoli gate, which is also a reversible three-bit gate for the classical computer. It may also be called the controlled-controlled-NOT or C^2NOT gate, for reasons that may easily be seen from the truth table shown in Fig. 9.5. There are two control inputs, C_1 and C_2, and the other input, A, swaps between 0 and 1 in the classical case, $|0\rangle$ and $|1\rangle$ in the quantum case, only if *both* the control qubits are 1 (or $|1\rangle$ in the quantum case).

Whereas classical irreversible computing may be carried out without going higher than double-bit gates, classical reversible computing requires a three-bit gate. The Fredkin gate and the Toffoli gate are alternatives. The latter is most often discussed, but for the billiard-ball computer Fredkin felt that the former was essential. This was because for the Toffoli gate (the truth table of which is shown in Fig. 9.5) the operations corresponding to the two bottom lines in the truth table do not conserve 1s, and, in the context of the billiard-ball computer, 1s are just billiard balls. The Fredkin gate was designed to avoid this problem.

Three-qubit gates, in particular the Toffoli gate, are used in quantum computation, but, unlike the classical case, they are not fundamental since they may be constructed from the $\pi/8$, phase, Hadamard and CNOT gates.

We will make one further comment about quantum logic gates, in fact about the CNOT gate. Let us examine what happens when it is applied to a state in which the first qubit is in state $(1/\sqrt{2})(|0\rangle + |1\rangle)$ and the second is in state $|0\rangle$. We will represent the combined state of the initial system as $(1/\sqrt{2})(|00\rangle + |10\rangle)$. This is clearly an unentangled state since the states of the two qubits may be stated, indeed have been stated, independently.

When the CNOT gate is applied to this initial state, each term in the sum must be handled separately. According to the truth table, the $|00\rangle$ remains as $|00\rangle$, but the $|10\rangle$ becomes $|11\rangle$. The combined state of the system is now $(1/\sqrt{2})(|00\rangle + |11\rangle)$. But now the states of the various qubits are *not* independent. If the first qubit is in state $|0\rangle$, so also is the second, but if the first is in $|1\rangle$, the second is again in the same state as the first. Results of measurement on the first qubit will be correlated with those of measurement on the second.

This is just the definition of an entangled state, so we see that the CNOT gate has the property of causing entanglement. This is a crucial role in quantum computation. Where quantum computation and quantum information handling are more powerful than the classical equivalents, it is because they take advantage of fundamentally quantum modes of behaviour. The most important of these is the ability to form superpositions and so to demonstrate interference, which would be important in all areas of quantum information. Entanglement would be only slightly less central; it is present and important in nearly all aspects of quantum information theory.

The gate which most clearly unlocks the power of the quantum superposition is the Hadamard gate. It replaces our initial states, $|0\rangle$ and $|1\rangle$, by linear combinations. One will not be surprised that both the Hadamard gate and the CNOT gate, or in some cases more powerful gates based on the same principles, are fundamental in all applications of quantum computation.

To conclude this section, we may ask the following question – where has the vast amount of parallel computing actually been performed (before measurement destroys most of its fruits)? It will be remembered from Chapter 8 that Deutsch is one of the leading adherents of the many-universes interpretation of quantum theory, and his answer [252, 469] is unequivocal – the different values of the function have been calculated in the different worlds of what he calls the multiverse.

Deutsch believes that the case for many universes is irrefutable directly from the structure of quantum theory, so logically the facts of quantum information can and need add nothing to make it firmer. Nevertheless he believes that quantum computation and also quantum cosmology are the areas of physics that have convinced others through their psychological impact. (In quantum cosmology one clearly cannot have an observer external to the whole Universe as would be demanded by Copenhagen.)

Deutsch [252] in fact issues a challenge to those who do not accept many worlds. In one particular application of quantum computing, Shor's algorithm for factorising large prime numbers, which is discussed in the section after next, Deutsch says that the computer resources used are of order 10^{500} times those which seem to be available (the number of atoms in the visible Universe being a mere 10^{80}). Deutsch then uses the explanation that the algorithm is marshalling the resources of 10^{500} universes. Of those who maintain belief in a single universe, he demands – how do you *explain* the computing power seemingly available?

Deutsch requires that physical theories provide explanation – not just prediction, which merely requires solving a few equations. In practice, though, the majority of researchers in quantum information theory, particularly those who moved directly into the area without having previously worked on more fundamental questions, seem happy just to use the quantum formalism, and are strikingly uninterested in deeper explanations. It should not be thought that studying quantum computation actually demands adherence to many universes. Steane [502], in fact, has replied to Deutsch by arguing that a quantum computer requires only one universe, and that it is misleading to argue that quantum computers perform more operations simultaneously than a single universe would allow.

Uses of quantum computation – the Deutsch algorithm

Deutsch's paper did not immediately attract great attention. This was partly because its presentation and even its title suggested a bias towards rather formal mathematical or even philosophical ideas, yet actually it required a good understanding both of quantum theory and of the theory of computation. Apart from Feynman and his circle, there were few who could have responded to the paper with great profit.

Perhaps more important, though, was that it was not clear that quantum computation would be useful. We have already seen the frustrating fact that, although the quantum computer appears to be doing a vast amount of useful work along parallel paths, the projection postulate allows us to pull out only the result obtained on one of these paths.

In his original paper, Deutsch did manage to come up with one application, one case where a quantum computer could beat any classical computer by a factor of two. This is always referred to as Deutsch's algorithm. It was immensely important from a formal point of view, for it demonstrated a violation of the original Church–Turing argument, which implied that no computer could do better than the best classical computer. Deutsch's argument was also extremely clever – beautifully simple in principle, with just enough complication in practice to make the argument rich and powerful.

Yet the formal importance and general interest of Deutsch's algorithm stood in stark contrast to the unfortunate fact that, try as Deutsch might, it was difficult for him to make a convincing case that the algorithm was of any real use. Additionally, though it could produce conclusions twice as fast as a classical computer, it was successful only on one run in two. Overall Deutsch's algorithm may ironically have played an unfortunate role, maybe serving to convince any physicist or computer scientist who might have been brought on board that, however interesting quantum computation might be in theory, it was extremely unlikely to be of any use in practice.

Before I explain the algorithm itself, I will attempt to give some general understanding of how we may seek to violate the previous argument regarding the effect of the measurement at the end of a computation. We have said that the measurement collapses down our set of answers to just a sample one of them. The trick is to avoid taking the whole of just one answer, and in a sense to sample some part of all of them, or combined properties of all of them, in a way cleverly designed to give the information we require.

In the Deutsch algorithm, the operative element is the Hadamard gate, which allows us to sample both $|0\rangle$ and $|1\rangle$ rather than one or the other. In more complicated algorithms for quantum computers, the Hadamard gate is replaced by a set of gates that perform a Fourier transform, which, as I shall proceed to explain, is more complicated than the transformation achieved by a Hadamard gate, but has a similar basic form. (To be a little more accurate, one might say that the Hadamard transformation is a very simple example of a Fourier transformation.)

We may acquire an idea of Fourier methods by attempting to describe a musical note played by a particular instrument. It is well known that a note that may be described, say, as a middle-C, that is to say as a plane wave with a single frequency, almost certainly actually has a more complicated structure; indeed that it is the general nature of the structure, the range of tones included for a particular instrument, that gives that instrument its special character.

How do we describe the note using either one parameter or a very small number of parameters? One idea might be to pick a time at random and use the value of the amplitude at that time. This is in some ways analogous to what might be called the brute-force approach to quantum computing; one just makes a measurement and takes whatever value turns up. In both situations the answer will be practically worthless; one is taking one answer at random from a set of such values about which one is learning nothing.

In the music case, the mathematics of the Fourier method allow one to ascertain the amplitude of the vibration exactly at middle-C, and then to find the amplitudes of the various harmonics that will be present. Since the different harmonics have different periodicities, this may be described as studying the various periodicities of the wave-form. With a very few parameters we are now able to describe reasonably well the whole wave-form, and we have achieved this essentially by sampling the whole wave-form judiciously. While the analogous case in quantum computing is somewhat more complex mathematically, the same general ideas apply.

We now turn to Deutsch's algorithm itself. Deutsch considered evaluation of what may be a complicated function of x, $f(x)$. The input x and the output $f(x)$ both take one of only two possible values, either 0 or 1. Clearly there are four possibilities: (a) $f(0)$ and $f(1)$ are both 0; (b) they are both 1; (c) $f(0)$ is 0 and $f(1)$ is 1; or (d) $f(0)$ is 1 and $f(1)$ is 0.

The question to be answered is as follows – are the two possible values of the output, $f(0)$ and $f(1)$, the same or not? To determine this classically, one must do two runs of the program, computing first $f(0)$ and then $f(1)$. The answer to the question is then obvious. What Deutsch's algorithm showed is that with a quantum computer one may answer the question with a single run. As already mentioned, though, only half the runs are successful; in the other half the program announces that it has failed. (Whether an individual run ends in success or failure is random.)

Deutsch attempted to generate interest in the algorithm by making it solve a real problem. Suppose that one has 24 hours to make a decision on whether to invest in the stock market. Suppose again that one will invest only if two indicators, which we may call $f(0)$ and $f(1)$, agree. Suppose yet again, though, that an individual computation of f takes 24 hours. Classically it will take 48 hours to obtain the result, too late to invest even if the two answers agree. With quantum computation, though, a single run is all that is required, and we just have time to make the investment decision.

The scenario is amusing but scarcely convincing enough to justify making the effort to build a quantum computer, even one with the small number of gates that would be required. Presumably it might be easier just to buy two classical computers! Nevertheless we will describe in some detail how Deutsch's algorithm works [455, 501], because, in this simplest possible application, it is possible to see clearly the jobs performed by the various quantum logic gates.

The central processing element in the circuit, as we may call it, will be called U. It has two inputs and, because it must, of course, be reversible, two outputs. U is called an f-controlled NOT gate, and it performs two tasks. If $|x\rangle$ is sent in as qubit 1, where the value of x may be 0 or 1, U first calculates $f(x)$, which also has one of the values 0 or 1; this process, as has been said, may be a long one. The input for qubit 2 is $|0\rangle$, but the second function of U is to change this if and only if $f(x)$ is 1. The value of f controls the gate, but qubit 1 does not change its value.

We now examine the various cases. If $x = 0$, then, if $f(0) = 0$, qubit 2 stays as $|0\rangle$, but if $f(0) = 1$, it changes to $|1\rangle$. In either case, the combined state of the system is $|0, f(0)\rangle$. In exactly the same fashion it can be seen that, if $x = 1$, then, irrespective of whether $f(1) = 0$ or 1, the final state of the system is $|1, f(1)\rangle$. The effect of U is to copy the value of $f(x)$ onto qubit 2.

But in fact what is sent as qubit 1 to U is neither $|0\rangle$ nor $|1\rangle$. While the input to the whole circuit is $|0\rangle$, this is first passed through a Hadamard gate, which has as output $(1/\sqrt{2})(|0\rangle + |1\rangle)$, and it is this that is sent on as qubit 1 to U. Since an input of $|0\rangle$ yields $|0, f(0)\rangle$ and an input of $|1\rangle$ yields $|1, f(1)\rangle$, the laws of quantum theory tell us that $(1/\sqrt{2})(|0\rangle + |)$ must yield $(1/\sqrt{2})(|0, f(0)\rangle + |1, f(1)\rangle)$.

The nature of this state is drastically different depending on whether or not $f(0) = f(1)$. If they are equal, the state is non-entangled. If, for example, they both equal 0 (case (a) above), the state is $(1/\sqrt{2})(|00\rangle + |10\rangle)$, which can be separated because qubit 2 is definitely $|0\rangle$, and qubit 1 is definitely $(1/\sqrt{2})(|0\rangle + |1\rangle)$. Thus the states of each qubit may be stated independently. Similarly, if $f(0)$ and $f(1)$ are both equal to 1 (case (b)), the combined state is $(1/\sqrt{2})(|01\rangle + |11\rangle)$, again non-entangled.

If, however, $f(0)$ and $f(1)$ are different, the combined state of the two qubits is entangled. If $f(0)$ is equal to 0 and $f(1)$ is equal to 1 (case (c) above) the combined state is $(1/\sqrt{2})(|00\rangle + |11\rangle)$. This is entangled because the states of each qubit are correlated; if qubit 1 were found to be $|0\rangle$, so would qubit 2, but if qubit 1 were found to be $|1\rangle$, again so would qubit 2. If $f(0)$ is equal to 1 and $f(1)$ is equal to 0 (case (d) above) the combined state is $(1/\sqrt{2})(|01\rangle + |10\rangle)$, again entangled because the states of the two qubits are inter-connected.

Already in the process we see the power of the Hadamard gate to produce superpositions of the basic states $|0\rangle$ and $|1\rangle$, and that of a controlled gate to produce entanglement. The next stages use further Hadamard gates, one for each qubit; here we may usefully speak not of forming superpositions but of causing the various paths through the circuit to interfere.

First qubit 2 is sent through a Hadamard gate. For case (a), $|0\rangle$ becomes $(1/\sqrt{2})(|0\rangle + |1\rangle)$, and the combined state of the system remains separable. After passing through the gate, each qubit is in the state $(1/\sqrt{2})(|0\rangle + |1\rangle)$. It is not actually necessary, but we may write the state of the combined system as $\frac{1}{2}(|00\rangle + |01\rangle + |10\rangle + |11\rangle)$. Case (b) is very similar; after passage through the gate, qubit 2 is in the state $(1/\sqrt{2})(|0\rangle - |1\rangle)$, qubit 1 is still $(1/\sqrt{2})(|0\rangle + |1\rangle)$; if we wish we may write the state of the combined system as $\frac{1}{2}(|00\rangle - |01\rangle + |10\rangle - |11\rangle)$.

For cases (c) and (d) the system remains entangled. On applying the appropriate transformation for each of $|0\rangle$ and $|1\rangle$ for qubit 2, we obtain for the state of the combined system $\frac{1}{2}(|00\rangle + |01\rangle + |10\rangle - |11\rangle)$ for case (c), and $\frac{1}{2}(|00\rangle - |01\rangle + |10\rangle + |11\rangle)$ for case (d).

We now measure the state of qubit 2. Since in each case there are two terms in the combined system with it equal to $|0\rangle$ and another two with it equal to $|1\rangle$, there is an equal probability, equal to $\frac{1}{2}$, of obtaining each answer. What we are interested in is how the state of qubit 1 depends on the result we obtain in measuring qubit 2. For the separable cases (a) and (b), the answer is obvious. Since the state is separable, the individual states of qubits 1 and 2 are independent, and, whichever result we obtain for qubit 2, qubit 1 will be left in its state $(1/\sqrt{2})(|0\rangle + |1\rangle)$.

Things are more tricky for the entangled cases (c) and (d), for which the final state of qubit 1 will in general depend on the result for qubit 2. First suppose that we measure qubit 2 to be in state $|0\rangle$. By von Neumann's projection postulate, after the measurement we retain only the terms in the state of the system in which qubit 2 is $|0\rangle$. In each case, these are $|00\rangle + |10\rangle$ inside the bracket, and since, of the $\frac{1}{2}$ factor, $1/\sqrt{2}$ belongs to each qubit, after qubit 2 has been measured to be $|0\rangle$ the state of qubit 1 is $(1/\sqrt{2})(|0\rangle + |1\rangle)$.

Also with probability $\frac{1}{2}$, though, we may measure qubit 2 as being in state $|1\rangle$. For the state of qubit 1 after the measurement, we take those terms in which qubit 2 is $|1\rangle$. For case (c), these are $|01\rangle - |11\rangle$, so the state of qubit 1 is $(1/\sqrt{2})(|0\rangle - |1\rangle)$. For case (d), they are $-|01\rangle + |11\rangle$, so the state of qubit 1 is $(1/\sqrt{2})(-|0\rangle + |1\rangle)$.

The results for cases (c) and (d) differ by a minus sign. However, a minus sign multiplying the *entire* state of a system is merely a change of phase and has no physical significance. Actually, if for case (b) we worked from the state of the combined system, we would end up with an extra minus sign over what was said above, but again it is physically irrelevant.

So, ignoring these signs, we may sum up the situation as follows. If the measurement of qubit 2 gives the result $|0\rangle$, which occurs with probability $\frac{1}{2}$, then in all cases qubit 1 will be left in the state $(1/\sqrt{2})(|0\rangle + |1\rangle)$. Since all cases are the same, clearly we cannot hope to distinguish between them, and, in this case, the method has failed.

However, if the measurement of qubit 2 gives the result $|1\rangle$, which also occurs with probability $\frac{1}{2}$, then qubit 1 will be left in the state $(1/\sqrt{2})(|0\rangle + |1\rangle)$ for cases (a) and (b), where $f(0)$ and $f(1)$ were equal, but in state $(1/\sqrt{2})(|0\rangle - |1\rangle)$ for cases (c) and (d), where $f(0)$ and $f(1)$ were different.

Clearly there is hope of distinguishing between these cases, and this is achieved by the passage of qubit 1 through yet another Hadamard gate. Cases (a) and (b) yield a final result of $|0\rangle$; cases (c) and (d) yield $|1\rangle$. This is precisely what we were aiming at!

At double the speed of a classical computer but only successful in half the cases it tries, Deutsch's algorithm was undoubtedly interesting, but scarcely awesome in its power. So it was important when in 1992 Deutsch, together with Richard Jozsa [503], extended the problem considerably. The input could now consist of n qubits, so the number of input functions was 2^n. One has to be assured that the 2^n values of f are either all the same, or balanced – in the sense that f takes each of its possible values 0 and 1 on exactly the same number of occasions.

To solve the problem classically all 2^n values of f must be calculated, but the Deutsch–Jozsa algorithm can solve the problem with one calculation of f. There is now a serious gain. The number 2^n increases exponentially with n, so Deutsch and Jozsa are able to claim an exponential gain on moving from classical to quantum computation. Theoretically this is immensely important, and it is just the claim we shall be meeting again in the following section. Still it must be admitted that the Deutsch–Jozsa problem, with its rather artificial demand for f to be balanced or constant, is even less realistic than the Deutsch problem.

Indeed, until 1994 the general consensus was that quantum computation was unlikely to be able to improve on classical computation for any serious problem. This was to change dramatically with the arrival of Shor's algorithm in 1994. This will be discussed in the following section, but to conclude this section, we will note that, in 1998, Artur Ekert and his group at Oxford [504] were able to improve Deutsch's original algorithm so that it gave a result every time rather than only with a probability of $\frac{1}{2}$.

Uses of quantum computation – the Shor and Grover algorithms

In 1994, Peter Shor [505] astounded those interested in quantum computation by demonstrating that a quantum computer could solve efficiently a problem – that of factorising large numbers – which was not only central in computer science, but also had immense applications for all concerned with the security of information – financiers, the military and so on.

It was Shor's paper that triggered the enormous interest in quantum computation that has lasted to the present day, encouraging many workers to move into the area, and many different types of organisation to fund the research reasonably liberally.

Why is factorisation a difficult problem, and what is meant by an efficient solution? To explain this, we start with multiplication; to be specific we will multiply together two prime numbers. If they both have n digits, the usual method (or algorithm) we all learned at school requires roughly n^2 basic operations. Actually there is a rather more efficient algorithm available [506], which for large n requires only $n \log n \log(\log n)$ operations (the logs are all to base e).

The important point is to study what happens to the number of operations as n increases; we have already discussed this approach for Feynman's quantum simulation and for the Deutsch–Jozsa algorithm. For either of the algorithms for multiplication, the number increases polynomially, that is to say as a power of n such as n^2 or n^3. This is a moderate increase in the demand for computational resources; problems of this nature are said to belong to complexity class P, standing for polynomial, and are regarded as tractable computationally. Indeed, it is straightforward to perform the multiplication $131 \times 227 = 29\,737$ on paper, or, of course, on a calculator or a computer.

However, if we are not shown the above sum, but are merely presented with the number 29 737 and asked to find the factors, things are much more difficult. (Note that 131 and 227 are both prime so there are no other factors of 29 737.) The standard method is to start at the lowest primes and work our way up, checking whether each prime is a factor of 29 737. We must go as far as the square root of 29 737, since clearly, if it does have factors, one of them must be smaller than that. This is rather tedious even for relatively small numbers.

Factorisation of a number with n digits requires about $10^{n/2}$ basic operations. The difference from the case of multiplication is that here the dependence on n is exponential, which means that the number of operations increases dramatically, far more speedily for large n than for any polynomial dependence. As for multiplication there are somewhat more efficient algorithms, but they don't change the basic nature of the problem.

Problems of this type are in the complexity class NP, which stands for non-deterministic polynomial. This implies that they could notionally be solved polynomially, but only on a hypothetical computer able to make choices that turned out to be correct at all branching points in the program.

Problems of NP class cannot be solved polynomially, at least on classical computers and with present algorithms, but they may be checked polynomially because to check them one has only to follow the branch downwards. For the example under consideration, checking a factorisation is just the same as performing the original multiplication, which we have seen may easily be accomplished in polynomial time.

Problems of NP class may be referred to as intractable computationally. Factorising a number of 250 digits might take the most powerful computer on Earth today about a million years. As we have seen from Moore's Law, computers rapidly increase in power. Also, prizes offered by magazines for factorising large numbers have encouraged co-operative action by a large number of computer owners [455], and it is even possible that an individual might hack into large numbers of computers and use their combined power.

Today numbers of rather more than 300 digits are thought to be impregnable; to factorise them would take all the computers in the world working together a time greater than the lifetime of the Universe. If computers increase dramatically in speed, one may always add on another 50 digits! All that we have said, though, ignores the possibility of quantum computers, as we shall indeed shortly see.

Why is this matter so important? The reason is that the inability to factorise large numbers is what makes the famous cryptography algorithm of Ronald Rivest, Adi Shamir and Len Adleman, the RSA algorithm, secure. Alice and Bob are the names always used for the two friends who desire to communicate without Eve, the eavesdropper, becoming aware of the contents of their message. The RSA algorithm, developed in 1977, allows them to do so via a public key, that is to say without sharing a private key.

Since it will be important later in the chapter when we study quantum cryptography, we will look briefly at the problem of key distribution, referring first to private keys. This is one of the safest ways of passing confidential information. If Alice and Bob both possess a particular key, then Alice may use the key and a standard algorithm to encrypt her message; she then sends it to Bob, who can decrypt it using the same key. If Eve intercepts the encrypted message, it is of no use to her because she does not possess the key.

What are the potential difficulties? The first is technical. The key will be of a certain number of binary digits. In principle it is possible for Eve to check all possibilities until one produces a readable decrypted message. The means of avoiding this possibility is to make the key sufficiently long that the number of possibilities is too large, and, in practice, 10^{18} is standard [507]. This is effectively the decision of the US government, which wishes the number to be large enough that commercial organisations can operate securely, but not so large that they themselves, with their superior computational resources, cannot decrypt messages as they choose. (In these matters there is necessarily a tension between the desire to maintain freedom of communication and that to catch criminals and terrorists.)

The second difficulty, as Simon Singh [507] reports, seems mundane, but has plagued cryptographers throughout history, and became the most demanding problem in the post-war era. It is that of key distribution. One point that has not been stressed so far is that any individual key must be used only once. A single encrypted message can give no information about the unencrypted message, but with two messages, as Singh explains, Eve may make sophisticated guesses at the nature of the key such that she can make sense of *both* messages. Once she has the key, she can obviously herself decrypt any message it has been used to encrypt.

Thus Alice and Bob must have in common a list of keys, each of which may be used only once; the famous term is the *one-time pad*. How are these keys to be distributed? Obviously the easiest possibility is if Alice and Bob may meet up to exchange keys. The other possibility is for couriers to distribute keys, but of course one needs to be assured both that the couriers are trustworthy, and that their passage around the world would not be inhibited by the representatives of Eve.

We are all familiar, perhaps from films, with military personnel delivering briefcases padlocked to their wrists. It is worth remembering, though, that the briefcase contains not, as one probably assumes when watching the film, the actual message, but merely the key. By the 1970s the sheer volume of material being transported by government, military and commercial organisations, together with the security necessary, was practically overwhelming. And some of the most famous spies in history have been caught through using a one-time pad more than once.

In this chapter we look at two possible solutions. Immediately we look briefly at public-key cryptography, which unfortunately may be undermined by quantum information theory. Later we look at quantum cryptography, which may have the power to recover the situation.

The RSA method of public-key cryptography exploits the fact that, ignoring the possible effects of quantum computation, it is effectively impossible to factorise the product of two very large prime numbers. Alice picks two such numbers, p and q, and their product, pq, which we will call N, is her public key, together with another number that she chooses, e. This public key is announced quite openly for the benefit of anybody who may wish to send her coded messages. The values of p and q themselves she keeps strictly secret.

The precise details of the method [455, 507] are a little detailed mathematically, and we will only sketch them. Bob, or anybody else who wants to send her a message, first changes the words of the message into binary digits, probably using the ASCII (the American Standard Code for Information Interchange), and thence into a decimal number M.

Bob now calculates C, the encrypted message to be sent to Alice, using the formula $C = M^e$ (mod N). This expression uses so-called modular arithmetic. To obtain x mod y, we take the remainder when x is divided by y, so, for example, 83 mod 15 is equal to 8. The importance of this step is that if Bob knows M then it is straightforward for him to calculate C, but the reverse does not apply.

The nature of modular arithmetic is such that, if Eve discovers C, then with the large sizes of numbers that would be used in realistic applications of this method, it is effectively impossible for her to deduce the value of M, the actual decrypted message. (Eve's problem may be spoken of as finding a *discrete logarithm*, a term we shall meet again briefly in connection with Shor's work.)

Alice, though, is in a more fortunate position. She knows not only N but also p and q, and, taking advantage of some very old but still subtle and powerful mathematics, she is able to find a value for another quantity d from the relation $de = 1 \bmod[(p - 1)(q - 1)]$. We may refer to d as her *private* key or *decryption key*, and she uses it to reverse Bob's encryption using the equation $M = C^d$ (mod N).

This gives her the message Bob sent. Note, though, that Eve's inability to obtain the message relies on her not being able to factorise N; if she could do so she would be in exactly the same position as Alice. It is this requirement that Shor's discovery threatens to undermine.

We now turn to Shor's discovery itself. Shor's paper [505, 508] was a theoretical *tour de force*, bringing together the problems of factorisation, period finding and obtaining discrete logarithms. All are related, and all were intractable computationally, which is to say that the computer resources required increased exponentially with the size of the problem. Shor was able to demonstrate a method by which they could all be computed in polynomial time.

For the factorisation problem, any calculation of the time required depends on details of the implementation. Richard Hughes [509] has estimated that, even with arrangements that are probably far from optimal, factorising a number of ℓ bits might require about 5ℓ qubits, and the time taken would be proportional to ℓ^3; in fact it would be about $25\ell^3$. Note the polynomial dependence on ℓ. A 512-bit number, some 155 decimal digits, which is about the largest factorisable at the present, might take about 30 seconds to factorise using Shor's algorithm. What is more worrying to those whose security relies on schemes such as RSA is that a number twice as long, often used today and regarded as completely impregnable, would take only about eight times as long to factorise, in other words a few minutes.

The Shor scheme is complicated and I shall only sketch some of its interesting aspects. I shall say nothing further about discrete logarithms, but start by explaining the connection between factorisation and finding periods. It is actually quite a subtle property of numbers.

Numbers we need to consider for factorisation are the product of two distinct odd numbers, so the smallest is 15, and we will start with this number, which we call N. We take x to be in turn 2, 3, 4, ... up to N - 1, which is 14, though in fact leaving out the factors of 15, which are 3 and 5. (In a real application, of course, we would not know these factors, since they are what we are trying to discover, but in such a case the possibility of accidentally hitting a factor is extremely small.)

Let us choose $x = 2$, and first write down the sequence $x, x^2, x^3, x^4, x^5, \ldots$ This is just 2, 4, 8, 16, 32, . . ., but let us now change this so that the numbers are not x^n but x^n (mod N). (Modular methods are described earlier in this section.) Our sequence is now 2, 4, 8, 1, 2, 4, 8, ... If we were to continue, the four numbers 2, 4, 8, 1 would repeat indefinitely. In other words, the values form a periodic function of period 4. We will say that this is the *order* of the function and call it f.

If we choose different values of x, we obtain similar results. For example, if $x = 7$, then the sequence is 7, 4, 13, 1, 7, . . ., and the period is again 4. While both those values of x give the same period, for other values the period will be different; for this particular case, for example, if x is 4 or 11, the order is 2.

A little algebra follows just for those who like it. Since $x^{f+1} = x \bmod N$, then $x^f - 1 = 0 \bmod N$. (We have divided both sides by x, and noted that x and N do not have common factors.) But x^f cannot equal 1, since x is greater than 1 and f is at least 1, so $x^f - 1$ must be a multiple of N.

Let us suppose first that $x^f - 1$ is actually equal to N. Then, using the very well-known identity $a^2 - b^2 = (a - b)(a + b)$, we can obtain $x^f - 1 = (x^{f/2} - 1)(x^{f/2} + 1)$ and hence $(x^{f/2} - 1)(x^{f/2} + 1) = N$. This relation is likely to be useful only if f is even, in which case it tells us that $x^{f/2} - 1$ and $x^{f/2} + 1$ are factors of N.

We consider the case $x = 2$, for which $f = 4$. Then $x^f - 1$ is exactly 15; $x^{f/2} - 1$ and $x^{f/2} + 1$ are 3 and 5, exactly the factors required. The case of $x = 1$, $f = 2$ is almost identical.

However if $x^f - 1$ is a multiple of N, then $x^{f/2-1}$ and $x^{f/2} + 1$ will usually be multiples of factors of N. Let us consider $x = 11$, $f = 2$. In this case $x^f - 1$ is 120, a multiple of 15, and $x^{f/2} - 1$ and $x^{f/2} + 1$ are 10 and 12, the first a multiple of 5, the second a multiple of 3.

How, though, can we obtain the actual factors from the numbers obtained, 10 and 12? We may note that 5 is the highest (or maximum) common factor, HCF or MCF, of 10 and 15; 3 is the HCF of 12 and 15. However, in a real application of this work with large numbers this would not be known; the factors of the number corresponding to 15 are what the method is trying to find out!

We may use, though, a well-established method of Euclid. To find the HCF of 10 and 15 we divide the larger number by the smaller and keep the remainder. So $15/10 = 1 + 5/10$. We then divide the 10 by the 5; this time there is no remainder so this confirms that 5 is the HCF of the numbers we started with. Similarly we may show that 3 is the HCF of 12 and 15.

This establishes the connection between factorisation and period finding. Unfortunately it must be recognised that there is a probability of the method failing. First the order f may be odd, and another possibility of failure is shown by choosing $x = 14$. For this case we can calculate $f = 2$, and $x^f - 1$ is 195, a multiple of 15 as we expect, but $x^{f/2} - 1$ and $x^{f/2} + 1$ are 13 and 15, which does not help us find factors of 15. Nevertheless it can be shown that the probability of success for an arbitrarily chosen value of x is greater than $\frac{3}{4}$, and this probability may be increased by trying a number of values of x.

While the quantum computer circuit required to implement the algorithm is complicated, it may be analysed [504] in a way analogous to the circuit for the Deutsch–Jozsa algorithm. The number N to be factorised is given. A number x is chosen, and it may be checked using Euclid's method that it has no common factor with N. Hadamard gates are used to set the first of two quantum registers into a superposition of states representing all numbers a as far as the limit set by the number of qubits available. The gate U in this case calculates $x^a \bmod N$, so that the second register holds the values of this for all values of a up to the same limit; the states of the two registers are entangled.

A Fourier transform is now applied to the first register by a series of logic gates, which I will not describe. Because of the entanglement, the Fourier transform produces the periodicity of the values of the function in the second register, and, if the values of the qubits on the first register are measured, there is a good probability, as explained above, that the factors of N may be discovered.

We now turn to the other important quantum algorithm, Grover's algorithm [510–512]. This relates to searching in a telephone directory for a name if you are given the corresponding telephone number. If there are N items, classically you must look through them one by one, and on average it will take $N/2$ steps. Grover's algorithm reduces this to roughly $\sqrt{N}$. Both are polynomial, so the algorithm does not, like the Shor algorithm, change the complexity class of the problem. Nevertheless the gain is substantial, and the areas of application are widespread.

As well as searching any form of directory, it could be used in searching for a short piece of text in an enormous number of computerised volumes. It could be used for searching through lists of keys in cryptography, assuming that an example of an unencrypted text and the encrypted version of the same text are available, or through various move strategies in chess. Indeed, for any problem where finding a solution is hard but checking one is easy, one can always resort to producing a complete list of possible solutions and then using Grover's algorithm.

To see how the algorithm works, we must remember that a quantum computer may examine all N items simultaneously. We shall see that we can also manipulate all N items simultaneously. The usual difficulty is that we can read only one item at a time.

How the Grover algorithm works is by manipulating the information in a way that discriminates very slightly but consistently in favour of the item we wish to extract; we shall call it $|i\rangle$. The algorithm starts in a standard way with a Hadamard gate, which gives us a superposition of all the items. We then apply an operator that picks out $|i\rangle$ from all the other items and changes its sign.

Three more operations follow, a Hadamard gate, then an operation that changes the signs of all the items *apart from* $|i\rangle$, and finally another Hadamard gate. Overall it is clear that item $|i\rangle$ has been treated a little differently from all the others.

Detailed analysis shows that the contribution of $|i\rangle$ is increased, and the whole operation may be repeated a large number of times. However, the number of times it is repeated must be just right! It has been said – Steane [454] attributes the remark to Kristen Fuchs – that applying Grover's algorithm is like baking a soufflé. The desired answer rises slowly but surely, but the process must be stopped at exactly the right time; otherwise the answer, like the soufflé, collapses again.

To put things a little more specifically, following the first Hadamard gate, all items are represented equally in the superposition produced. Since there are N items, the laws of quantum theory tell us that the coefficient of each must be $\sqrt{1/N}$, a very small number of course. It is mathematically convenient to write this as the sine of an angle, say $\sin\theta$, where θ is in radians. (Since, in fact, for a very small angle, $\sin\theta$ and the value of θ in radians are almost exactly the same, we may just say θ is equal to $\sqrt{1/N}$.)

We shall put the coefficient of $|i\rangle$ equal to $\sin\alpha$. So, at the commencement of the process, α is equal to θ. Following the first use of the series of Grover operations, α increases to 3θ, and the coefficient of $|i\rangle$ increases to $\sin(3\theta)$. The coefficients of all the other items must decrease slightly to balance, but we need not dwell on this. Instead we note that each further use of the operations may be shown to increase α a little further, to 5θ, 7θ and so on.

How many times should we perform the operations? Readers may be aware that $\pi/2$ radians is equal to $90°$, so $\sin(\pi/2)$ is equal to 1. If we increase α this far, the coefficient of $|i\rangle$ will have reached 1, and we are *certain* to obtain the answer we desire. To balance this increase, the coefficients of all the other possibilities have at this stage decreased to zero. Since the angle must increase by almost exactly $\pi/2$, and the increase at each stage is $2\sqrt{1/N}$, the number of series of operations should be the first divided by the second, which is $(\pi/4)\sqrt{N}$. We note the $\sqrt{N}$ dependence.

If more operations are performed, α increases further. As it moves on from $\pi/2$ or $90°$, $\sin\alpha$ decreases from 1, and the probability of obtaining the answer $|i\rangle$ decreases, so it is essential that the number of sets of operations is just right.

Decoherence and quantum error correction

It cannot be over-emphasised just how fragile the operation of a quantum computer must be. There are several reasons for this. First there is the very nature of the qubit as compared with the bit. A physical system representing a bit on a classical computer is analogous to a light switch [454]; it requires only two positions of equilibrium, 0 and 1. If the system moves from one of these, there is a strong restoring force to take it back; there is also dissipation of energy, so it may come to rest at the equilibrium position rather than oscillating indefinitely. (These positive features are essentially those of digital computers and digital recording devices, and explain why they are more reliable than their analog equivalents and have displaced them.)

Turning now to the quantum case, the very nature of the qubit means that the states $|0\rangle$ and $|1\rangle$ are not physically special; any superposition of $|0\rangle$ and $|1\rangle$ has just as much physical meaning. In addition, there can be no mechanism for dissipation of energy, since schemes for quantum computation must be reversible. It may be said that quantum computers have some of the rather unsatisfactory features of analog computers.

Schemes for quantum computation rely heavily on the maintenance of superpositions of states, particularly including those of entangled states. As discussed in Chapter 8 in some detail, interaction with the environment will inevitably cause decoherence. A state of the form $(a|0\rangle + b|1\rangle)|e\rangle$, where $|e\rangle$ is a state of the environment, may become $a|0\rangle|e_0\rangle + b|1\rangle|e_1\rangle$, where $|e_0\rangle$ and $|e_1\rangle$ are distinct states of the environment.

Whereas it is an essential feature of quantum computation to arrange that $|0\rangle$ and $|1\rangle$ may be made to interfere, because $|e_0\rangle$ and $|e_1\rangle$ are quite different states of the environment, which is macroscopic, not only can they themselves not interfere, but also their presence prevents $|0\rangle$ and $|1\rangle$ from interfering. Although, as stressed in Chapter 8, there is actually no collapse of the wave-function of the entire system, qubit plus environment, if one considers the qubit alone it behaves very much like a mixture of $|0\rangle$ and $|1\rangle$. Inevitably this degrades the performance of the computer.

The amount of decoherence may be reduced but is impossible to eliminate altogether. The controlled-NOT gate is particularly susceptible, because it makes use of coupling between qubits, and so inevitably brings in other couplings as well. Plenio and Knight [513] have estimated that, every time this gate is used, the probability of decoherence may be at least 10^{-6}.

While this may sound reasonably low, it must be remembered that establishing the quantum computer as genuinely superior to a classical computer in its own areas of use, that is to say factorising a number of perhaps 200 digits, will require thousands of qubits and billions of gates. Thus the amount of decoherence is still too large by a factor of about 10^7. As Steane [454] says, it is easier to imagine increasing the speed of a classical computer by 10^7 than reducing the decoherence of a quantum computer by the same factor.

It is scarcely surprising that there have been extreme pessimists about the possibilities of quantum computation ever being useful. They have included Haroche and Raimond [514] and Landauer [515]. Before deciding that they were *too* pessimistic, we should perhaps wait for the machines to be produced, but the reason why many think that this was the case is the large amount of successful work that has been performed on codes for quantum error correction, by Shor [516] and by Steane [517] independently, and by many others [454].

Codes for quantum error correction were broadly based on analogous codes for correcting errors in classical information, though we shall see that the quantum case presents very severe extra problems. In his lectures, Feynman [483] gave a readable account of the classical methods. They were important in the early days of classical computers, when valves were used and the computers were prone to error, but they became much less so when transistors were introduced and errors became far less common. But the codes have always been exceptionally important for devices transmitting information, since the channels must always be assumed to be noisy.

The simplest check on a message is to add a parity bit. If there is an odd number of 1s, the parity bit is 1; otherwise it is 0. So the message 0011010 becomes 00110101. If there is a single error in transmitting the message, whether in one of the original bits or in the parity bit, there will be a mismatch between the two parts of the message, and it must be re-sent.

This technique can cope only with a single error. A combination of two errors, whether both in the original message, or one in the original message and one in the parity bit, will not be detected. If there is a significant possibility of this occurring, additional check bits must be included. Another limitation in the above is that we do not know where the error has occurred, so we cannot automatically correct it; we must just resend the message.

The following scheme enables us to find and correct a single error. We imagine a number of lines of text, each with the same number of bits. To the message we add a parity operator for each line, one for each column and also one for the entire message. It is easy to see that a single error, whether in the actual message or in the parity bits, may be found and corrected. In some cases it may be possible to correct two or more errors, but not if they occur in the same line or column of the message.

Another useful scheme in classical error correcting, which is also the basis of Steane's scheme of quantum error correction, is the Hamming code [457, 518]. The {7, 4, 3} Hamming code [454] represents each of the 16 possible 4-bit messages by a 7-bit codeword, so that each codeword differs from all others in at least three places. For example, 0000 is represented by 0000000, 0001 by 1010101 and so on.

One of the codewords closest to that for 0000 is that for 1001, which is 0101010. The codewords for 0000 and 1001 differ in three places. If each has one error, 0000000 might become 0100000, and 0101010 might become 0100010, but they still differ in a single place. Provided that there is indeed only the one error, it is quite clear what each incorrect codeword should actually be.

We now turn to quantum error codes themselves. A problem over and above classical correction is that we must be careful not to perform a measurement on any qubit we are actually using for computation, because we will disturb it.

Yet another problem of quantum computation as compared with classical computation is the so-called no-cloning theorem. In the classical case, one may always take a number of copies of part of a particular message or computation, and trust that, even if an error occurs working with one of the copies, the majority will give correct results. For the quantum case, the no-cloning theorem says that making such copies is impossible.

This theorem, which was proved independently by Wooters and Zurek [519], and by Dieks [520], tells us that it is not possible to find an operator U that will clone an arbitrary quantum state presented to it. The proof is simple. If we imagine such a U, it must cause the reactions $U|\alpha, 0\rangle \to |\alpha, \alpha\rangle$ and also $U|\beta, 0\rangle \to |\beta, \beta\rangle$, where $|\alpha\rangle$ and $|\beta\rangle$ are any two states of any system, and the state of the second qubit, initially $|0\rangle$, becomes $|\alpha\rangle$ or $|\beta\rangle$, a clone of the first.

Now imagine the same operator applied to the second qubit in state $|0\rangle$ as before, but the first in state $(1/\sqrt{2})(|\alpha\rangle + |\beta\rangle)$. The laws of quantum theory tell us that we can add together the two previous equations, multiply by $1/\sqrt{2}$, and obtain for the resulting combined state $(1/\sqrt{2})(|\alpha, \alpha\rangle + |\beta, \beta\rangle)$. But this is not a cloning of the original state. A cloning would require both qubits to end up in $(1/\sqrt{2})(|\alpha\rangle + |\beta\rangle)$; the combined state would be $\frac{1}{2}(|\alpha, \alpha\rangle + |\beta, \alpha\rangle + |\alpha, \beta\rangle + |\beta, \beta\rangle)$.

This shows us that imagining an operator such as U leads to inconsistencies, or, in other words, there can be no cloning operator.

Quantum error codes attempt to solve both problems – the restrictions on measurement and on cloning – by introducing extra qubits, called the *ancilla*, which is a Latin word for maidservant. These extra qubits aid the computation process but they do not contribute directly to the final answer. So they may be measured; also in some ways they play a role similar to the one that might be played by copies of the original qubit if it were possible to make them.

For a single qubit, errors may be complicated in form, but they can all be built up out of three basic types. These are

$$\alpha|0\rangle + \beta|1\rangle \rightarrow \alpha|1\rangle + \beta|0\rangle,$$

$$\alpha|0\rangle + \beta|1\rangle \rightarrow \alpha|0\rangle - \beta|1\rangle,$$

$$\alpha|0\rangle + \beta|1\rangle \rightarrow \alpha|1\rangle - \beta|0\rangle.$$

The first two are called bit flip and phase flip, respectively, and the third is a combination of them. Alternatively they may be called X, Z and Y classes of errors.

Shor's original scheme required a total of nine qubits for each qubit of information. This is not the minimum for coping with a single error; Steane's scheme used only seven, and in fact the number can be reduced to five [521]. We shall sketch Shor's original scheme, though, which is perhaps the simplest [455].

First we will see how to correct for bit flips, which requires only three qubits, the one carrying the information and an ancilla containing two more. Initially the qubits of the ancilla are in state $|0\rangle$. In an encoder, two controlled-NOT gates are used to entangle each ancilla qubit with the information qubit. The ancilla qubits are flipped if but only if the information qubit is $|0\rangle$, so, if the state of the information qubit is $\alpha|0\rangle + \beta|1\rangle$, the state produced for the combined system is $\alpha|000\rangle + \beta|111\rangle$. This state is entangled, as will be expected from controlled-NOT gates.

At this stage we do not actually do any computing; we just allow the qubits to propagate, and we study errors that may occur and how they may be corrected. The correction procedure will be to pass the qubits through a decoder, which, just like the encoder, consists of two controlled-NOT gates. Again each ancilla qubit flips if the information qubit is $|1\rangle$ but not otherwise.

If no errors have occurred, the decoder will change $\alpha|000\rangle + \beta|111\rangle$ back to $\alpha|000\rangle + \beta|100\rangle$, which is separable; the states of the information qubit may be written as $\alpha|0\rangle + \beta|1\rangle$, and the ancilla qubits are both $|0\rangle$. The decoder has precisely undone the effect of the encoder.

If the second qubit has experienced a bit flip during the transmission process, the state of the qubits will have become $\alpha|010\rangle + \beta|101\rangle$, which will be decoded as $\alpha|010\rangle + \beta|110\rangle$. Again this is separable; the states of the three qubits may be written as $\alpha|0\rangle + \beta|1\rangle$, $|1\rangle$ and $|0\rangle$, respectively. If it is the third qubit that has experienced a bit flip, the situation is directly analogous to the previous one.

If it is the first or information qubit that has undergone a bit flip, though, things are different. The state of the qubits becomes $\alpha|100\rangle + \beta|011\rangle$, which is decoded as $\alpha|111\rangle + \beta|011\rangle$. Again the system is separable; the first qubit is in state $\alpha|1\rangle + \beta|0\rangle$, and the other qubits are in state $|1\rangle$.

I have stressed that in all cases the state of the system is not entangled, because the next state is to measure the output from the ancillary qubits. If there were entanglement between the qubits of the ancilla and the information qubit, a measurement of the first would affect the second, which is exactly what the ancilla is there to avoid.

We focus our attention, though, on the state of the information qubit. This is exactly the same as before the transmission of the qubits for all cases except the last, for which there has been a bit flip from $\alpha|0\rangle + \beta|1\rangle$ to $\alpha|1\rangle + \beta|0\rangle$. We have seen that this case is the only one for which both qubits of the ancilla have ended up as $|0\rangle$.

The course of action is clear. The state of the ancilla qubits is measured; if both are $|0\rangle$, then a bit flip is applied to the information qubit. This means that in all cases the final state of the information qubit is $\alpha|0\rangle + \beta|1\rangle$, exactly as we require. This has been achieved without measuring the information qubit itself, which would be disastrous, since it would be collapsed either to $|0\rangle$ or to $|1\rangle$.

As we have said, this procedure corrects only for bit flips, not for phase flips. However, Shor was able to show how the latter could be corrected for. The technique is very closely related to the one just described; the only change is that a Hadamard gate is added to each output of the encoder and each input of the decoder. The laws of quantum mechanics then mean that phase flips are corrected, three qubits again being required. In Shor's general scheme, both corrections are applied and nine qubits are required in all.

I have described this scheme in some detail to give a flavour of how quantum error correction actually works, admittedly in a very simple case. We have, of course, only scratched the surface of the topic, and there are massive questions remaining. For a start we have considered only errors occurring in the transmission of information, not in the computation process itself. The latter may be handled in the area called *fault-tolerant computation* [501].

The codes discussed, using between five and nine qubits, correct for only a single error. In any workable quantum computer the rate of errors is almost certain to be such that we must correct for many more errors. We may indeed do so, but the number of qubits required will be considerably larger – 23 to correct for three simultaneous errors, 55 to correct for five, and 87 to correct for seven [455].

As well as massively increasing the scale of the computer and the computational effort required, we need to be sure that it is worthwhile. After all, all the additional gates added to help correct for errors will introduce their own. It might be imagined that they might only make matters worse. However, it can be shown that, provided that the error rate for each individual component is low enough, something like 10^{-5}, error correction can in theory be carried out successfully and we thus retain the possibility of the construction of quantum computers genuinely out-performing classical computers in specific areas.

This does not mean that their construction will be easy! With error correction, the total number of operations required might increase by a factor of thousands. The crucial gain, though, is that the probability of decoherence which may be tolerated every time a gate is used will increase from 10^{-13}, which may be shown to be impossible for ever on simple physical grounds, to that factor of 10^{-5}, which number Steane [454] describes as 'daunting but possible'.

To conclude this section we will briefly mention another possible way of controlling decoherence. In the previous chapter we met the quantum Zeno effect, which has the property of inhibiting, or in the limit suppressing, certain forms of evolution of a quantum state. It is natural to examine whether the effect could be used to suppress precisely the decoherence that threatens the working of the quantum computer, and many studies of how this might be achieved have been performed [522, 523].

Creating the quantum computer: general criteria

As we have seen, creating the quantum computer that will really demonstrate the power of quantum computation by factorising large numbers is an immense task. There is, though, a first stage in which quantum computers may perform the vastly easier though still extremely demanding tasks of demonstrating Deutsch's algorithm, factorising 15, which is the smallest number that is applicable for Shor's algorithm, or using Grover's algorithm for small numbers of items. This stage requires only two or three qubits and perhaps a few hundred gates.

Work has been performed using a very large number of techniques, including nuclear magnetic resonance, the ion trap, optical photons, cavity quantum electrodynamics, polymers, superconductors, quantum dots and neutral atoms in an optical lattice. In this section we shall concentrate on the first two on the list. A reasonable amount of success has been achieved in carrying out the first stage. The problem in moving even to address the second stage is that in general the techniques most useful for the first stage are not the ones most suitable for extension to a large number of qubits and gates, *scaling up* as it is called. Consideration is being given to this stage, although its realisation will be well in the future. (The titles of successive books edited by Samuel Braunstein [524, 525] are quite informative.)

Before looking at individual techniques, we shall review a set of five basic criteria for an effective quantum computer, which have been listed by David DiVincenzo [526] and have become extremely well known.

The first requirement is the natural one for a scalable physical system with well-characterised qubits. The second is for the ability to set the qubits to $|0\rangle$. This is obviously required at the beginning of any computation, but quantum error correction also needs a continuous supply of such qubits. This may be achieved for some implementations by cooling, or by measuring and projecting into various states, all of which apart from $|0\rangle$ must then be removed.

The third criterion is that the decoherence time must be much longer than the gate operation time. The discussion of the previous section suggests that by 'much longer', we should mean something like by a factor of 10^5. Note that, were it not for quantum error correction, this criterion would have to refer to the total computation time rather than the gate operation time.

The fourth criterion is for a universal set of gates. This is easy to demand, but much more difficult to supply! Although the number of gates actually required is small, just single-qubit gates and the controlled-NOT double-qubit gate, one still requires interactions to be switched on and off at precise times. In practice, if there is an interaction provided by Nature between the objects constituting the qubits, it will not be straightforward to switch it off; if there is no such interaction, a so-called 'bus qubit' is required to interact with each qubit in turn, but would seem certain to be a vehicle for decoherence.

The fifth criterion is the requirement of a capability for measurement of qubits. Real measurements will rarely be 100% efficient, and often practical quantum computation may be achieved with less, perhaps by repeating a computation a number of times.

DiVincenzo asks which technology will be successful in meeting these criteria, but suggests that it is impossible for anybody to answer this question, and that it may be counterproductive even to ask it. Rather physicists should work on as great a variety of schemes as possible, and make progress as opportunities and ideas present themselves.

We proceed to outline two of the most successful of the many schemes so far studied, nuclear magnetic resonance and the ion trap.

Creating the quantum computer: nuclear magnetic resonance

Nuclear magnetic resonance or NMR is one of the techniques for quantum computation [501, 527] with which most progress has been made, although, as we shall see, there are a couple of stings in the tail.

The reason why progress with NMR has been fast is that many of the manipulations of the spin-state required for quantum computation are the same as those which have been carried out routinely in NMR for many years.

The NMR technique was invented by Felix Bloch and Edward Purcell in 1945, using the experience physicists had gained with radio waves in the recently finished war. The technique uses the fact that some nuclei, just like the electron, have spin angular momenta and associated magnetic moments. In fact any neutron or proton has a spin of $\hbar/2$, exactly the same as an electron. However, in a pair of protons or a pair of neutrons the directions of the two spins are opposite, so the total spin is zero. So nuclei with even numbers of protons and of neutrons have no total spin, but all other nuclei have a non-zero total spin, which is an integer multiple of $\hbar/2$, and thus they also have a magnetic moment.

(The basic unit of nuclear magnetic moment, the nuclear magneton, is smaller than the basic unit of electron magnetic moment that we met in Chapter 4, the Bohr magneton, by a factor equal to the ratio of the proton mass to the electron mass.)

In NMR a strong magnetic field is applied. We may describe the behaviour of the nuclear system in two ways, one broadly classical, the other quantum-mechanical, and the two are intimately connected. The classical way is to say that the nuclear spins precess about the direction of the magnetic field, rather like a top spinning fast about its own axis, but with its axis also rotating slowly about a direction fixed in space. This precession is at a frequency, f, which is characteristic of the particular isotope of the particular atomic species. Frequencies for different nuclei lie in the radio-frequency range of the electromagnetic spectrum.

For the quantum-mechanical description, let us focus on a nuclear species with spin $\hbar/2$. An important example is a nucleus of (ordinary) hydrogen, which contains of course just a single proton and no neutrons. This will have two energy-levels, corresponding to the two orientations of the nuclear spin, and the energy difference between these levels is just the product of Planck's constant, h, and f, the precession frequency we met in the previous paragraph. The most elementary quantum theory then tells us that spins will make transitions between the two levels, and energy will be absorbed, if the frequency of the exciting radiation is exactly f.

Boltzmann's distribution, which we met in Chapters 2 and 4, tells us that, if we have a large number of nuclei, there will be fewer in the upper energy-level than in the lower energy-level. More nuclei will have their spins and hence their magnetic moments pointing in the direction of the field than have them pointing in the opposite direction. This means that the system will have a net magnetic moment in the direction of the applied magnetic field.

But it is essential for the way NMR quantum computation is forced to work that, because the energy difference is extremely small, the imbalance is also very small. The difference between the numbers of spins pointing in the two directions is tiny, perhaps one in a million of the total number of spins present.

It is clear that an obvious use of NMR is to detect in a sample the presence of a particular species of nucleus. If the sample is exposed to a range of frequencies of radio waves, energy will be absorbed at the characteristic frequencies of the various species of nuclei present. The word 'resonance' refers to a very wide range of processes throughout physics where an applied signal that may vary in frequency causes an extremely large output when its frequency is at or around a particular frequency characteristic of the system, known as the resonance frequency.

However, NMR can achieve far more than this. So far we have been thinking about the frequency of an isolated nucleus. In practice a particular nucleus is in a magnetic field, which is the sum of the large applied field and small additional fields produced by the magnetic moments of its neighbours. Thus nuclei of the same species but in different chemical environments are in slightly different magnetic fields, and thus have slightly different characteristic frequencies.

A standard example is ethyl alcohol. The chemical formula for this is CH_3CH_2OH, and there are three types of hydrogen differently sited chemically, those in the CH_3, those in the CH_2 and that in the hydroxyl group OH. When the frequency distribution of absorption of energy is studied, there are indeed three peaks at slightly different frequencies and with strengths in the ratio $3:2:1$.

This approach is the basis of the use of NMR in medicine, for which it is re-christened MRI – magnetic resonance imaging – to avoid frightening users with the dreaded N-word. In MRI, the applied magnetic field is arranged to vary across the human body, and the signal produced gives detailed information on the chemical environments of protons through the body, including any important aspects of the local biochemistry.

We now turn to quantum computing, in which a hydrogen nucleus – a spin-$\frac{1}{2}$ particle – is used as a qubit; this means that a molecule with two hydrogen atoms may represent two qubits. Many of the operations required for quantum computation may be investigated using a conventional NMR spectrometer.

Fig. 9.6. (a) Energies of the states of the two spins without and with coupling. (b) Resonance frequencies of the two spins without and with coupling. The dotted transition corresponds to the controlled-NOT gate (From [506]).

To understand this, let us look at the energy-level diagram for the pair of spins. Figure 9.6(a) shows the energies in the absence of dipole–dipole interaction. The spins are assumed to be in different chemical environments but to behave independently. Thus the energy required in order to jump from $|00\rangle$ to $|10\rangle$ and that to jump from $|01\rangle$ to $|11\rangle$ are the same, hf_1, and relate only to the first spin. Similarly the energies required in order to jump from $|00$ to $|01\rangle$ and from $|10\rangle$ to $|11\rangle$ are the same, hf_2, and relate only to the second spin. (The transitions marked in Fig. 9.6(a) show that the only transitions likely to occur are those for which only one spin changes its state.)

Figure 9.6(b) shows the energies with the interaction between spins included. This interaction gives a positive contribution to the energy if the spins are in the same direction, that is to say for $|11\rangle$ and $|00\rangle$, but an equal and opposite negative contribution if the spins are in opposite directions, for $|10\rangle$ and $|01\rangle$. It is convenient to call this energy $hf_0/2$, so the energies of the transitions become $h(f_1 + f_0)$, $h(f_1 - f_0)$, $h(f_2 + f_0)$ and $h(f_2 - f_0)$. To obtain the frequencies for the transitions, one just removes the h, and, except if there should be a coincidence, all the frequencies will be different.

Now imagine radiation at one of these frequencies, for example $f_2 + f_0$, applied to the system. The only type of transition it will cause is that from $|10\rangle$ to $|11\rangle$. In the language of quantum computation, this is precisely a controlled-NOT gate; spin 2 changes from $|0\rangle$ to $|1\rangle$ only if spin 1 is in state $|0\rangle$. DiVincenzo's fourth criterion, availability of quantum gates, is extremely well met by NMR.

However, there is a major problem with quantum computation via NMR. It is totally impossible to prepare or measure an individual set of qubits. Instead an entire macroscopic system, liquid or perhaps solid, is used, consisting of perhaps 10^{20} molecules. Preparation just requires that, as we have already seen, at any particular temperature there is a surplus of spins in the lower energy level. Qubits are easy to obtain, as already explained, and readout, again from the whole macroscopic system, is also reasonably straightforward. There is also a long coherence time, of the order of a second, because the system is quite self-contained, with only weak links to the outside. However, the typical gate time is also long, in the millisecond range, so the number of gates is limited to about 100.

Overall, NMR quantum computation performs well on the DiVincenzo criteria, and this is why the most progress has probably been made with this technique. Probably the greatest achievement has been that of Isaac Chuang's group [528, 529] in factorising the number 15 using a quantum computer. The work required about 300 independently controlled radio-frequency pulses to manipulate seven qubits. A recent general account of the technique has been given by Jonathan Jones [530].

Now for the difficulties of quantum computation by NMR. First there is scalability. As the number of qubits increases, more and more quantum states become available for the system, but the number useful for quantum computation stays constant. The signal strength therefore decreases as 2^{-n}, where n is the number of qubits. How far the technique could go is questionable, but, if we are optimistic, perhaps to 50 qubits, which is just moving into the region where quantum computers can out-perform classical ones.

The second problem at the moment is the suggestion [531] that the computation carried out in these experiments might not be quantum-mechanical at all. Just because NMR quantum computation works with very small differences between the populations of the various energy-levels, it may be argued that the state of the system can be decomposed into separable states. In other words, the experiments may be described without using the idea of entanglement, and the states may be modelled classically, although the classical models may be rather contrived [532].

Taking into account all the points, both positive and negative, it seems that, while NMR quantum computation may play an extremely useful role in clarifying some of the ways forward, examining the difficulties encountered, and discovering how they may be overcome, it will not be the technique that produces a really useful quantum computer.

Creating the quantum computer: the ion trap

We met the ion trap in the previous chapter while studying the quantum Zeno effect. This device has been used with great success to study individual ions exposed to static or oscillating electric fields [533, 534, 148], and it is very natural that the same technique should be the basis of one of the most successful approaches to practical quantum computation.

In a sense, the very existence of the ion trap confounded Bohr and believers in the Copenhagen orthodoxy. It was an article of belief for these people that it would never be possible to study individual atomic systems, and of course, in the 1920s and 1930s, this was very much the experimental fact. Thus awkward questions about how individual systems might behave, at a measurement for example, could be avoided, and attention could be directed towards the expectation values of physical observables and the relatively benign properties of ensembles.

However, from the 1950s on, physicists developed techniques to trap single ions. One might imagine that it would be possible to produce an electric-field configuration with a global minimum at a particular point where the ion might sit. A very old result called Earnshaw's theorem says that this is not possible, and most ion traps use an electric potential of quadrupolar geometry, supplemented with a magnetic field or a radio-frequency potential [533, 534].

As experimental sophistication has developed, and in almost deliberate defiance of Bohr, it has become possible to trap a single ion for periods of weeks; some researchers become so attached to a particular ion that they give it a pet name! This experimental success resulted in Hans Dehmelt and Wolfgang Paul being awarded the Nobel Prize for Physics in 1989.

A component of this development of enormous importance has been that of laser cooling. An ion when initially trapped will have a temperature of a few thousand degrees Kelvin. Cooling it means that the possibility of its escaping is much diminished, and also its properties become easier to determine in the absence of thermal motion. Laser cooling of ions (and atoms) works through light acting mechanically on the ion or atom. A clever arrangement of beams results in the interaction always reducing its speed. This technique may bring the temperature of an ion down to about 10^{-4} K.

Laser cooling is the first stage in one of the most dramatic achievements in physics over recent years, the production of Bose–Einstein condensation. In Chapter 4, we met Bose–Einstein statistics. A strange prediction of these statistics is that, at low enough temperatures, nearly all the particles in a system of bosons will be in the ground-state. This state of affairs was achieved in dilute gases of alkali-metal atoms in 1997 [535]. Actually, a further reduction of the temperature to about 10^{-7} K was required, and this was arranged by forced evaporation, rather as a cup of tea will cool as it evaporates, since it loses the faster molecules preferentially.

Further Nobel Prizes were awarded: to Steven Chu, Claude Cohen-Tannoudji and William Phillips for laser cooling in 1997; and to Eric Cornell, Wolfgang Ketterle and Carl Wieman for achieving the Bose–Einstein condensation in 2001.

Here we are concerned with use of the ion trap and laser cooling as an implementation of quantum computation. In this method, ions are confined in a linear trap at a pressure of about 10^{-8} Pa. The ions are separated by 2×10^{-5} m by their electrostatic repulsion. Thus the ions are lined up rather like the elements in a highly technologically advanced abacus.

A laser beam is split by beam-splitters and acousto-optic modulators into several beam-pairs, and one beam-pair is used to illuminate each ion. For each ion, two states, $|g\rangle$ and $|e\rangle$ (for ground- and excited states), are used, so each ion represents a single qubit. Each laser beam-pair drives transitions between the two states, thus allowing a single-qubit gate to be applied to any ion, using transitions such as $|g\rangle \rightarrow |e\rangle$ and $|e\rangle \rightarrow |g\rangle$.

Also required is the two-qubit controlled-NOT gate, which is much less straightforward to implement. In fact it was the ingenious proposal [536] of Juan Cirac and Peter Zoller in 1995 of how to achieve this that really kick-started the development of the ion-trap method of quantum computation. Indeed, one may go further than this and suggest that, like the publication of Shor's algorithm, this paper was actually one of the main stimulations for the intense generation of interest in quantum computation itself around 1994–5.

In this scheme, the laser performs a second task: it puts the string of ions into motion. To be precise, it puts a single quantum of vibrational or sound energy into the string. (Physicists would say that it puts a single *phonon* into the string, a phonon being just a single quantum of sound energy, broadly equivalent to a photon, a single quantum of light energy.) This phonon allows interaction between a particular ion and its nearest neighbour, which is required for the controlled-NOT gate.

The scheme consists of four operations: (i) a 180° pulse on ion m; (ii) a 360° pulse on its nearest neighbour, ion n; (iii) a second 180° pulse on ion m; and (iv) a 90° pulse on ion n. In stage (i), a phonon will be excited if ion m is initially in state $|e\rangle$, but in any case it will disappear in stage (iii). The only change between the beginning of the process and its end is that, for ion n, states $|g\rangle$ and $|e\rangle$ will be interchanged if but only if ion m starts in state $|e\rangle$. This is exactly what is required for a controlled-NOT gate.

Progress on implementing the ion-trap quantum computer has been slow but steady. Much of it has been made by David Wineland and his group at Boulder, Colorado. As early as 1995, they [537] were able to implement a controlled-NOT gate, though the two quantum bits were stored in the internal and external degrees of freedom of a single trapped ion. They were also able to produce the required phonon mode for two trapped $^9Be^+$ ions [538], and to produce entangled states of two trapped ions in a deterministic fashion [539]. They referred to the latter achievement as providing entangled states *on demand*. More recently, Rainer Blatt's group in Innsbruck [540] reported the realisation of a Cirac–Zoller controlled-NOT gate implemented with two $^{40}Ca^+$ ions.

While the basic requirements for a quantum computer have been demonstrated, the question of scalability is extremely important. It would seem possible to reach perhaps ten qubits in a straightforward way, but there are experimental and theoretical obstacles to going much further. The Boulder group [541–544] has developed the idea of linking together a number of small ion-trap systems by quantum communication. With massive parallel gate operation between these individual systems, there seems at least a possibility of creating a scalable quantum computer.

Quantum cryptography

In our discussion of Shor's algorithm, the basic ideas of quantum cryptography were sketched. We learned that, if a quantum computer were to be built and Shor's algorithm implemented, public-key cryptography would be undermined, since it relies on the effective impossibility of factorising large numbers.

By an uncanny coincidence, however, quantum information theory also provides the possibility of solving the major problems of *private*-key cryptography. It will be remembered that private-key cryptography is unassailable in principle, but suffers from the fact that key distribution, while straightforward in theory, in practice presents massive administrative difficulties. Quantum theory is able to remove these difficulties; indeed, although the area is traditionally referred to as quantum cryptography, really it is just quantum key distribution [545–548].

The general ideas were introduced principally by Bennett and Giles Brassard and collaborators [549, 550] in the early 1980s, though Stephen Wiesner had invented similar ideas as early as 1970, but found it impossible to publish them. The most well-known variant is the so-called BB84 protocol introduced by Bennett and Brassard [550].

To explain what is involved, I shall briefly review some of the basic concepts of quantum measurement in the appropriate context of polarised light. We may imagine an analyser of polarised light oriented so that it may distinguish between photons with vertical and horizontal polarisations. If a stream of photons, each polarised either in the vertical or in the horizontal direction, impinges on the analyser, an accurate record of the polarisation may be produced for each case.

However, let us now consider the case in which a photon with direction of polarisation bisecting the horizontal and vertical is incident on the analyser. There are in fact two possible directions, which we may call $+45°$ and $-45°$. For such a photon, the analyser will record either a horizontal or a vertical polarisation at random; for a stream of such photons, the numbers of horizontal and vertical registrations will be approximately equal.

Of course we may alternatively consider an analyser that can distinguish between photons polarised in the $+45°$ and $-45°$ directions, but this will not be able to register appropriately photons polarised vertically or horizontally.

In the BB84 protocol, Alice and Bob (the universal heroine and hero of quantum communication) are attempting to create a shared set of private keys. They may communicate using a quantum channel such as an optical fibre, but also use a public channel, which may just be a telephone connection.

Alice has four polarisers available, which polarise photons in the horizontal, vertical, $+45°$ and $-45°$ directions respectively. Bob has two analysers, one that can distinguish between vertically and horizontally polarised photons, and a second that distinguishes between photons polarised at $+45°$ and $-45°$.

Alice sends a stream of photons to Bob, in each case using a polariser selected at random and recording her choice. Bob detects each photon using one of his analysers, again selected at random. He also records both his choice of analyser and the result he obtains.

It is clear that, in 50% of the cases, there are matches between Alice and Bob, but in the other 50% mismatches. In the latter, Alice may have sent a vertically polarised photon, but Bob has used an analyser distinguishing between the $+45°$ and $-45°$ directions. In such a case, he has a 50% chance of measuring either result, so the result actually obtained is meaningless.

To sort out matches from mismatches and using the public channel, Bob sends Alice a list of the analysers he used in each case; she responds, also on the public channel, with a list of the numbers of the various cases for which polariser and analyser were compatible. For these events, Alice and Bob now share common knowledge in the form of Alice's polariser settings and Bob's results. This may be called the *raw quantum transmission* or RQT, and may be used to generate a shared key, at least provided that no eavesdropper has intervened.

In discussions of quantum communication the eavesdropper is invariably referred to as Eve. It is essential that her presence is detected, because, if Alice and Bob are aware that she has gained even partial knowledge of the key, it must not be used. To check this, Alice and Bob now use the public channel to compare a section of the RQT, a section that must obviously then be discarded.

Let us examine how Eve will actually deal with the signal if she does intercept it on its way to Bob. We shall consider in detail only the most straightforward strategy she might adopt. She will intercept and study each photon with her own analyser, and the direction of this can be selected only at random. Let us consider only the cases for which Alice's and Bob's choices were compatible, so the event has become part of the RQT.

In 50% of these cases, Eve's choice of analyser will be compatible with the choices of Alice and Bob. Her measurement result agrees with Alice's choice of polariser. She must send on to Bob a photon corresponding to this result, which he then analyses in the normal way. His result will agree with Alice's records, and of course also with those of Eve. Eve's actions have had no effect on the RQT.

However, in 50% of the cases, Eve's polariser will be incompatible with the choices of Alice and Bob. Alice may have sent a vertically polarised photon, but Eve uses the analyser that distinguishes between the $+45°$ and $-45°$ directions. Her result will be random. Whichever result she obtains, she must send on to Bob a photon corresponding to the measurement she has made. Since we are assuming that this is an event ending up in the RQT, he is using the analyser that distinguishes between vertical and horizontal, and his result again will be random. There is a 50% probability that his measurement indicates a vertical polarisation, which is what Alice actually sent, but a 50% probability of his measuring a horizontal polarisation, which would be incorrect.

So, overall, in 50% of the cases, Eve makes the right choice of analyser and remains undetected. In 25% of the cases she makes the wrong choice, but by good luck (for her) Bob's reading is unaffected. But in 25% of cases, Bob's result will have been altered by the presence of Eve, and in the checked section of the RQT there will be an error. Thus, if there is an eavesdropper using this strategy,

Alice and Bob will expect that there will be mismatches in about a quarter of the events listed in the RQT. If so, the whole RQT must be abandoned.

There are more sophisticated eavesdropping strategies [551–553], but, for noiseless channels, the eavesdropper will always leave some detectable trace, so, in the absence of such, Alice and Bob know they have a secure key. For channels with noise, Alice and Bob can still assess how much information Eve may have obtained.

Provided that this is not too large, they may then use error correction to remove errors, along the lines of the section earlier in this chapter, and then *privacy amplification*, a series of operations that may cut down Eve's potential knowledge to minuscule proportions, though only at the expense of drastically reducing the size of the usable key. The whole process may be known as *key distillation* [554].

It should be noticed that the possibility of maintaining security depends crucially on the no-cloning theorem. If she had a cloning device, Eve could clone the initial state she intercepts on its way from Bob, and send it directly on to Alice. Thus her presence could never be detected. She could also clone a large number of copies of the original state for her own benefit, and so determine the direction of its polarisation. If, for example, use of the vertical/horizontal polariser on a number of cloned copies always gives the result as horizontal polarisation, but the +45°/−45° polariser gives either answer randomly, it is clear that the original state was polarised in the horizontal direction.

It may have occurred to the reader that, rather unusually for applications of quantum foundations or quantum information theory, no use has been made of EPR or the ideas of Bell. However, a completely different protocol was proposed by Artur Ekert [555] in 1991, and this protocol is based totally on EPR and Bell's theorem.

In this protocol, one of an EPR pair of particles is sent to Alice and the other to Bob. They each make a measurement along the x-, y- or z-axis; their choices are independent and random. Later they separate out those results for the cases where they have made the same choice of measurement, and thus form their key (rather similarly to BB84). The remaining measurements are then used to test Bell's inequality. If violation is found, they may conclude that there has been no eavesdropper, but if the inequality is upheld, they may conclude either that there has been an eavesdropper, or, of course, that hidden variables have been found! Indeed, any eavesdropper is in a sense simulating a hidden variable by collapsing probabilities into actualities.

We now turn to practical implementation of the BB84 scheme. The first successful experiments were carried out by Bennett and Brassard in 1989 [556]. Weak polarised pulses were transmitted through free space over a distance of about 30 cm. The bit rate was slow and system errors were around 4%, but the key distillation was performed efficiently. As Brassard later reported [456], any potential Eve would have had a field day; the devices used to put the photons in different polarisation states gave very different noises depending on the state chosen.

Nevertheless, despite the practical limitations, this was an exceedingly important achievement for quantum information theory. Deutsch commented [456] that this was the first device produced of any type whose capabilities exceeded those of a Turing machine.

This makes the point that, whereas useful implementation of the quantum computer is almost certainly decades away, the requirements of quantum cryptography are comparatively limited. It is comparatively easy to make the scheme work in a laboratory, and the real challenge is to develop it so it works in a practical and challenging environment.

One of the main participating groups in this endeavour has been that of Gisin, which has extended the work on Bell's inequalities described in Chapter 7 in a fairly natural way. In 1995, they [557] generated a quantum key shared by users 23 km apart across Lake Geneva. The connector was standard Swiss Telecom optical fibre, and the standard telecommunication wavelength of 1300 nm was used. The key was encoded in very weak laser pulses, each pulse containing on average only 0.12 photons. This group has continued to explore a variety of approaches [557–559]. Other groups [547, 560] have also succeeded in performing quantum key distribution along many kilometres of optical fibre.

Richard Hughes' group [561, 562] at Los Alamos has been able to demonstrate quantum key distribution in free space over a distance of about 0.5 km in broad daylight. Problems arising from a high background of photons from the Sun and a turbulent atmosphere were overcome by a careful choice of experimental parameters. The atmosphere has a high transmission window for photons around 770 nm and has little effect on the polarisation states of photons, and the periods involved in atmospheric turbulence are sufficiently long that its effect can be compensated for.

This is a very exciting development since it implies that perfectly secure communication between the Earth and a satellite may be possible. Quantum key distribution can also provide secure communication, for example, over distances within the financial area of any big city. This is one area of quantum information theory that is already ripe for exploitation. All that is required is for those responsible to pluck up the courage to embrace the new technology.

Quantum teleportation

Quantum teleportation is the third main branch of today's quantum information theory. The achievement of quantum computation and quantum cryptography (or quantum key distribution) is and will be to take areas of today's classical science and technology, and to show how quantum techniques enable things to be done much more successfully and efficiently. In contrast, the very idea of quantum teleportation makes it clear that it is not a development of classical techniques, or even an entirely new way of doing something that is possible classically, but an entirely new field of study, totally dependent on novel quantum technology.

Teleportation – the process of disassembling a person at one point of space, A, and reconstituting him or her, unchanged in any way, at a remote location, B, without the subject having actually travelled from A to B – is, of course, a staple of science fiction, a highly convenient device for speeding up the action. Quantum teleportation achieves much the same thing, but, at least so far, only for the state of a single elementary particle, such as an electron.

In science fiction, teleportation obeys the no-cloning restriction (at least if there are no mishaps with the technology!). The subject indeed disappears at point A and re-appears at point B. However, it would be common for the process at least to appear to violate the requirements of locality and special relativity. The disappearance and reconstitution may sometimes be simultaneous. More often though, perhaps, one follows immediately from the other. Disappearance at A is followed immediately by the reconstitution at B, despite the fact that A and B may be at opposite ends of the Universe, so the time it might take light to travel from A to B might be hours or days.

Quantum teleportation also obeys the no-cloning theorem, but in addition it obeys special relativity. We saw in Chapter 7 that quantum theory is in conflict with locality, but it retains parameter independence and so does not conflict with relativity; information cannot be sent from A to B at a speed greater than that of light. We shall see that quantum teleportation, like quantum key distribution, requires use of a public classical channel as well as the quantum channel, and this makes it quite obvious that special relativity cannot be threatened.

The method was invented by a group of very well-known quantum information scientists [563] in 1993. In the technique, Alice has control of the state to be teleported, a spin-$\frac{1}{2}$ particle, which we will take to be an electron. She may know the state of the particle, since she may, for example, have seen it emerge from a particular beam of a Stern–Gerlach device with its magnetic field set at a particular angle. But she does not *need* to know the state. What she is teleporting is not her knowledge of the state, but the state itself.

In contrast, although at the end of the process Bob is in possession of an electron whose state is identical to that of Alice's electron before the process begins, he does *not* know this state, and, of course, any attempt on his part to investigate the state would almost certainly change it.

To explain quantum teleportation, we need one further technical element of quantum theory, the so-called Bell states. We have already met what we may call the EPR–Bohm state which we can write as $(1/\sqrt{2})(|+_1 -_2\rangle - |-_1 +_2\rangle)$. I will spell out in detail what this means. $|+\rangle$ and $|-\rangle$ denote the two spin-states of the electron. $|+\rangle$ might alternatively be written as $|s_z = +\frac{1}{2}\rangle$ and $|-\rangle$ as $|s_z = -\frac{1}{2}\rangle$.

The state as written is for two electrons we number as 1 and 2. So the state $|+_1 -_2\rangle$ would imply that electron 1 is in the state $|+\rangle$ and electron 2 in the state $|+\rangle$. Similarly $|-_1 +_2\rangle$ would imply that electron 1 is in state $|-\rangle$ and electron 2 in state $|+\rangle$. Of course the EPR–Bohm state is neither of these, but a superposition of both. It is thus precisely an entangled state, and this was the entire point of the original EPR argument. (The $1/\sqrt{2}$ factor is for normalisation.)

This state is spoken of as being maximally entangled. In general, discussing the amount of entanglement in a complicated system is very difficult. There is no obvious way to define it, though many people have had different though related ideas. But it is clear that, for the single-spin system, this EPR–Bohm state is completely entangled.

However, it is only one of four such maximally entangled states for this system, and these states have been called the Bell states [564]. For convenience we will write it as $|B_{b12}\rangle$ (with B for Bell, b because it is the second one in the usual order of writing, and 1 and 2 for the two particles).

The first one is usually written as $|B_{a12}\rangle$ and is $(1/\sqrt{2})(|+_1 -_2\rangle + |-_1 +_2\rangle)$. Notice that the only difference from $|B_{b12}\rangle$ is that there is a plus sign rather than a minus sign inside the second set of parentheses. Technically this may be referred to as a relative phase factor. It does not really change much about the basic nature of the state, but it is very important in any detailed calculations.

The other Bell states may be written as $|B_{c12}\rangle$ and $|B_{d12}\rangle$, or explicitly as $(1/\sqrt{2})(|+_1 +_2\rangle + |-_1 -_2\rangle)$ and $(1/\sqrt{2})(|+_1 +_2\rangle - |-_1 -_2\rangle)$. These are similar to each other, but rather different from $|B_{a12}\rangle$ and $|B_{b12}\rangle$, because, for them, the two spins will always be in the same state, whether it is $|+\rangle$ or $|-\rangle$, whereas for $|B_{a12}\rangle$ and $|B_{b12}\rangle$ the two spins must be in different states.

We may describe the four Bell states as a *complete* set of maximally entangled states for the two spins, in the sense that any state at all of the spins may be written mathematically as a sum of the four Bell states with particular values of the coefficients multiplying the various states.

Now we come to the actual method of quantum teleportation. The particle whose state is to be teleported is in the charge of Alice. We will call it particle 1. Its state may be completely general and we will call it $|\phi_1\rangle$. Particles 2 and 3 are put into an EPR–Bohm state, which is $|B_{b23}\rangle$, so the initial state of the total system is $|\phi_1 B_{b23}\rangle$. Of this EPR–Bohm pair, particle 2 is sent to Alice and particle 3 to Bob.

Alice therefore has particles 1 and 2, and she elects to make a Bell-state measurement of these two spins. According to the prescription for measurement we have used throughout the book, this means that she projects the states of particles 1 and 2 into one of the four Bell states we listed above, and records which state she has obtained. We will need a little maths to examine what happens. Our initial state $|\phi_1 B_{b23}\rangle$ may be written as

$$|\phi_1 B_{b23}\rangle = \frac{1}{2}\left(-|B_{b12}\phi_3\rangle + |B_{a12}\phi_3'\rangle + |B_{c12}\phi_3''\rangle + |B_{d12}\phi_3'''\rangle\right).$$

I will explain what this means. Let us concentrate first on particles 1 and 2, the particles which Alice is manipulating. The factor outside the parentheses on the right-hand side of the equation is $\frac{1}{2}$, so this tells us that there is a probability of $(\frac{1}{2})^2$ or $\frac{1}{4}$ of Alice ending up with her particles in each of the four Bell states when she performs her measurement. Particles 1 and 2 may end up in state $|B_{b12}\rangle$; the probability of this happening is $\frac{1}{4}$, and, if this happens, Alice will say that she has performed a Bell-state measurement on particles 1 and 2, and they are in state $|B_{b12}\rangle$. The same goes with the same probability for states $|B_{a12}\rangle$, $|B_{c12}\rangle$ and $|B_{d12}\rangle$.

But when Alice's measurement puts particles 1 and 2 into, for example, $|B_{a12}\rangle$, the equation above tells us that she also affects the state of particle 3, although this is in the possession of Bob. This, of course, is exactly what one means by saying that the system is entangled.

If Alice obtains the result $|\phi_{b12}\rangle$, then the equation tells us that particle 3 has gone into state $|\phi_3\rangle$. This is saying that particle 3 is in state $|\phi\rangle$, exactly the state that particle 1 was in before the Bell-state measurement. This looks good for teleportation. (Readers may spot an awkward-looking minus sign in the equation, but again this is only a change of phase and is irrelevant in this context.)

Unfortunately, if Alice obtains any of the other results, Bob's particle 3 is left in different states, $|\phi_3'\rangle$, $|\phi_3''\rangle$, or $|\phi_3'''\rangle$. However, the good news is that each of these is closely related to $|\phi_3\rangle$. In each case, to get back to $|\phi_3\rangle$, Bob has to perform a simple rotation of 180°, about the z-, y- or x-axis, respectively.

But, without further information, Bob does not know which rotation to perform, or whether to leave the particle alone. This is where the classical channel comes in. Using this channel, Alice has to inform Bob which result she obtained in her Bell-state measurement. Since her results are in one-to-one correspondence with the different actions he must take, this immediately gives him the information he needs. He performs the required rotation, or if appropriate just does nothing, and is left with particle 3 in state $|\phi_3\rangle$, exactly the state in which particle 1 was at the beginning of the process. There is no cloning, of course, because Alice's particle is no longer in this state; particles 1 and 2 are left in the Bell state into which her measurement put them.

The quantum teleportation scheme is complicated, but, within five years, several groups, those of Zeilinger [203], Francesco De Martini [565] and Jeff Kimble [566], had performed experiments that they all claimed had demonstrated the technique.

Bouwmeester and colleagues [203] in Zeilinger's group used polarisation states of photons. The experiment used many of the same techniques as in the same group's study of GHZ systems described in Chapter 7. The method requires production and measurement of entangled states. Production is by parametric down-conversion giving photons 2 and 3, and the same pulse creates another pair, one of which is photon 1, while the other is a trigger indicating that a photon to be teleported is on its way.

Photons 1 and 2 are to be projected into a Bell state. Their individual times of arrival within a given pulse are sufficiently imprecise that Alice cannot tell which is which, so technically they become indistinguishable, which is an essential aspect of the method. These photons are superposed at a beam-splitter and proceed to detectors. In 25% of the cases, two different detectors, f_1 and f_2, register, indicating that state $|B_{b12}\rangle$ has been formed. (This is the only state utilised in the investigation under discussion; extra polarisers would be required to utilise $|B_{a12}\rangle$, while, for $|B_{c12}\rangle$ and $|B_{d12}\rangle$, quantum logic gates would be needed.)

In the experiment, the direction of polarisation of photon 1 was $+45°$, and this is to be teleported. Photon 3 may be investigated by a polarising beam-splitter, followed by detection in detector d_2 if its direction of polarisation is $+45°$, in detector d_1 if it is $-45°$. That quantum teleportation has occurred is indicated by the fact that when f_1 and f_2 register, the third photon is detected at d_2, never at d_1.

In the experiment of [565], in contrast, only two photons are used. While this departs from the prescribed scheme in detail, it makes the experiment simpler, and also makes it possible to use all four Bell states. The EPR state is realised by entanglement of path rather than of polarisation, and the two measurements required are on the path and the polarisation of the same photon.

In [566], optical coherent states are teleported. A higher fidelity is reported than is possible without the use of entanglement. The authors describe this as the first realisation of quantum teleportation where every state entering the device is teleported.

Over more recent years much further work has been performed; here I mention only some recent work performed by Gisin's group [567], which is on the same lines as the work of this group on Bell's inequalities and quantum cryptography which I have discussed already. In these experiments, there is a demonstration of quantum teleportation between laboratories separated by 55 m physically, but by 2 km of standard telecommunications fibre. A possible application is to help to extend quantum cryptography to larger distances.

Should useful quantum computers ever be built, quantum teleportation will be an essential component, since it will convey information from one component to another. It is fascinating to ask whether the technique could ever be extended to much larger objects, even to human beings. Sam Braunstein, however, has estimated that it would take a hundred million centuries to transmit the amount of information required to teleport a human being along a single channel [456].

10

Bohr or Einstein?

In this last chapter I shall attempt a very brief assessment of the Bohr–Einstein debate. Who won it in the view of the audience? Who actually had the better case? I will also ask two rather more specific questions. The first is 'Have the contributions of John Bell favoured the position of either of the protagonists?' The second is 'Has the dramatic arrival and progress of quantum information theory anything to tell us about the Bohr–Einstein debate?'

To help to answer these questions, let me first re-assemble a few points, putting together in turn an argument *for* each of the protagonists *against* a rather hostile critic. First Bohr, and I take as his critic John Bell himself. Bell was prepared, in fact, to give Bohr great credit for what Bell called the *pragmatic* approach to quantum theory, which I discussed earlier. This approach says that the world must be divided into 'classical' and 'quantum' parts, with an arbitrarily placed cut between them; that the macroscopic measuring device must be described in classical terms, but that we should not expect to picture the 'quantum' system in physical terms, and should be content just to possess rules of calculation that work well. But Bell regarded Bohr's philosophy of what lay behind pragmatism, *complementarity*, as ill-defined, unsatisfactory and bizarre.

To reply, on Bohr's behalf, to Bell, I would argue that complementarity is not necessarily any of these things. Regarded as a practical means of regulating discussion so that any *conceivable* experiment could be discussed, but conceptual contradictions and paradoxes avoided, it was reasonably straightforward and very successful. The first two rounds of the Bohr–Einstein debate demonstrated just how powerful Bohr's ideas were, and how beautifully his abstract and conceptual arguments matched detailed analysis of individual thought-experiments.

I would add another point in support of Bohr. I suspect that Bell did not realise how completely the questions he himself found of such interest and

413

concern stemmed from Bohr. At a time when, with very few exceptions, physicists were following Heisenberg in taking the view that finding correct mathematical answers was all-important, that no strictly physical discussion was possible, necessary, or even desirable, it was Bohr who insisted that the connection between unanalysable quantum system and permanent experimental results *must* be addressed. Bell may not have liked Bohr's answers, but he could have given him more credit for recognising the importance of the questions.

Indeed, one could perhaps imagine a different Bohr, for whom complementarity was immensely useful, but not a philosophy to be raised practically to the status of an unchallengeable dogma; for whom the divide between classical and quantum was doubtless necessary, at least for the time being, but was not a matter of principle, and was, in the final analysis, open to deeper explanation; a Bohr who could have genuinely welcomed theories such as those of Bohm and GRW as helping to clarify his own ideas. It would be difficult to criticise this other Bohr at all!

I'm even tempted to suggest that the rigidity of the Bohr position was more the responsibility of von Neumann and Rosenfeld than of Bohr himself . . . But that would not be correct. Bohr supported von Neumann whole-heartedly, and I don't really think he would have clashed with Rosenfeld. For all Bohr's great achievements, which I have tried to highlight in this book, I have to admit that, as a result of the dominance of his own arguments, others were stifled, that rather dubious positions on EPR and the quantum/classical divide were not effectively challenged for a considerable period, and that, as a result, science was harmed.

Now I turn to Einstein, and criticisms of his ideas on quantum theory by Pais [3]. While Pais was in general a very great admirer of Einstein, he comments that 'Bohr . . . could have said: "Einstein is a great man, I am fond of him, but in regard to quantum theory he is out to lunch".' Of EPR, Pais says that 'There are a few who opine that Einstein concluded with a bang, while most consider it a whimper.'

The position adopted, and, I hope, demonstrated, in this book has been that, on the contrary, Einstein's arguments in EPR were audacious, lucid and effective. They left Bohr's position immeasurably weaker, and they brought such notions as *realism* and *locality* to centre-stage. What is disappointing is that Einstein's *positive* achievement was so insubstantial. The fact that his own work in the years after 1935 was not fruitful was scarcely likely to encourage others to take much interest in his challenge to Bohr. It was particularly disappointing that he could see no merit at all in the hidden-variable model of Bohm.

So to our questions. Who convinced the community of physicists? Obviously in any timescale up to 50 years, the answer must be Bohr. This was important because it did allow physicists to achieve enormous advances in applications of quantum theory, secure in the knowledge, or at least belief, that it had a firm logical under-pinning.

More recently those physicists interested in the foundations of quantum theory have paid increasing attention to aspects of Einstein's work: to EPR, to discussion of determinism, locality and realism. Indeed, Bell described himself as a follower of Einstein, and Selleri and his fellow 'realists' may base their programme, at least in part, on Einstein's views, but very few would be interested in Einstein's own work on the unified field theory. The only way in which one might convince oneself that Einstein had really captured his audience would be by taking the view that, after several decades of Bohr's total dominance, *any* questioning of that point of view constitutes success for Einstein.

Who actually won the debate – in strictly logical terms? During the first two rounds of the debate, Bohr showed brilliantly the very considerable strength of complementarity, but his and its lack of flexibility put him on thin ground in tackling EPR. Of course even Einstein admitted that Bohr's position on EPR was logically tenable given his assumptions (though these assumptions, particularly wholeness, in the form of denial of independent existence, were not likely to appeal to most who studied his position seriously). But the quantum/classical dichotomy was, in a sense, and on a longer timescale, a more fundamental difficulty. However useful it was, it could not really be defended from a logical point of view.

Bohr's position was considerably helped by the experimental limitations of his own time. Since then, two major experimental advances have been made. The first is to experiments on individual systems, using, for example, the ion trap and the neutron interferometer. The suggestion here is not so much that Bohr's position will be found to be wrong, but that it will seem more and more unconvincing. The second advance is to working with macroscopic systems. Whatever the results of the experiments based on Leggett's work, one must surely suspect that, *in the end*, the gap between microscopic and macroscopic will be bridged, in defiance of Bohr, for whom macroscopic must be logically independent of microscopic. Thus it is difficult to see Bohr as a long-term winner of the debate in a strictly logical sense.

But neither can Einstein be regarded as a winner. At best he put up arguments that severely damaged Bohr's position, and stamped his authority on the debate by re-introducing the ideas of realism and locality, but he never put forward a clear and thorough position that could be considered seriously as winning the argument.

Let us now turn to the work of Bell. Bell started as the follower of Einstein, but ended up popularly known as 'the man who proved Einstein wrong'. Indeed,

Bell's theorem is often taken as vindicating totally Bohr's position. This though is quite wrong. Complementarity was built on the central idea that there could be no hidden variables. Yet the very existence of the Bohm interpretation demonstrated a contradiction of that article of faith, as Bell stressed repeatedly. The fact that *local* hidden variables in quantum theory were ruled out should really have been regarded as scant consolation for Bohr.

Einstein, of course, would have liked to have retained locality *and* realism. There is no doubt that his vision of physics was dealt quite a savage blow by Bell's theorem and the subsequent experimental work. In Chapter 7 I discussed how he might have adjusted his view of the physical world to cope with these new facts.

So from this point of view Bell's work was somewhat destructive for the positions of both Einstein and Bohr. Yet we may think about things from a different point of view, asking not whose views were more correct or less correct, but whose approach and suggestions stimulated and encouraged thoughtful consideration of the problems of quantum theory, the possibility of interesting and constructive ideas, and perhaps eventually important new physics. From this point of view, Einstein's approach must be judged far superior to that of Bohr.

Einstein stimulated the work of Bohm, and Bohm and Einstein both stimulated the work of Bell. Eventually even the most diehard opponent of hidden variables was forced to admit that interesting and important ideas about realism and locality had emerged, ideas that were even experimentally testable.

Even more so, this applies to the rise of quantum information theory. From the strictly logical position, quantum information theory relies solely on quantum theory itself, not on Bohr's or Einstein's or any other approach to it. Yet, in practice, scientific ideas do not arise from a conceptual vacuum. Studies on realism and completeness, the idea of entanglement, all the arguments and analysis of Bell's theorem were, in practice, essential requirements for the emergence of quantum information theory. It is difficult to see how this theory could have been produced without the contributions of EPR, Schrödinger, Bell and others over several decades. It has also been stressed that the experimental advances made in the course of investigating Bell's inequalities were central in moving on to implementation of quantum information theory.

Similarly, one does not have to be a strong advocate of many worlds to appreciate that the position of David Deutsch may have had a constructive role in his appreciation of the importance of quantum computation.

I have said that, at least at one level and in the short term, the general espousing of Bohr's position in the 1920s and 1930s was pragmatically useful, for it freed physicists to study the many applications of quantum theory. At a higher level, though, and in a rather longer term, it was anything but the most useful approach. It denied those who might have been interested the opportunity to think deeply, to progress further in understanding the theory, and just possibly to reach important new conclusions.

One last question – for whose work on the foundations of quantum theory should we be more grateful – Bohr or Einstein? If one restricts oneself to the scientific content, one might rate the honours even. Bohr certainly introduced many important concepts that will always remain important in thinking about quantum theory. But the steady critique of Bohr's position by Einstein also brought out essential elements of the theory.

It is when we move outside the strictly scientific area that the Bohrian perspective seems, as least in retrospect, badly flawed, and we may take from this an important general message. It can never be acceptable in science to stifle thought and inhibit discussion, as the advocates of Copenhagen certainly did. Einstein of all people deserved better from the scientific community than the treatment he actually received. As science moves on we should not make the same mistake again.

References

[1] A. Pais, 'Subtle Is the Lord . . . ': The Science and the Life of Albert Einstein (Clarendon, Oxford, 1982)

[2] A. Fölsing, Albert Einstein: A Biography (Viking, London, 1997)

[3] A. Pais, Niels Bohr's Times, in Physics, Philosophy and Polity (Clarendon, Oxford, 1991)

[4] N. Blaedel, Harmony and Unity: The Life of Niels Bohr (Springer-Verlag, Berlin, 1988)

[5] M. Born (ed.), The Born–Einstein Letters (Macmillan, London, 2005)

[6] W. Moore, Schrödinger: Life and Thought (Cambridge University Press, Cambridge, 1989)

[7] M. Gell-Mann, The Nature of the Physical World (Wiley, New York, 1979)

[8] T. S. Kuhn, The Structure of Scientific Revolutions (Chicago University Press, Chicago, 1962)

[9] T. S. Kuhn, The Copernican Revolution (Harvard University Press, Cambridge, Massachusetts, 1957)

[10] J. J. Langford, Galileo, Science and the Church (University of Michigan Press, Ann Arbor, 1966)

[11] J. Renn (ed.), Galileo in Context (Cambridge University Press, Cambridge, 2001)

[12] A. R. Hall, Isaac Newton, Adventurer in Thought (Blackwell, Oxford, 1992)

[13] P. Fara, Newton: The Making of a Genius (Macmillan, London, 2002)

[14] A. J. Berry, Henry Cavendish: His Life and Scientific Work (Hutchinson, London, 1960)

[15] M. Berry, Principles of Cosmology and Gravitation (Institute of Physics, Bristol, 1989)

[16] E. J. Squires, Conscious Mind in the Physical World (Adam Hilger, Bristol, 1990)

[17] I. Stewart, Does God Play Dice?: The Mathematics of Chaos (Blackwell, Oxford, 1989)

[18] J. Gribbin, Deep Simplicity: Chaos, Complexity and the Emergence of Life (Allen Lane, London, 2004)

[19] L. P. Williams, Michael Faraday: A Biography (Chapman and Hall, London, 1965)

[20] G. Cantor, Michael Faraday, Sandemanian and Scientist: A Study of Science and Religion in the Nineteenth Century (Macmillan, London, 1991)

[21] C. W. F. Everitt, James Clerk Maxwell (Scribner, New York, 1975)

[22] P. M. Harman, The Natural Philosophy of James Clerk Maxwell (Cambridge University Press, Cambridge, 1998)

[23] S. G. Brush, The Kind of Motion We Call Heat (North-Holland, Amsterdam, 1976)

[24] P. V. Coveney and R. Highfield, *The Arrow of Time* (Allen, London, 1990)

[25] D. Lindley, *Boltzmann's Atom: The Great Debate That Launched a Revolution in Physics* (Free Press, New York, 2001)

[26] A. J. Ihde, *The Development of Modern Chemistry* (Dover, New York, 1984)

[27] A. Pais, *Inward Bound: Of Matter and Forces in the Physical World* (Clarendon, Oxford, 1986)

[28] J. D. Watson, *The Double Helix* (Weidenfeld and Nicolson, London, 1968)

[29] B. R. Wheaton, *The Tiger and the Shark: Empirical Roots of Wave–Particle Dualism* (Cambridge University Press, Cambridge, 1983)

[30] A. Einstein, *The Meaning of Relativity* (Princeton University Press, Princeton, New Jersey, 2004)

[31] E. F. Taylor and J. A. Wheeler, *Spacetime Physics* (Freeman, New York, 2nd edn, 1993)

[32] T. S. Kuhn, *Black-Body Theory and the Quantum Discontinuity 1894–1912* (Clarendon, Oxford, 1978)

[33] M. J. Klein, Max Planck and the beginning of the quantum theory, *Archive for the History of Exact Sciences* **1**, 459–79 (1962)

[34] H. Kangro, *History of Planck's Radiation Law* (Taylor and Francis, London, 1976)

[35] K. Mendelssohn, *The World of Walther Nernst, the Rise and Fall of German Science* (Macmillan, London, 1973)

[36] F. K. Richtmyer, E. H. Kennard and J. N. Cooper, *Introduction to Modern Physics* (McGraw-Hill, New York, 1969)

[37] T. S. Kuhn, J. L. Heilborn, P. L. Forman and L. Allen, *Sources for History of Quantum Physics: An Inventory and Report* (American Philosophical Society, Philadelphia, 1967): http://amphilsoc.org/library/guides/ahqp/

[38] J. L. Heilbron, Rutherford–Bohr atom, *American Journal of Physics* **49**, 223–31 (1981)

[39] M. Jammer, *The Conceptual Development of Quantum Mechanics* (McGraw-Hill, New York, 1966)

[40] A. Sommerfeld, *Atomic Structure and Spectral Lines* (Methuen, London, 1923)

[41] B. L. van der Waerden, *Sources of Quantum Mechanics* (North-Holland, Amsterdam, 1967)

[42] R. H. Stuewer, *The Compton Effect* (Neale Watson, New York, 1975)

[43] J. C. Slater, *Solid State and Molecular Theory: A Scientific Biography* (Wiley, New York, 1975)

[44] D.C. Cassidy, *Uncertainty: The Life and Science of Werner Heisenberg* (Freeman, New York, 1992)

[45] H. Kragh, *Dirac: A Scientific Biography* (Cambridge University Press, Cambridge, 1990)

[46] P. Forman, Weimar culture, causality and quantum theory, *Historical Studies in the Physical Sciences* **3**, 1–116 (1971)

[47] M. Beller, Born's probabilistic interpretation: a case study of 'concepts in flux', *Studies in the History and Philosophy of Science* **21**, 563–88 (1990)

[48] S. Weinberg, *Dreams of a Final Theory: The Search for the Fundamental Laws of Nature* (Vintage, London, 1993)

[49] S. W. Hawking, *A Brief History of Time* (Guild, London, 1990)

[50] L. Hoddeson, E. Braun, J. Teichmann and S. Weart (eds.), *Out of the Crystal Maze: Chapters from the History of Solid-State Physics* (Oxford University Press, Oxford, 1992)

[51] M. Finnis, *Interatomic Forces in Condensed Matter* (Oxford University Press, Oxford, 2003)

[52] S. Körner (ed.), *Observation and Interpretation in the Philosophy of Physics* (Dover, New York, 1962)

[53] P. C. W. Davies and J. R. Brown (eds.), *The Ghost in the Atom* (Cambridge University Press, Cambridge, 1986)

[54] L. I. Schiff, *Quantum Mechanics* (McGraw-Hill, New York, 1949)

[55] H. J. Folse, *The Philosophy of Niels Bohr* (North-Holland, Amsterdam, 1985)

[56] D. Murdoch, *Niels Bohr's Philosophy of Physics* (Cambridge University Press, Cambridge, 1987)

[57] J. Honner, *The Description of Nature: Niels Bohr and the Philosophy of Quantum Physics* (Clarendon, Oxford, 1987)

[58] J. Faye, *Niels Bohr: His Heritage and Legacy. An Anti-Realist View of Quantum Mechanics* (Kluwer, Dordrecht, 1991)

[59] S. Petruccioli, *Atoms, Molecules and Paradoxes: Niels Bohr and the Construction of a New Physics* (Cambridge University Press, Cambridge, 1993)

[60] J. Faye and H. T. Folse (eds.), *Niels Bohr and Contemporary Philosophy* (Kluwer, Dordrecht, 1994)

[61] A. Plotnisky, *Complementarity and Anti-Epistemology after Bohr and Derrida* (Duke University Press, Durham, North Carolina, 1994)

[62] J. Ottaviani and L. Purvis, *Suspended in Language: Niels Bohr's Life, Discoveries and the Century He Shaped* (GT Labs, Ann Arbor, Michigan, 2004)

[63] N. Bohr, *Niels Bohr's Philosophical Writings* (Ox Bow Press, Woodbridge, 4 Vols, 1987–99)

[64] R. G. Colodny (ed.), *Frontiers of Science and Philosophy* (George Allen and Unwin, London, 1964)

[65] M. Beller, The birth of Bohr's complementarity: the context and the dialogues, *Studies in the History and Philosophy of Science* **23**, 147–80 (1992)

[66] J. Kalchar (ed.), *Niels Bohr: Collected Works*, Vol. 6 (North-Holland, Amsterdam, 1985)

[67] M. Jammer, *The Philosophy of Quantum Mechanics* (Wiley, New York, 1974)

[68] N. Bohr, *Atomic Physics and Human Knowledge* (Wiley, New York, 1958)

[69] K. Popper, Quantum mechanics without the observer, in *Quantum Theory and Reality* (M. Bunge, ed.) (Springer, New York, 1967), pp. 7–44.

[70] M. Bunge, The turn of the tide, in *Quantum Theory and Reality* (M. Bunge, ed.) (Springer, New York, 1967), pp. 1–6

[71] J. Baggott, *The Meaning of Quantum Theory* (Oxford University Press, Oxford, 1992)

[72] H. P. Stapp, The Copenhagen interpretation, *American Journal of Physics* **40**, 1098–116 (1972)

[73] P. A. Schilpp (ed.), *Albert Einstein: Philosopher-Scientist* (Library of the Living Philosophers, Evanston, Illinois, 1949)

[74] S. Rozental (ed.), *Niels Bohr: His Life and Work As Seen by Friends and Colleagues* (North-Holland, Amsterdam, 1967)

[75] F. Capra, *The Tao of Physics: An Exploration of the Parallels between Modern Physics and Eastern Mysticism* (Flamingo, London, 1975)

[76] J. S. Bell, *Speakable and Unspeakable in Quantum Mechanics* (Cambridge University Press, Cambridge, 2nd edn, 2004)

[77] J. de Boer, E. Dal and O. Ulfbeck (eds.), *The Lesson of Quantum Theory* (Elsevier, Amsterdam, 1986)

[78] J. Bernstein, *Quantum Profiles* (Princeton University Press, Princeton, New Jersey, 1991)

[79] N. R. Hanson, *Patterns in Discovery* (Cambridge University Press, Cambridge, 1958)

[80] P. K. Feyerabend, *Against Method: Outline of an Anarchistic Theory of Knowledge* (New Left Books, London, 1975)

[81] P. K. Feyerabend, *Science in a Free Society* (New Left Books, London, 1978)

[82] N. R. Hanson, Copenhagen interpretation of quantum theory, *American Journal of Physics* **27**, 1–15 (1959)

[83] P. A. M. Dirac, *The Principles of Quantum Mechanics* (Clarendon, Oxford, 1930)

[84] N. Bohr, *Essays 1958–62 on Atomic Physics and Human Knowledge* (Wiley, New York, 1963)

[85] A. Thwaite (ed.), *Larkin at Sixty* (Faber, London, 1982)

[86] A. Motion, *Philip Larkin: A Writer's Life* (Faber, London, 1993)

[87] J. von Neumann, *Mathematical Foundations of Quantum Mechanics* (Princeton University Press, Princeton, New Jersey, 1955)

[88] L. E. Ballentine, The statistical interpretation of quantum mechanics, *Reviews of Modern Physics* **42**, 358–81 (1970)

[89] L. E. Ballentine, Resource letter IQM-2: foundations of quantum mechanics since the Bell inequalities, *American Journal of Physics* **55**, 785–92 (1987)

[90] N. Macrae, *John von Neumann: The Scientific Genius Who Pioneered the Modern Computer, Game Theory, Nuclear Deterrence and Much More* (American Mathematical Society, Providence, Rhode Island, 1999)

[91] J. A. Wheeler and W. H. Zurek (eds.), *Quantum Theory and Measurement* (Princeton University Press, Princeton, New Jersey, 1983)

[92] L. J. Good (ed.), *The Scientist Speculates – An Anthology of Partly-Baked Ideas* (Heinemann, London, 1961)

[93] J. Mehra, *The Solvay Conferences on Physics* (Reidel, Dordrecht, 1975)

[94] A. Fine, *The Shaky Game* (University of Chicago Press, Chicago, 1986)

[95] R. Guy and R. Deltete, Fine, Einstein and ensembles, *Foundations of Physics* **20**, 943–65 (1990)

[96] A. Fine, Einstein and ensembles: response, *Foundations of Physics* **20**, 967–89 (1990)

[97] D. Home and M. A. B. Whitaker, Ensemble interpretations of quantum mechanics: a modern perspective, *Physics Reports* **210**, 223–317 (1992)

[98] L. E. Ballentine, Einstein's interpretation of quantum mechanics, *American Journal of Physics* **40**, 1763–71 (1972)

[99] A. Einstein, Physics and reality, *Journal of the Franklin Institute* **221**, 349–82 (1936)

[100] E. Schrödinger, Die gegenwärtige Situation in der Quantenmechanik [The present situation in quantum mechanics], *Naturwissenschaften* **23**, 807–12, 823–8, 844–9 (1935) (Translation by J. D. Trimmer in [91], pp. 152–67)

[101] E. P. Wigner, On hidden variables and quantum mechanical probabilities, *American Journal of Physics* **38**, 1005–9 (1970)

[102] R. Penrose and C. Isham (eds.), *Quantum Concepts in Space and Time* (Oxford University Press, Oxford, 1986)

[103] A. Einstein, B. Podolsky and N. Rosen, Can quantum-mechanical description of physical reality be considered complete?, *Physical Review* **47**, 777–80 (1935)

[104] P. Lahti and P. Mittelstaedt (eds.), *Symposium on the Foundations of Modern Physics: Fifty Years of the Einstein–Podolsky–Rosen Gedankenexperiment* (World Scientific, Singapore, 1985)

[105] M. Dickson, Bohr on Bell: a proposed reading of Bohr and its implications for Bell's theorem, in *Non-locality and Modality* (T. Placek and J. Butterfield, eds.) (Kluwer, Dordrecht, 2002), pp. 19–32

[106] M. A. B. Whitaker, The EPR paper and Bohr's response: a re-assessment, *Foundations of Physics* **34**, 1305–1340 (2004)

[107] D. Bohm, *Quantum Theory* (Prentice-Hall, Englewood Cliffs, New Jersey, 1951)

[108] D. Bohm, Interview with D. Home, *Science Today*, November, 1986, p. 25

[109] N. Bohr, Can quantum-mechanical description of physical reality be considered complete?, *Physical Review* **48**, 696–702 (1935)

[110] P. K. Feyerabend, Complementarity, *Proceedings of the Aristotelian Society* (Supplementary Volume) **32**, 75–104 (1958)

[111] J. T. Cushing, A Bohmian response to Bohr's complementarity, in [60], pp. 57–75

[112] M. Beller and A. Fine, Bohr's response to EPR, in [60], pp. 1–31.

[113] M. Beller, *Quantum Dialogue* (University of Chicago Press, Chicago, 1999)

[114] H. Halvorson and R. Clifton, Reconsidering Bohr's reply to EPR, in *Non-locality and Modality* (T. Placek and J. Butterfield, eds.) (Kluwer, Dordrecht, 2002), pp. 3–18

[115] A. Einstein and L. Infeld, *The Evolution of Physics* (Cambridge University Press, Cambridge, 1938)

[116] P. Speziali (ed.) *Albert Einstein–Michele Besso Correspondence* (Hermann, Paris, 1972)

[117] P. C. W. Davies, *The Forces of Nature* (Cambridge University Press, Cambridge, 2nd edn, 1986)

[118] S. Eliezer and Y. Eliezer, *Fourth State of Matter: An Introduction to the Physics of Plasma* (Adam Hilger, Bristol, 1989)

[119] F. Close, *Too Hot to Handle: The Story of the Race for Cold Fusion* (Allen, London, 1990)

[120] S. Raimes, *The Wave Mechanics of Electrons in Metals* (North-Holland, Amsterdam, 1961)

[121] A Landé, *Quantum Mechanics* (Pitman, London, 1951)

[122] D. Pines, David Bohm 1917–92, *Physics World* **6** (3), 67 (1993)

[123] D. Bohm, *Causality and Chance in Modern Physics* (Routledge and Kegan Paul, London, 1957)

[124] D. Bohm, *Wholeness and the Implicate Order* (Routledge and Kegan Paul, London, 1980)

[125] D. Bohm and B. J. Hiley, *The Undivided Universe: An Ontological Interpretation of Quantum Theory* (Routledge and Kegan Paul, London, 1993)

[126] D. Bohm, *Thought As a System* (Routledge, London, 1994)

[127] B. J. Hiley and F. D. Peat (eds.), *Quantum Implications: Essays in Honour of David Bohm* (Routledge and Kegan Paul, London, 1987)

[128] P. Holland, *The Quantum Theory of Motion* (Cambridge University Press, Cambridge, 1993)

[129] D. Bohm, A suggested interpretation of the quantum theory in terms of 'hidden variables', I and II, *Physical Review* **85**, 166–93 (1952)

[130] W. Heisenberg, *Physics and Philosophy* (Allen and Unwin, London, 1959)

[131] A. George (ed.), *Louis de Broglie, physicien et penseur* (Albin-Michel, Paris, 1953)

[132] M. A. B. Whitaker, John Bell and the most profound discovery of science, *Physics World* **11** (12), 29–34 (1998)

[133] M. A. B. Whitaker, John Bell in Belfast: early years and education, in [215], pp. 7–20

[134] A. Shimony, John S. Bell: Some reminiscences and reflections, in [215], pp. 51–60

[135] R. Jackiw and A. Shimony, The depth and breadth of John Bell's physics, *Physics in Perspective* **4**, 78–116 (2002)

[136] M. Born, *Natural Philosophy of Cause and Chance* (Clarendon, Oxford, 1949)

[137] H. P. Stapp, Are superluminal connections necessary?, *Il Nuovo Cimento* **40B**, 191–205 (1977)

[138] N. D. Mermin, Is the Moon there when nobody looks?, *Physics Today* **38** (4), 38–47 (1985)

[139] J. F. Clauser, M. A. Horne, A. Shimony and R. A. Holt, Proposed experiment to test hidden-variable theories, *Physical Review Letters* **23**, 880–4 (1969)

[140] F. Selleri, *Quantum Paradoxes and Physical Reality* (Kluwer, Dordrecht, 1990)

[141] S. J. Freedman and J. F. Clauser, Experimental test of local hidden-variable theories, *Physical Review Letters* **28**, 938–41 (1972)

[142] J. F. Clauser, Experimental investigation of a polarization correlation anomaly, *Physical Review Letters* **36**, 1223–6 (1976)

[143] E. S. Fry and R. C. Thompson, Experimental test of local hidden-variable theories, *Physical Review A* **37**, 465–8 (1976)

[144] R. A. Holt, *Atomic Cascade Experiments* (PhD Thesis, Harvard University, 1973)

[145] J. F. Clauser and A. Shimony, Bell's Theorem: experimental tests and implications, *Reports on Progress in Physics* **41**, 1881–1927 (1978)

[146] A. Aspect, P. Grangier and G. Roger, Experimental realization of Einstein–Podolsky–Rosen *gedankenexperiment*: a new violation of Bell's inequalities, *Physical Review Letters* **49**, 91–4 (1982)

[147] A. Aspect, J. Dalibard and G. Roger, Experimental tests of Bell's inequalities using time-varying analysers, *Physical Review Letters* **49**, 1804–7 (1982)

[148] M. A. B. Whitaker, Theory and experiment in the foundations of quantum theory, *Progress in Quantum Electronics* **24**, 1–106 (2000)

[149] A. Zeilinger, Testing Bell's inequalities with periodic switching, *Physics Letters A* **118**, 1–2 (1986)

[150] P. M. Pearle, Hidden-variable example based upon data rejection, *Physical Review D* **2**, 1418–25 (1970)

[151] P. H. Eberhard, Background level and counter efficiencies required for a loophole-free Einstein–Podolsky–Rosen experiment, *Physical Review A* **47**, R747–50 (1993)

[152] Y. H. Shih and C. O. Alley, New type of Einstein–Podolsky–Rosen–Bohm experiment using pairs of light quanta produced by optical parametric down conversion, *Physical Review Letters* **61**, 2921–4 (1988)

[153] Z. Y. Ou and L. Mandel, Violation of Bell's inequality and classical probability in a two-photon correlation experiment, *Physical Review Letters* **61**, 50–3 (1988)

[154] P. G. Kwiat, K. Mattle, H. Weinfurter, A. Zeilinger, A. V. Sergienko and Y. H. Shih, New high-intensity source of polarization-entangled photon pairs, *Physical Review Letters* **75**, 4337–41 (1995)

[155] G. Weihs, T. Jennewein, C. Simon, H. Weinfurter and A. Zeilinger, Violation of Bell's inequalities under strict Einstein locality conditions, *Physical Review Letters* **81**, 5039–43 (1998)

[156] J. G. Rarity and P. R. Tapster, Experimental violation of Bell's inequality based on phase and momentum, *Physical Review Letters* **64**, 2495–8 (1990)

[157] M. A. Horne, A. Shimony and A. Zeilinger, Two particle interferometry, *Physical Review Letters* **62**, 2209–12 (1989)

[158] J. D. Franson, Bell inequality for position and time, *Physical Review Letters* **62**, 2205–8 (1989)

[159] P. G. Kwiat, A. M. Steinberg and R. Y. Chiao, High-visibility interference in a Bell-inequality experiment for energy and time, *Physical Review A* **47**, R2472–5 (1993)

[160] S. Aerts, P. Kwiat, J. A. Larsson and M. Zukowski, Two-photon Franson-type experiments and local realism, *Physical Review Letters* **83**, 2872–5 (1999)

[161] P. Samuelsson, E. V. Sukhorukov and M. Büttiker, Orbital entanglement and violation of Bell's inequalities in mesoscopic conductors, *Physical Review Letters* **91**, 157002 (2003)

[162] W. Tittel, J. Brendel, B. Gisin, T. Herzog, H. Zbinden and N. Gisin, Experimental demonstration of quantum-correlations over more than 10 kilometers, *Physical Review A* **57**, 3229–32 (1998)

[163] W. Tittel, J. Brendel, H. Zbinden and N. Gisin, Violation of Bell's inequalities by photons more than 10 km apart, *Physical Review Letters* **81**, 3563–6 (1998)

[164] W. Tittel, J. Brendel, N. Gisin and H. Zbinden, Long-distance Bell-type tests using energy–time entangled photons, *Physical Review A* **59**, 4150–63 (1999)

[165] H.-D. Zeh, On the interpretation of measurement in quantum theory, *Foundations of Physics* **1**, 69–76 (1970)

[166] W. H. Zurek, Decoherence and the transition from quantum to classical, *Physics Today* **44** (10), 36–44 (1991)

[167] W. H. Furry, Note on the quantum-mechanical theory of measurement, *Physical Review* **49**, 393–9 (1936)

[168] P. Pearle, Might god toss coins?, *Foundations of Physics* **12**, 249–63 (1982)

[169] G.-C. Ghirardi, A. Rimini and T. Weber, Uniform dynamics for microscopic and macroscopic systems, *Physical Review D* **34**, 470–91 (1986)

[170] P. G. Kwiat, P. H. Eberhard, A. M. Steinberg and R. Y. Chiao, Proposal for a loophole-free Bell inequality experiment, *Physical Review A* **49**, 3209–20 (1994)

[171] S. F. Huelga, M. Ferrero and E. Santos, Loophole-free test of the Bell inequality, *Physical Review A* **51**, 5008–11 (1995)

[172] E. Santos, Entropy inequalities and Bell inequalities for two-qubit systems, *Physical Review A* **69**, 022305 (2004)

[173] E. Santos, Proposed optical tests of Bell's inequalities not resting upon the fair sampling assumption, *Physical Letters A* **327**, 33–7 (2004)

[174] T. K. Lo and A. Shimony, Proposed molecular test of local hidden-variable theories, *Physical Review A* **23**, 3003–12 (1981)

[175] E. S. Fry, T. Walther and S. Li, Proposal for a loophole free test of the Bell inequalities, *Physical Review A* **52**, 4381–95 (1995)

[176] B. J. Oliver and C. R. Stroud, Bell's inequalities for Rydberg atoms, *Journal of the Optical Society of America B* **4**, 1426–8 (1987)

[177] M. Freyberger, P. K. Aravind, M. A. Horne and A. Shimony, Proposed test of Bell's inequality without a detection loophole by using entangled Rydberg atoms, *Physical Review A* **53**, 1232–44 (1996)

[178] E. Hagley, X. Maître, G. Nogues, C. Wunderlich, M. Brune, J. M. Raimond and S. Haroche, Generation of Einstein–Podolsky–Rosen pairs of atoms, *Physical Review Letters* **79**, 1–5 (1997)

[179] A. Beige, W. J. Munro and P. L. Knight, Bell's inequality test with entangled atoms, *Physical Review A* **62**, 052102 (2000)

[180] C. Brif and A. Mann, Bell's inequality test with entangled atoms, *Europhysics Lettes* **49**, 1–7 (2000)

[181] M. Lamehi-Rachti and W. Mittig, Quantum mechanics and hidden variables: a test of Bell's inequalities by the measurement of spin correlation in low-energy proton–proton scattering, *Physical Review D* **14**, 2543–55 (1976)

[182] A. Datta and D. Home, Is Bell-type inequality violated for $K_0\bar{K}_0$, $B_0\bar{B}_0$ systems?, *Foundations of Physics Letters* **4**, 165–78 (1991)

[183] F. Selleri, Incompatibility between local realism and quantum mechanics for pairs of neutral kaons, *Physical Review A* **56**, 3493–506 (1997)

[184] A. Bramon and M. Nowakowski, Bell inequalities for entangled neutral kaons, *Physical Review Letters* **83**, 1–5 (1999)

[185] R. A. Bertlmann and B. C. Hiesmayr, Bell's inequalities for entangled kaons and their unitary time evolution, *Physical Review A* **63**, 062112 (2001)

[186] M. Genovese, Entanglement properties of kaons and tests of hidden-variable models, *Physical Review A* **69**, 022103 (2004)

[187] A. Go (Belle collaboration), Observation of Bell inequality violation in B mesons, *Journal of Modern Optics* **51**, 991–8 (2004)

[188] R. A. Bertlmann, A. Bramon, G. Garbarino and B. C. Hiesmayr, Violation of a Bell inequality in particle physics experimentally verified?, *Physics Letters A* **332**, 355–60 (2004)

[189] D. H. Perkins, *Introduction to High Energy Physics* (Cambridge University Press, Cambridge, 4th edn, 2000)

[190] J. S. Bell and R. A. Bertlmann, Magic moments, *Nuclear Physics B* **177**, 218–36 (1981)

[191] R. A. Bertlmann, Magic moments: a collaboration with John Bell, in [215], pp. 29–47

[192] A. Aspect, Bell's inequality test: more ideal than ever, *Nature* **398**, 189–90 (1999)

[193] M. A. Rowe, D. Klepinski, V. Meyer, C. A. Sackett, W. M. Itano, C. Monroe and D. J. Wineland, Experimental violation of a Bell's inequality with efficient detection, *Nature* **409**, 791–4 (2001)

[194] P. Grangier, Count them all, *Nature* **409**, 774–5 (2001)

[195] M. Barbieri, F. De Martini, G. DiNepi and P. Mataloni, Towards a test of non-locality without 'supplementary assumptions', *Physics Letters A* **334**, 23–9 (2005)

[196] J. Barrett, D. Collins, L. Hardy, A. Kent and S. Popescu, Quantum nonlocality, Bell inequalities and the memory loophole, *Physical Review A* **66**, 042111 (2002)

[197] H. Nha and H. J. Carmichael, Proposed tests of quantum nonlocality for continuous variables, *Physical Review Letters* **93**, 020401 (2004)

[198] R. García-Patrón, J. Fiurášek, N. J. Cerf, J. Wenger, R. Tualle-Brouri and P. Grangier, Proposal for a loophole-free Bell test using homodyne detection, *Physical Review Letters* **93**, 130409 (2004)

[199] D. M. Greenberger, M. A. Horne and A. Zeilinger, Going beyond Bell's theorem, in *Bell's Theorem, Quantum Theory and Conceptations of the Universe* (M. Kafatos, ed.) (Kluwer, Dordrecht, 1989)

[200] D. M. Greenberger, M. A. Horne, A. Shimony and A. Zeilinger, Bell's theorem without inequalities, *American Journal of Physics* **58**, 1131–43 (1990)

[201] N. D. Mermin, Quantum mysteries revisited, *American Journal of Physics* **58**, 731–4 (1990)

[202] D. Bouwmeester, J.-W. Pan, M. Daniell, H. Weinfurter and A. Zeilinger, Observation of three-photon Greenberger–Horne–Zeilinger entanglement, *Physical Review Letters* **82**, 1345–9 (1999)

[203] D. Bouwmeester, J.-W. Pan, K. Mattle, M. Eibl, H. Weinfurter and A. Zeilinger, Experimental quantum teleportation, *Nature* **390**, 575–9 (1997)

[204] J.-W. Pan, D. Bouwmeester, H. Weinfurter and A. Zeilinger, Experimental entanglement swapping: entangling photons that never interacted, *Physical Review Letters* **80**, 3891–4 (1998)

[205] L. Hardy, Quantum mechanics, local realistic theories and Lorentz-invariant realistic theories, *Physical Review Letters* **68**, 2981–4 (1992)

[206] L. Hardy, Spooky action at a distance in quantum mechanics, *Contemporary Physics* **39**, 419–29 (1998)

[207] A. Elitzur and L. Vaidman, Quantum mechanical interaction-free measurements, *Foundations of Physics* **23**, 987–97 (1993)

[208] L. Hardy, Nonlocality for two particles without inequalities for almost all entangled states, *Physical Review Letters* **71**, 1665–8 (1993)

[209] G. Di Giuseppe, F. De Martini and D. Boschi, Experimental test of the violation of local realism in quantum mechanics without Bell inequalities, *Physical Review A* **56**, 176–81 (1997)

[210] D. Boschi, S. Branca, F. De Martini and L. Hardy, Ladder proof of nonlocality without inequalities: theoretical and experimental results, *Physical Review Letters* **79**, 2755–8 (1997)

[211] A. van der Merwe, F. Selleri and G. Tarozzi (eds.), *Bell's Theorem and the Foundations of Modern Physics* (World Scientific, Singapore, 1992)

[212] C. H. Thompson, The chaotic bell: an intuitive analogy for EPR experiments, *Foundations of Physics Letters* **9**, 357–82 (1996)

[213] C. H. Thompson, Subtraction of accidentals and the validity of Bell tests, *Galilean Electrodynamics* **14**, 43–50 (2003)

[214] J. T. Cushing and E. McMullin (eds.), *Philosophical Consequences of Quantum Theory: Reflections on Bell's Theorem* (University of Notre Dame Press, Notre Dame, Indiana 1989)

[215] R. Bertlmann and A. Zeilinger (eds.), *Quantum (Un)speakables: From Bell to Quantum Information* (Springer-Verlag, Berlin, 2002)

[216] N. Herbert, *Quantum Reality: Beyond the New Physics* (Anchor, New York, 1987)

[217] M. Redhead, *Incompleteness, Nonlocality and Realism, a Prolegomenon to the Philosophy of Quantum Mechanics* (Oxford University Press, Oxford, 1987)

[218] W. M. Dickson, *Quantum Chance and Non-locality: Probability and Non-locality in the Interpretations of Quantum Mechanics* (Cambridge University Press, Cambridge, 1998)

[219] J. Bub, *Interpreting the Quantum World* (Cambridge University Press, Cambridge, 2000)

[220] T. Maudlin, *Quantum Non-locality and Relativity: Metaphysical Aspects of Modern Physics* (Blackwell, Oxford, 2nd edn, 2002)

[221] J. Baggott, *Beyond Measure: Modern Physics, Philosophy and the Meaning of Quantum Theory* (Oxford University Press, Oxford, 2004)

[222] G.-C. Ghirardi, *Sneaking a Look at God's Cards: Unravelling the Mysteries of Quantum Mechanics* (Princeton University Press, Princeton, New Jersey, 2nd edn, 2005)

[223] J. Gribbin, The man who proved Einstein was wrong, *New Scientist* **128**, 43–5 (1990)

[224] R. H. Romer, John S. Bell (1928–90), the man who proved Einstein was right, *American Journal of Physics* **59**, 299–300 (1991)

[225] J. T. Cushing, *Quantum Mechanics: Historical Contingency and the Copenhagen Hegemony* (University of Chicago Press, Chicago, 1994)

[226] D. Howard, Review of *The Shaky Game* by A. Fine, *Synthese* **86**, 123–41 (1991)

[227] C. Philippidis, C. Dewdney and B. J. Hiley, Quantum interference and the quantum potential, *Il Nuovo Cimento* **52B**, 15–28 (1979)

[228] C. Dewdney and B. J. Hiley, A quantum potential description of one-dimensional time-dependent scattering from square barriers and square wells, *Foundations of Physics* **12**, 27–48 (1982)

[229] C. R. Leavens, Time of arrival in quantum and Bohmian mechanics, *Physical Review A* **58**, 840–7 (1998)

[230] J. C. Muga, R. Sala Mayato and I. L. Egusquiza (eds.), *Time in Quantum Mechanics* (Springer-Verlag, Berlin, 2002)

[231] H. Everett, *The Theory of the Universal Wave Function* (PhD Thesis, Princeton University, 1957)

[232] H. Everett, 'Relative state' formulation of quantum mechanics, *Reviews of Modern Physics* **29**, 454–62 (1957)

[233] J. A. Wheeler, Assessment of Everett's 'relative state' formulation of quantum theory, *Reviews of Modern Physics* **29**, 463–5 (1957)

[234] B. de Witt and N. Graham (eds.) *The Many-Worlds Interpretation of Quantum Mechanics* (Princeton University Press, Princeton, New Jersey, 1973)

[235] E. J. Squires, Many views of one world, *European Journal of Physics* **8**, 171–3 (1987)

[236] D. Albert and B. Loewer, Interpreting the many-worlds interpretation, *Synthese* **77**, 195–213 (1988)

[237] M. Lockwood, *Mind, Brain and the Quantum* (Blackwell, Oxford, 1989)

[238] M. Lockwood, 'Many minds' interpretations of quantum mechanics, *British Journal for the Philosophy of Science* **47**, 159–88 (1996)

[239] H. P. Stapp, *Mind, Matter and Quantum Mechanics* (Springer-Verlag, Berlin, 2nd edn, 2004)

[240] J. Barrett, *The Quantum Mechanics of Minds and Worlds* (Oxford University Press, Oxford, 1999)

[241] M. J. Donald, Quantum theory and the brain, *Proceedings of the Royal Society A* **427**, 43–93 (1990)

[242] M. A. B. Whitaker, The relative states and many worlds interpretations of quantum mechanics and the EPR problem, *Journal of Physics A* **18**, 253–64 (1985)

[243] M. A. B. Whitaker, Many worlds, many minds, many views, *Revue Internationale de Philosophie* **54**, 369–91 (2000)

[244] J. B. Hartle and S. W. Hawking, Wavefunction of the universe, *Physical Review D* **28**, 2960–75 (1983)

[245] B. J. Carr and M. J. Rees, The anthropic principle and the structure of the physical world, *Nature* **278**, 605–12 (1979)

[246] J. D. Barrow and F. J. Tipler, *The Anthropic Cosmological Principle* (Clarendon, Oxford, 1986)

[247] N. Bostrom, *Anthropic Bias: Observation Selection Effects in Science and Philosophy* (Routledge, New York, 2002)

[248] J. Polkinghorne, *Belief in God in an Age of Science* (Yale University Press, New Haven, Connecticut, 1998)

[249] C. W. Misner, K. S. Thorne and J. A. Wheeler, *Gravitation* (Freeman, San Francisco, California, 1973)

[250] M. A. B. Whitaker, On Hacking's criticism of the Wheeler anthropic principle, *Mind* **97**, 259–64 (1988)

[251] D. Deutsch, Quantum theory as a universal physical theory, *International Journal of Theoretical Physics* **24**, 1–41 (1985)

[252] D. Deutsch, *The Fabric of Reality* (Allen Lane, London, 1997)

[253] L. E. Ballentine, Can the statistical postulate of quantum theory be derived? – a critique of the many-universes interpretation, *Foundations of Physics* **3**, 229–40 (1973)

[254] H. P. Stapp, Bell's theorem and the foundations of quantum physics, *American Journal of Physics* **53**, 306–17 (1985)

[255] A. Sudbery, This damned quantum jumping, *Studies in the History and Philosophy of Modern Physics* **34**, 387–411 (2002)

[256] A. Kent, Against many-world interpretations, *International Journal of Theoretical Physics A* **5**, 1745–62 (1990)

[257] L. Vaidman, On schizophrenic experiences of the neutron or why we should believe in the many-worlds interpretation of quantum theory, *International Studies in the Philosophy of Science* **12**, 245–61 (1998)

[258] L. Vaidman, The many-worlds interpretation of quantum mechanics, *Stanford Encyclopaedia of Philosophy* (2002): http://plato.stanford.edu/entries/qm-manyworlds/

[259] D. Deutsch, Quantum theory of probability and decisions, *Proceedings of the Royal Society A* **455**, 3129–37 (1999)

[260] S. Saunders, Time, decoherence and quantum mechanics, *Synthese* **102**, 235–66 (1995)

[261] S. Saunders, Time, quantum mechanics and tense, *Synthese* **107**, 19–53 (1996)

[262] S. Saunders, Time, quantum mechanics and probability, *Synthese* **114**, 373–404 (1998)

[263] S. Saunders, Derivation of the Born rule from operational assumptions, *Proceedings of the Royal Society A* **460**, 1–18 (2004)

[264] S. Saunders, What is probability?, in *Quo Vadis Quantum Mechanics?* (A. Elitzur, S. Dolev and N. Kolenda, eds.) (Springer-Verlag, Berlin, 2005)

[265] D. Wallace, Worlds in the Everett interpretation, *Studies in the History and Philosophy of Modern Physics* **33**, 637–61 (2002)

[266] D. Wallace, Everett and structure, *Studies in the History and Philosophy of Modern Physics* **34**, 87–105 (2003)

[267] D. Wallace, Everettian rationality: defending Deutsch's approach to probability in the Everett interpretation, *Studies in the History and Philosophy of Modern Physics* **34**, 415–39 (2003)

[268] H. Greaves, Understanding Deutsch's probability in a deterministic universe, *Studies in the History and Philosophy of Modern Physics* **35**, 423–56 (2004)

[269] D. C. Dennett, Real patterns, *Journal of Philosophy* **87**, 27–51 (1991)

[270] B. de Finetti, *Theory of Probability* (Wiley, New York, 1974)

[271] K. V. Laurikainen and C. Montonen (eds.), *Symposia on the Foundations of Modern Physics 1992* (World Scientific, Singapore, 1993)

[272] B. d'Espagnat, *Veiled Reality: An Analysis of Present-Day Quantum Mechanical Concepts* (Addison-Wesley, Reading, Massachusetts, 1995)

[273] B. d'Espagnat, *Conceptual Foundations of Quantum Mechanics* (Benjamin, Reading, Massachusetts, 2nd edn, 1976)

[274] D. Home and M. A. B. Whitaker, The inadequacy of effective incoherence interpretations of quantum theory as demonstrated by analysis of EPR measurements, *Physics Letters A* **211**, 5–11 (1996)

[275] L. D. Landau and E. M. Lifshitz, *Quantum Mechanics* (Pergamon, Oxford, 2nd edn, 1977)

[276] K. Gottfried, *Quantum Mechanics* (Benjamin, New York, 1966)

[277] N. G. van Kampen, 10 theorems about quantum-mechanical measurements, *Physica A* **153**, 97–113 (1988)

[278] K. Gottfried, Does quantum mechanics carry the seeds of its own destruction? *Physics World* **4** (10), 34–40 (1991)

[279] N. G. van Kampen, Quantum criticism, *Physics World* **3** (10), 20 (1990)

[280] N. G. van Kampen, Mystery of quantum measurement, *Physics World* **4** (12), 16–17 (1991)

[281] R. Peierls, In defence of 'measurement', *Physics World* **4** (1), 19–20 (1991)

[282] E. J. Squires, Quantum challenge, *Physics World* **5** (1), 18 (1992)

[283] J. Ellis and D. Amati, *Quantum Reflections* (Cambridge University Press, Cambridge, 2000)

[284] M. Bell, K. Gottfried and M. Veltmann (eds.), *Quantum Mechanics, High Energy Physics and Accelerators* (World Scientific, Singapore, 1995). [The papers on quantum mechanics were republished on their own in 2001 as *John S. Bell on the Foundations of Quantum Mechanics* with the same editors and publisher.]

[285] K. Gottfried, Inferring the statistical interpretation of quantum mechanics from the classical limit, *Nature* **405**, 533–6 (2000)

[286] K. Gottfried and T.-M. Yan, *Quantum Mechanics: Fundamentals* (Springer, New York, 2nd edn, 2003)

[287] A. Whitaker, review of [286], *Journal of Physics A: Mathematical and General Physics* **37**, 3073–6 (2004)

[288] N. G. van Kampen, Macroscopic systems in quantum mechanics, *Physica A* **194**, 542–50 (1993)

[289] N. G. van Kampen, Note on the 2-slit experiment, *Journal of Statistical Physics* **77**, 345–50 (1994)

[290] M. A. B. Whitaker, On a model of wave-function collapse, *Physica A* **255**, 455–66 (1998)

[291] L. E. Ballentine, What do we learn about quantum mechanics from the theory of measurement?, *International Journal of Theoretical Physics* **27**, 211–8 (1988)

[292] D. Giulini, E. Joos, C. Kiefer, J. Kupsch, I.-O. Stamatescu and H.-D. Zeh, *Decoherence and the Appearance of a Classical World in Quantum Theory* (Springer-Verlag, Berlin, 2nd edn, 2003)

[293] J. L. Anderson; G. C. Ghirardi, R. Grassi and P. Pearle; N. Gisin; D. Z. Albert and G. Feinberg; P. R. Holland; V. Ambegaokar; K. J. Epstein; and W. H. Zurek, Negotiating the tricky border between quantum and classical, *Physics Today* **46** (4), 13–51, 81–90 (1993)

[294] P. Atkins, *Galileo's Finger: The Great Ideas of Science* (Oxford University Press, Oxford, 2003)

[295] R. B. Griffiths, Consistent histories and the interpretation of quantum mechanics, *Journal of Statistical Physics* **36**, 219–72 (1984)

[296] R. B. Griffiths, Correlations in separated quantum systems: a consistent history analysis of the EPR problem, *American Journal of Physics* **55**, 11–17 (1987)

[297] R. B. Griffiths, Consistent interpretation of quantum mechanics using quantum trajectories, *Physical Review Letters* **70**, 2201–4 (1993)

[298] R. Omnès, Consistent interpretations of quantum mechanics, *Reviews of Modern Physics* **64**, 339–82 (1992)

[299] R. Omnès, *Interpretation of Quantum Mechanics* (Princeton University Press, Princeton, New Jersey, 1994)

[300] R. Omnès, *Understanding Quantum Mechanics* (Princeton University Press, Princeton, New Jersey, 1999)

[301] R. Omnès, *Quantum Philosophy: Understanding and Interpreting Contemporary Science* (Princeton University Press, Princeton, New Jersey, 1999)

[302] S. Kobayashi, H. Ezawa, Y. Murayama and S. Nomura (eds.), *Proceedings of the Third International Symposium on the Foundations of Quantum Mechanics in the Light of New Technology* (Physical Society of Japan, Tokyo, 1990)

[303] M. Gell-Mann and J. B. Hartle, Classical equations for quantum systems, *Physical Review D* **47**, 3345–82 (1993)

[304] L. Diósi and B. Lukács (eds.) *Stochastic Evolution of Quantum States in Open Systems and Measurement Processes* (World Scientific, Singapore, 1994)

[305] F. Dowker and A. Kent, On the consistent histories approach to quantum mechanics, *Journal of Statistical Physics* **82**, 1575–646 (1996)

[306] R. B. Griffiths, Bohmian mechanics and consistent histories, *Physics Letters A* **261**, 227–34 (1999)

[307] J. B. Hartle, Bohmian histories and decoherent histories, *Physical Review A* **69**, 042111 (2004)

[308] N. Yamada, Speakable and unspeakable in the tunnelling time problem, *Physical Review Letters* **83**, 3350–3 (1999)

[309] D. Sokolovski, Nature of quantum measurement of quantities not represented by Hermitian operators, *Physical Review A* **66**, 032107 (2002)

[310] I. L. Egusquiza and J. G. Muga, Consistent histories, the quantum Zeno effect, and time of arrival, *Physical Review A* **62**, 032103 (2000)

[311] R. B. Griffiths, Consistent histories and quantum delayed choice, *Fortschritte der Physik* **46**, 741–8 (1998)

[312] R. B. Griffiths, *Consistent Quantum Theory* (Cambridge University Press, Cambridge, 2003)

[313] N. Gisin and I. C. Percival, The quantum-state diffusion model applied to open systems, *Journal of Physics A* **25**, 5677–91 (1992)

[314] N. Gisin and I. C. Percival, Quantum state diffusion, localization and quantum dispersion entropy, *Journal of Physics A* **26**, 2233–43 (1993)

[315] N. Gisin and I. C. Percival, The quantum state diffusion picture of physical processes, *Journal of Physics A* **26**, 2245–60 (1993)

[316] N. Gisin, Quantum measurements and stochastic processes, *Physical Review Letters* **52**, 1657–60 (1984)

[317] L. Diósi, Quantum stochastic processes as models for state vector reduction, *Journal of Physics A* **21**, 2885–98 (1988)

[318] L. Diósi, N. Gisin, J. Halliwell and I. C. Percival, Decoherent histories and quantum state diffusion *Physical Review Letters* **74**, 203–7 (1995)

[319] H. D. Zeh, There are no quantum jumps, nor are there particles, *Physics Letters A* **172**, 189–92 (1993)

[320] N. Gisin and I. C. Percival, Stochastic wave equations versus parallel world components, *Physics Letters A* **175**, 144–5 (1993)

[321] L. Diósi, N. Gisin and W. T. Strunz, Non-Markovian quantum state diffusion, *Physical Review A* **58**, 3068–79 (1998)

[322] T. Yu, L. Diósi, N. Gisin and W. T. Strunz, Non-Markovian quantum state diffusion: perturbation approach, *Physical Review A* **60**, 91–103 (1999)

[323] J. T. Stockburger and H. Grabert, Non-Markovian quantum state diffusion, *Chemical Physics* **268**, 249–56 (2001)

[324] H. P. Breuer and F. Petruccione, Relativistic formulation of quantum-state diffusion, *Journal of Physics A: Mathematical and General Physics* **31**, 33–52 (1998)

[325] L. Diósi, Relativistic formulation of quantum-state diffusion, *Journal of Physics A: Mathematical and General Physics* **31**, 9601–3 (1998)

[326] H. P. Breuer and F. Petruccione, Reply to 'Relativistic formulation of quantum-state diffusion', *Journal of Physics A: Mathematical and General Physics* **31**, 9605–12 (1998)

[327] I. C. Percival, Primary state diffusion, *Proceedings of the Royal Society of London A* **447**, 189–209 (1994)

[328] I. C. Percival, *Quantum State Diffusion* (Cambridge University Press, Cambridge, 1998)

[329] T. P. Spiller, Quantum state diffusion [Review of Ref. 328], *Studies in the History and Philosophy of Modern Physics* **33**, 707–16 (2002)

[330] N. D. Mermin, Whose knowledge?, in [215], pp. 271–80

[331] T. A. Brun, J. Finkelstein and N. D. Mermin, How much state assignments can differ, *Physical Review A* **65**, 032315 (2002)

[332] Č. Brukner and A. Zeilinger, Operationally invariant information in quantum measurements, *Physical Review Letters* **83**, 3354–7 (1999)

[333] Č. Brukner and A. Zeilinger, Information and fundamental elements of the structure of quantum theory, in *Time, Quantum, Information* (L. Castell and O. Ischbek, eds.) (Springer-Verlag, Berlin, 2003)

[334] Č. Brukner and A. Zeilinger, Quantum physics as a science of information, in *Quo Vadis Quantum Mechanics* (A. Elitzur, S. Dolev and N. Kolenda, eds.) (Springer-Verlag, Berlin, 2005)

[335] C. F. von Weizsäcker, *Aufbau der Physik* (Carl Hanser, Munich, 1958)

[336] J. A. Wheeler, Law without law, in *Quantum Theory and Measurement* (T. A. Wheeler and W. H. Zurek, eds.) (Princeton University Press, Princeton, New Jersey, 1983), pp. 182–2

[337] E. Nelson, *Dynamical Theories of Brownian Motion* (Princeton University Press, Princeton, New Jersey, 1967)

[338] E. Nelson, *Quantum Fluctuations* (Princeton University Press, Princeton, New Jersey, 1985)

[339] L. de la Peña and A. M. Cetto, Stochastic theory for classical and quantum mechanical systems, *Foundations of Physics* **5**, 355–70 (1975)

[340] L. de la Peña and A. M. Cetto, Does quantum mechanics accept a stochastic support?, *Foundations of Physics* **12**, 1017–37 (1982)

[341] L. de la Peña and A. M. Cetto, Why Schrödinger's equation?, *International Journal of Quantum Theory* **12**, 23–37 (1977)

[342] T. W. Marshall, Random electrodynamics, *Proceedings of the Royal Society A* **276**, 475–91 (1963)

[343] G. Tarozzi and A. van der Merwe (eds.), *Open Questions in Quantum Physics* (Reidel, Dordrecht, 1985)

[344] E. Santos, Comment on 'Source of vacuum electromagnetic zero-point energy', *Physical Review A* **44**, 3383–4 (1991)

[345] P. Garbaczewski and J. P. Vigier, Quantum dynamics from the Brownian recoil principle, *Physical Review A* **46**, 4634–8 (1992)

[346] T. C. Wallstrom, Inequivalence between the Schrödinger equation and the Madelung hydrodynamic equation, *Physical Review A* **49**, 1613–17 (1994)

[347] G. Bacciagaluppi, Nelsonian mechanics revisited, *Foundations of Physics Letters* **12**, 1–16 (1999)

[348] P. Pearle, Stochastic dynamical reduction theories and superluminal communication, *Physical Review D* **33**, 2240–52 (1986)

[349] E. J. Squires, *The Mystery of the Quantum World* (Adam Hilger, Bristol, 1986)

[350] G. C. Ghirardi, R. Grassi and P. Pearle, Relativistic dynamic reduction models – general framework and examples, *Foundations of Physics* **20**, 1271–1316 (1990)

[351] P. Pearle, Relativistic collapse model with tachyonic features, *Physical Review A* **59**, 80–101 (1999)

[352] O. Nicrosini and A. Rimini, Relativistic spontaneous localization: a proposal, *Foundations of Physics* **33**, 1061–84 (2003)

[353] G. C. Ghirardi, P. Pearle and A. Rimini, Markov processes in Hilbert space and continuous localization of systems of identical particles, *Physical Review A* **42**, 78–89 (1990)

[354] A. Fine, M. Forbes and L. Wessels (eds.), *Proceedings of the 1990 Biennial Meeting of the Philosophy of Science Association* (Philosophy of Science Association, East Lansing, Michigan, 1990)

[355] P. J. Lewis, Quantum mechanics, orthogonality, and counting, *British Journal for the Philosophy of Science* **48**, 313–28 (1997)

[356] P. J. Lewis, Counting marbles: a reply to my critics, *British Journal for the Philosophy of Science* **54**, 165–70 (2003)

[357] R. Clifton and B. Monton, Losing your marbles in wavefunction collapse theories, *British Journal for the Philosophy of Science* **50**, 697–717 (1999)

[358] R. Clifton and B. Monton, Counting marbles with 'accessible' mass density: a reply to Bassi and Ghirardi, *British Journal for the Philosophy of Science* **52**, 125–30 (2000)

[359] G. Ghirardi and A. Bassi, Do dynamical reduction models imply that arithmetic does not apply to ordinary macroscopic objects?, *British Journal for the Philosophy of Science* **50**, 49–64 (1999)

[360] A. Bassi and G. Ghirardi, More about dynamical reduction and the enumeration principle, *British Journal for the Philosophy of Science* **50**, 719–34 (1999)

[361] A. Bassi and G. Ghirardi, Counting marbles: reply to Clifton and Monton, *British Journal for the Philosophy of Science* **52**, 125–30 (2001)

[362] A. Bassi and G. Ghirardi, Dynamical reduction models, *Physics Reports* **379**, 257–426 (2003)

[363] A. Cordero, Are GRW tails as bad as they say?, *Philosophy of Science* **66**, S59–71 (1999)

[364] D. Parker, Finding your marbles in wavefunction collapse theories, *Studies in the History and Philosophy of Modern Physics* **34**, 607–20 (2003)

[365] D. Z. Albert and L. Vaidman, On a proposed postulate of state-reduction, *Physics Letters A* **139**, 1–4 (1988)

[366] D. Z. Albert, *Quantum Mechanics and Experience* (Harvard University Press, Cambridge, Massachusetts, 1992)

[367] F. Aicardi, A. Borsellino, G. C. Ghirardi and R. Grassi, Dynamical models for state-vector reduction: do they ensure that measurements have outcomes?, *Foundations of Physics Letters* **4**, 109–28 (1991)

[368] P. Pearle and E. J. Squires, Bound state excitation, neutron decay experiments and models of wavefunction collapse, *Physical Review Letters* **73**, 1–5 (1994)

[369] B. Collett, P. Pearle, F. Avignone and S. Nussinov, Constraint on collapse models by limit on spontaneous X-ray emission in Ge, *Foundations of Physics*, **25**, 1399–1412 (1995)

[370] P. Pearle, J. Ring, J. I. Collar and F. T. Avignone, The CSL collapse model and spontaneous radiation: an update, *Foundations of Physics* **29**, 465–80 (1999)

[371] A. I. M. Rae, Can GRW theory be tested by experiments on SQUIDs?, *Journal of Physics A: Mathematical and General Physics* **23**, L57–60 (1990)

[372] A. M. Buffa, O. Nicrosini and A. Rimini, Dissipation and reduction effects of spontaneous localization on superconducting states, *Foundations of Physics Letters* **8**, 105–25 (1995)

[373] M. Tinkham, *Introduction to Superconductivity* (McGraw-Hill, New York, 2nd edn, 1996)

[374] O. Serot, N. Carjan and D. Strottman, Transient behaviour in quantum tunnelling: time dependent approach to alpha decay, *Nuclear Physics A* **569**, 562–74 (1994)

[375] P. T. Greenland, Seeking non-exponential decay, *Nature* **335**, 298 (1988)

[376] T. Jittoh, S. Matsumoto, J. Sato, Y. Sato and K. Takeda, Nonexponential decay of an unstable quantum system: small-Q-value s-wave decay, *Physical Review A* **71**, 012109 (2005)

[377] E. B. Norman, S. B. Gazes, S. G. Crane and D. A. Bennett, Test of the exponential decay law at short and long times, *Physical Review Letters*, **60**, 2246–9 (1988)

[378] E. B. Norman, B. Sur, K. T. Lesko, R. M. Larimer, D. J. Depaolo and T. L. Owens, An improved test of the exponential decay law, *Physics Letters B* **357**, 521–5 (1995)

[379] S. R. Wilkinson, C. F. Bharucha and six others, Experimental evidence for non-exponential decay in quantum tunnelling, *Nature* **387**, 575–7 (1997)

[380] N. G. Kelkar, M. Nowakowski and K. P. Khemchandani, Hidden evidence of nonexponential nuclear decay, *Physical Review C* **70**, 024601 (2004)

[381] R. H. Dicke, Interaction-free quantum mechanics — a paradox, *American Journal of Physics* **49**, 925–30 (1981)

[382] D. Home and M. A. B. Whitaker, Negative-result experiments, and the requirement of wavefunction collapse, *Journal of Physics A* **25**, 2387–94 (1992)

[383] A. Degasperis, L. Fonda and G. C. Ghirardi, Does the lifetime of an unstable system depend on the measuring apparatus?, *Il Nuovo Cimento* **21A**, 471–84 (1974)

[384] L. Fonda, G. C. Ghirardi, A. Rimini and T. Weber, On the quantum foundations of the experimental decay law, *Il Nuovo Cimento* **15A**, 689–704 (1973)

[385] B. Misra and E. C. G. Sudarshan, The Zeno's paradox in quantum theory, *Journal of Mathematical Physics* **18**, 756–63 (1977)

[386] C. B. Chiu, E. C. G. Sudarshan and B. Misra, Time evolution of unstable quantum states and a resolution of Zeno's paradox, *Physical Review D* **16**, 520–9 (1977)

[387] L. E. Ballentine, *Quantum Mechanics* (Prentice-Hall, Englewood Cliffs, New Jersey, 1990)

[388] D. Home and M. A. B. Whitaker, Reflections on the quantum Zeno paradox, *Journal of Physics A* **19**, 1847–54 (1986)

[389] D. Home and M. A. B. Whitaker, The many-worlds and relative states interpretations of quantum mechanics and the quantum Zeno paradox, *Journal of Physics A* **20**, 3339–45 (1987)

[390] M. A. B. Whitaker, On Squires' many-views interpretation of quantum theory, *European Journal of Physics* **10**, 73–4 (1989)

[391] D. Home and M. A. B. Whitaker, A critical re-examination of the quantum Zeno paradox, *Journal of Physics A* **25**, 657–64 (1992)

[392] S. Inagaki, M. Namiki and T. Tajiri, Possible observation of the quantum Zeno effect by means of neutron spin-flipping, *Physics Letters A* **166**, 5–12 (1992)

[393] D. Home and M. A. B. Whitaker, A unified framework for quantum Zeno processes, *Physics Letters A* **173**, 327–31 (1993)

[394] S. Pascazio, M. Namiki, G. Badurek and H. Rauch, Quantum Zeno effect with neutron spin, *Physics Letters A* **179**, 155–60 (1993)

[395] W. M. Itano, D. J. Heinzen, J. J. Bollinger and D. J. Wineland, Quantum Zeno effect, *Physical Review A* **41**, 2295–300 (1990)

[396] R. J. Cook, What are quantum jumps?, *Physica Scripta* **T21**, 49–51 (1988)

[397] D. J. Wineland and W. M. Itano, Laser cooling, *Physics Today* **40** (6), 34–40 (1987)

[398] W. M. Itano, J. C. Bergquist and D. J. Wineland, Laser spectroscopy of trapped atomic ions, *Science* **237**, 612–17 (1987)

[399] A. Peres and A. Ron, Incomplete collapse and partial quantum Zeno effect, *Physical Review A* **42**, 5720–22 (1990)

[400] T. Petrovsky, S. Tasaki and I. Prigogine, Quantum Zeno effect, *Physics Letters A* **151**, 109–13 (1990)

[401] H. Nakazato, M. Namiki and S. Pascazio, Temporal behaviour of quantum mechanical systems, *International Journal of Modern Physics B* **10**, 247–95 (1996)

[402] D. Home and M. A. B. Whitaker, A conceptual analysis of quantum Zeno: paradox, measurement and experiment, *Annals of Physics* **258**, 237–85 (1997)

[403] P. Facchi and S. Pascazio, Quantum Zeno and inverse quantum Zeno effects, *Progress in Optics* **42**, 147–217 (2001)

[404] P. Facchi and S. Pascazio, Quantum Zeno subspaces, *Physical Review Letters* **89**, 080401 (2002)

[405] H. Nakazato, T. Takazawa and K. Yuasa, Purification through Zeno-like measurements, *Physical Review Letters* **90**, 060401 (2003)

[406] K. Koshino and A. Shimizu, Quantum Zeno effect for exponentially decaying systems, *Physical Review Letters* **92**, 030401 (2004)

[407] P. Facchi, D. A. Lidar and S. Pascazio, Unification of dynamical decoupling and the quantum Zeno effect, *Physical Review A* **69**, 032314 (2004)

[408] P. Facchi, H. Nakazoto and S. Pascazio, From the quantum Zeno to the inverse quantum Zeno effect, *Physical Review Letters* **86**, 2699–703 (2001)

[409] S. M. Roy, Quantum Zeno and anti-Zeno paradoxes, *Pramāṇa* **56**, 169–78 (2001)

[410] M. C. Fischer, B. Gutierrez-Medina and M. G. Raizen, Observation of the quantum Zeno and anti-Zeno effects in an unstable system, *Physical Review Letters* **87**, 040202 (2001)

[411] A. Luis, Zeno and anti-Zeno effects in two-level systems, *Physical Review A* **67**, 062113 (2003)

[412] J. Ruseckas and B. Kaulakys, General expression for the quantum Zeno and anti-Zeno effects, *Physical Review A* **69**, 032104 (2004)

[413] H. P. Stapp, Quantum theory and the role of mind in physics, *Foundations of Physics* **31**, 1465–99 (2001)

[414] H. Rauch, W. Treimer and U. Bonse, Test of a single crystal neutron interferometer, *Physics Letters A* **47**, 369–71 (1974)

[415] U. Bonse and M. Hart, An X-ray interferometer, *Applied Physics Letters* **6**, 155–6 (1965)

[416] S. A. Werner, Neutron interferometry, *Physics Today* **33** (12), 24–30 (1980)

[417] D. M. Greenberger, The neutron interferometer as a device for illustrating the strange behaviour of quantum systems, *Reviews of Modern Physics* **55**, 875–905 (1983)

[418] H. Rauch, Neutron interferometric tests of quantum mechanics, *Contemporary Physics* **27**, 345–60 (1986)

[419] H. Rauch, Neutron interferometry, *Science* **262**, 1384–5 (1993)

[420] J. Summhammer, G. Badurek, H. Rauch and U. Kischko, Explicit experimental verification of quantum spin-state superposition, *Physics Letters A* **90**, 110–2 (1982)

[421] C. Dewdney, The quantum potential approach to neutron interferometry experiments, *Physica B* **151**, 160–70 (1988)

[422] H. Rauch and J. P. Vigier, Proposed neutron interferometry test of Einstein's 'Einweg' assumption in the Bohr–Einstein controversy, *Physics Letters A* **151**, 269–75 (1990)

[423] T. Unnerstall, A comment on the Rauch–Vigier experiments on neutron interferometry, *Physics Letters A* **151**, 263–8 (1990)

[424] P. B. Lerner, Comment on [422], *Physics Letters A* **157**, 309–10 (1991)

[425] H. Rauch and J. P. Vigier, Reply to comment on [422], *Physics Letters A* **157**, 311–13 (1991)

[426] S. A. Werner, R. Clothier, H. Kaiser, H. Rauch and H. Wölwitsch, Spectral filtering in neutron interferometry, *Physical Review Letters* **67**, 683–6 (1991)

[427] J. P. Vigier, Possible test of the reality of superluminal phase waves and particle phase space motions in the Einstein–de Broglie–Bohm causal stochastic interpretation of quantum mechanics, *Foundations of Physics* **24**, 61–83 (1994)

[428] M. Zawisky, M. Baron and R. Loidl, Three-beam interference and which-way information in neutron interferometry, *Physical Review A* **66**, 063608 (2002)

[429] A. G. Wagh, Neutron interferometry was the first to correlate complementarity and 'which-way' information, *Physics Letters A* **259**, 81–2 (1999)

[430] H. Rauch and S. Werner, *Neutron Interferometry: Lessons in Experimental Quantum Mechanics* (Oxford University Press, Oxford, 2000)

[431] M. Hafner and J. Summhammer, Experiment on interaction-free measurement in neutron interferometry, *Physics Letters A* **235**, 563–8 (1997)

[432] Y. Hasegawa, R. Loidl, G. Badurek, M. Baron and H. Rauch, Violation of a Bell-like inequality in single-neutron interferometer experiments: quantum contextuality, *Journal of Modern Optics* **51**, 967–72 (2004)

[433] S. Basu, S. Bandyopadhyay, G. Kar and D. Home, Bell's inequality for a single spin-$\frac{1}{2}$ particle and quantum contextuality, *Physics Letters A* **279**, 281–6 (2001)

[434] A. J. Leggett, Macroscopic quantum systems and the quantum theory of measurement, *Progress of Theoretical Physics* (Supplement) **69**, 80–100 (1980)

[435] A. J. Leggett, Macroscopic quantum tunnelling, *Physics Today* **37**, S15–16 (1984)

[436] A. J. Leggett, Schrödinger's cat and her laboratory cousins, *Contemporary Physics* **25**, 583–98 (1984)

[437] A. J. Leggett and A. Garg, Quantum mechanics versus macroscopic realism: is the flux there when nobody looks?, *Physical Review Letters* **54**, 857–60 (1985)

[438] A. J. Leggett, Macroscopic quantum tunnelling and related matters, *Japanese Journal of Applied Physics* **26**, Supplement 3, 1986–93 (1987)

[439] A. J. Leggett, Experimental approaches to the quantum measurement paradox, *Foundations of Physics* **18**, 939–52 (1988)

[440] A. J. Leggett, Some thought-experiments involving macrosystems as illustrations of various interpretations of quantum mechanics, *Foundations of Physics* **29**, 445–56 (1999)

[441] A. J. Leggett, The significance of the MQC experiment, *Journal of Superconductivity* **12**, 683–7 (1999)

[442] A. J. Leggett, Testing the limits of quantum mechanics: motivation, state of play, prospect, *Journal of Physics: Condensed Matter* **14**, R415–51 (2002)

[443] A. J. Leggett, Probing quantum mechanics towards the everyday world: where do we stand?, *Physica Scripta* **T102**, 69–73 (2002)

[444] A. J. Leggett, The quantum measurement problem, *Science* **307**, 871–2 (2005)

[445] J. R. Friedman, V. Patel, W. Chen, S. K. Tolpygo and J. E. Lukens, Quantum superposition of distinct macroscopic states, *Nature* **406**, 43–6 (2000)

[446] C. H. van der Waal, A. C. J. ter Haar and six others, Quantum superposition of macroscopic persistent-current states, *Science* **290**, 773–7 (2000)

[447] M. Arndt, O. Nairz, J. Vos-Andreae, C. Keller, G. van der Zouw and A. Zeilinger, Wave–particle duality of C_{60} molecules, *Nature* **401**, 680–2 (1999)

[448] O. Nairz, M. Arndt and A. Zeilinger, Quantum interference experiments with large molecules, *American Journal of Physics* **71**, 319–25 (2003)

[449] D. D. Awschalom, D. P. DiVincenzo and J. F. Smyth, Macroscopic quantum effects in nanometer-scale magnets, *Science* **258**, 414–21 (1992)

[450] B. Julsgaard, A. Kozhekin and E. S. Polzik, Experimental long-lived entanglement of two macroscopic objects, *Nature* **413**, 400–3 (2001)

[451] A. Elby and S. Foster, Why SQUID experiments can rule out non-invasive measurability, *Physics Letters A* **166**, 17–23 (1992)

[452] L. Hardy, D. Home, E. J. Squires and M. A. B. Whitaker, Realism and the quantum-mechanical two-state oscillator, *Physical Review A* **45**, 4267–70 (1992)

[453] D. Home and M. A. B. Whitaker, Quantum Zeno effect: relevance for local realism, macroscopic realism and noninvasive measurability at the macroscopic level, *Physics Letters A* **239**, 6–12 (1998)

[454] A. M. Steane, Quantum computing, *Reports on Progress in Physics* **61**, 117–73 (1998)

[455] J. R. Brown, *Quest for the Quantum Computer* (Simon and Schuster, Riverside, New Jersey, 2001) [hardback: *Minds, Machines and the Multiverse: The Quest for the Quantum Computer* (2000)]

[456] C. P. Williams and S. H. Clearwater, *Ultimate Zero and One* (Copernicus, New York, 2000)

[457] R. W. Hamming, *Coding and Informtion Theory* (Prentice-Hall, Englewood Cliffs, New Jersey, 2nd edn, 1986).

[458] C. E. Shannon, A mathematical theory of communication, *Bell Systems Technical Journal* **27**, 379–423, 623–56 (1948)

[459] A. Hodges, *Alan Turing: The Enigma* (Random House, London, 1992)

[460] F. H. Hinsley and A. Stripp (eds.), *Code Breakers: The Inside Story of Bletchley Park* (Oxford University Press, Oxford, 1993)

[461] A. M. Turing, On computable numbers, with an application to the Entscheidungsproblem, *Proceedings of the London Mathematical Society* **42**, 230–65 (1936–7)

[462] A. Church, An unsolvable problem of elementary number theory, *American Journal of Mathematics* **58**, 345–63 (1936)

[463] A. Church, A note on the *Entscheidungsproblem*, *Journal of Symbolic Logic* **1**, 40–1 (1936) [correction on pp. 101–2]

[464] E. Post, Finite combinatory processes — formulation I, *Journal of Symbolic Logic* **1**, 103–5 (1936)

[465] G. T. Q. Hoare, Post, Turing and a 'kind of miracle' in mathematical logic, *Mathematics Today* **40**, 132–7 (2004)

[466] R. Landauer, Information is physical, *Physics Today* **44** (5), 23–9 (1991)

[467] A. J. G. Hey (ed.), *Feynman and Computation* (Perseus, Reading, Massachusetts, 1999)

[468] R. Landauer, Information is inevitably physical, in [467], pp. 77–92

[469] D. Deutsch, Quantum theory, the Church–Turing principle and the universal quantum computer, *Proceedings of the Royal Society A* **400**, 97–117 (1985)

[470] R. Penrose, *The Emperor's New Mind: Concerning Computers, Minds and the Laws of Physics* (Oxford University Press, Oxford, 1989)

[471] R. Penrose, *Shadows of the Mind: A Search for the Missing Science of Consciousness* (Oxford University Press, Oxford, 1994)

[472] R. Penrose, *The Road to Reality: A Complete Guide to the Laws of the Universe* (Cape, London, 2005)

[473] S. Hawking and R. Penrose, *The Nature of Space and Time* (Princeton University Press, Princeton, New Jersey, 1996)

[474] J. Mehra, *The Beat of a Different Drum: The Life and Science of Richard Feynman* (Oxford University Press, Oxford, 1994)

[475] J. Gleick, *Genius: The Life and Science of Richard Feynman* (Pantheon, New York, 1992)

[476] J. Gribbin and M. Gribbin, *Richard Feynman: A Life in Science* (Viking, London, 1997)

[477] R. P. Feynman, R. B. Leighton and M. Sands, *The Feynman Lectures on Physics* (3 Vols, Addison-Wesley, Reading, Massachusetts, 1965)

[478] R. P. Feynman, *Surely You're Joking, Mr. Feynman!* (Norton, New York, 1985)

[479] R. P. Feynman, *What Do You Care What Other People Think?* (Norton, New York, 1988)

[480] R. P. Feynman, There's plenty of room at the bottom, in [467], pp. 63–76

[481] E. Regis, *Nano!* (Bantam, New York, 1995)

[482] M. Minsky, Richard Feynman and cellular vacuum, in [467], pp. 117–30

[483] R. P. Feynman, *Feynman Lectures on Computation* (A. J. G. Hey and R. W. Allen, eds.) (Addison-Wesley, Reading, Massachusetts, 1996)

[484] A. J. G. Hey, Richard Feynman and computation, *Contemporary Physics* **40**, 257–65 (1999)

[485] R. P. Feynman, Simulating physics with computers, in [467], pp. 133–53; reprinted from *International Journal of Theoretical Physics* **21**, 467–88 (1982)

[486] P. Benioff, Quantum mechanical Hamiltonian models of Turing machines, *Journal of Statistical Physics* **29**, 515–46 (1982)

[487] R. P. Feynman, Quantum mechanical computers, in [483], pp. 185–211; reprinted from *Optics News* (February 1985), pp. 11–20; *Foundations of Physics* **16**, 507–32 (1986)

[488] C. H. Bennett, The thermodynamics of computation – a review, *International Journal of Theoretical Physics* **21**, 905–40 (1982); also in [483], pp. 283–318

[489] C. H. Bennett and R. Landauer, The fundamental limits of computation, *Scientific American* **253**, 38–46 (November 1985)

[490] C. H. Bennett, Demons, engines and the Second Law, *Scientific American* **257**, 88–96 (November 1987)

[491] H. S. Leff and A. F. Rex (eds.), *Maxwell's Demon 2: Entropy, Classical and Quantum Information, Computing* (Institute of Physics, Bristol, 2003) [A first edition, *Maxwell's Demon: Entropy, Information and Computing* (Adam Hilger, Bristol; Princeton University Press, Princeton, 1990) contains a different selection, including more early papers]

[492] L. Szilard, Über die Entropieverminderung in einem thermodynamischen System bei Eingriffen intelligenter Wesen, *Zeitschrift für Physik* **53**, 840–56 (1929) [Translated by A. Rapopor and M. Knol as On the decrease in entropy in a thermodynamic system by the intervention of intelligent beings, *Behavioral Science* **9**, 301–10 (1964); also in [91] on 539–48.]

[493] J. von Neumann, *Theory of Self-reproducing Automata* (A. W. Burks, ed.) (University of Illinois Press, Urbana, 1966)

[494] R. Landauer, Irreversibility and heat generation in the computing process, *IBM Journal of Research and Development* **5**, 183–91 (1961); also in [491], pp. 148–56

[495] C. H. Bennett, Logical reversibility of computation, *IBM Journal of Research and Development* **17**, 525–32 (1973)

[496] C. H. Bennett, Notes on the theory of reversible computation, *IBM Journal of Research and Development* **32**, 16–23 (1988); also in [491], pp. 327–34

[497] O. Penrose, *Foundations of Statistical Mechanics* (Pergamon, Oxford, 1970)

[498] E. Fredkin and T. Toffoli, Conservative logic, *International Journal of Theoretical Physics* **21**, 219–53 (1982)

[499] J. J. Hopfield, Feynman and computation, in [467], pp. 3–6

[500] B. Schumacher, Quantum coding, *Physical Review A* **51**, 2738–47 (1995)

[501] J. Stolze and D. Suter, *Quantum Computing: A Course from Theory to Experiment* (Wiley-VCH, Weinheim, 2004)

[502] A. M. Steane, A quantum computer only needs one universe, *Studies in the History and Philosophy of Modern Physics* **34**, 469–78 (2003)

[503] D. Deutsch and R. Jozsa, Rapid solution of problems by quantum computation, *Proceedings of the Royal Society A* **439**, 552–8 (1992)

[504] R. Cleve, A. Ekert, C. Macchiavello and M. Mosca, Quantum algorithms revisited, *Proceedings of the Royal Society A* **453**, 339–54 (1998)

[505] P. W. Shor, Polynomial-time algorithms for prime factorisation and discrete logarithms on a quantum computer, in S. Goldwasser (ed.), *Proceedings of the 35th Annual Symposium on Foundations of Computer Science* (IEEE, Piscataway, New Jersey, 1994), pp. 124–49; also in *SIAM Journal of Computing* **26**, 1484–509 (1997).

[506] A. Barenco, Quantum physics and computers, *Contemporary Physics* **37**, 375–89 (1996)

[507] S. Singh, *The Code Book: The Science of Secrecy from Ancient Egypt to Quantum Cryptography* (Fourth Estate, London, 1999)

[508] A. Ekert and R. Jozsa, Quantum computation and Shor's factoring algorithm, *Reviews of Modern Physics* **68**, 733–53 (1996)

[509] R. J. Hughes, Quantum computation, in [467], pp. 191–221

[510] L. K. Grover, A fast quantum mechanical algorithm for database search, in *Proceedings of the 28th Annual Symposium on the Theory of Computation* (ACM Press, New York, 1996), pp. 212–9

[511] L. K. Grover, Quantum mechanics helps in searching for a needle in a haystack, *Physical Review Letters* **79**, 325–8 (1997)

[512] G. Brassard, Searching a quantum phonebook, *Science* **275**, 627–8 (1997)

[513] M. B. Plenio and P. L. Knight, Realistic lower bounds for the factorization time of large numbers on a quantum computer, *Physical Review A* **53**, 2986–90 (1996)

[514] S. Haroche and J.-M. Raimond, Quantum computing: dream or nightmare?, *Physics Today* **49** (8), 51–2 (1996)

[515] R. Landauer, Is quantum mechanics useful?, *Philosophical Transactions of the Royal Society A* **353**, 367–76 (1995)

[516] P. W. Shor, Scheme for reducing decoherence in quantum computer memory, *Physical Review A* **52**, R2493–6 (1995)

[517] A. M. Steane, Error correcting codes in quantum theory, *Physical Review Letters* **77**, 793–7 (1996)

[518] R. W. Hamming, Error detection and error correcting codes, *Bell Systems Technical Journal* **29**, 147–60 (1950)

[519] W. H. Wooters and W. H. Zurek, A single quantum cannot be cloned, *Nature* **299**, 802–3 (1982)

[520] D. Dieks, Communication by EPR devices, *Physics Letters A* **92**, 271 (1982)

[521] D. P. DiVincenzo and B. Terhal, Decoherence: the obstacle to quantum computation, *Physics World* **11** (3), 53–7 (1998)

[522] A. Luis, Quantum state preparation and control via the Zeno effect, *Physical Review A* **63**, 052112 (2001)

[523] P. Facchi, S. Tasaki, S. Pascazio, H. Nakazato, A. Tokuse and D. A. Lidar, Control of decoherence: analysis and comparison of three different strategies, *Physical Review A* **71**, 022302 (2005)

[524] S. L. Braunstein (ed.), *Quantum Computing: Where Do We Want to Go Tomorrow* (Wiley VCH, Berlin, 1999)

[525] S. L. Braunstein and H.-K. Lo (eds.), *Scalable Quantum Computers: Paving the Way to Realization* (Wiley-VCH, Berlin, 2001)

[526] D. P. DiVincenzo, The physical implementation of quantum computation, in [525], pp. 1–13

[527] N. A. Gershenfeld and I. L. Chuang, Bulk spin-resonance quantum computation, *Science* **275**, 350–6 (1997)

[528] L. M. K. Vanderspyen, M. Steffen, G. Breyta, C. S. Yannoni, M. H. Sherwood and I. L. Chuang, Experimental realisation of Shor's quantum factoring algorithm using nuclear magnetic resonance, *Nature* **414**, 883–7 (2001)

[529] J. A. Jones, Quantum computers get real, *Physics World* **15** (4), 21–2 (2002)

[530] J. A. Jones, Robust quantum information processing with techniques from liquid state NMR, *Philosophical Transactions of the Royal Society A* **61**, 1429–40 (2003)

[531] S. L. Braunstein, C. M. Caves, R. Jozsa, N. Linden, S. Popescu and R. Schack, Separability of very noisy mixed states and implications for NMR quantum computing, *Physical Review Letters* **83**, 1054–7 (1999)

[532] J. A. Jones, NMR quantum communication: a critical evaluation, in [525], pp. 139–54

[533] G. Z. K. Horvath, R. C. Thompson and P. L. Knight, Fundamental physics with trapped ions, *Contemporary Physics* **38**, 25–48 (1997)

[534] D. Leibfried, R. Blatt, C. Monroe and D. Wineland, Quantum dynamics of single trapped ions, *Reviews of Modern Physics* **75**, 281–324 (2003)

[535] C. Townsend, W. Ketterle and S. Stringari, Bose–Einstein condensation, *Physics World* **10** (3), 29–34 (1997)

[536] J. I. Cirac and P. Zoller, Quantum computation with cold trapped ions, *Physical Review Letters* **74**, 4091–4 (1995)

[537] C. Monroe, D. M. Meekhof, B. E. King, W. M. Itano and D. J. Wineland, Demonstration of a fundamental logic gate, *Physical Review Letters* **75**, 4714–17 (1995)

[538] B. E. King, C. S. Wood and six others, Cooling the collective motion of trapped ions to initialize a quantum register, *Physical Review Letters* **81**, 1525–8 (1998)

[539] Q. A. Turchette, C. S. Wood and six others, Deterministic entanglement of two trapped ions, *Physical Review Letters* **81**, 3631–4 (1998)

[540] F. Schmidt-Kaler, H. Haffner and ten others, Realization of the Cirac–Zoller controlled-NOT quantum gate, *Nature* **422**, 408–11 (2003)

[541] D. Kielpinski, C. Monroe and D. J. Wineland, Architecture for a large-scale ion-trap quantum computer, *Nature* **417**, 709–11 (2002)

[542] D. Leibfried, B. Demarco and twelve others, Quantum information with trapped ions at NIST, *Journal of Modern Optics* **50**, 1115–29 (2003)

[543] D. J. Wineland, M. Barrett and ten others, Quantum information processing with trapped ions, *Philosophical Transactions of the Royal Society A* **361**, 1349–61 (2003)

[544] T. Schaetz, D. Leibfried and eight others, Towards a scalable quantum computer/simulator based on trapped ions, *Applied Physics B – Lasers and Optics* **79**, 979–86 (2004)

[545] W. Tittel, G. Ribordy and N. Gisin, Quantum cryptography, *Physics World* **11** (3), 41–5 (1998)

[546] N. Gisin, G. G. Ribordy, W. Tittel and H. Zbinden, Quantum cryptography, *Reviews of Modern Physics* **74**, 145–95 (2002)

[547] R. J. Hughes, D. M. Alde, P. Dyer, G. G. Luther, G. L. Morgan and M. Schauer, Quantum cryptography, *Contemporary Physics* **36**, 149–65 (1995)

[548] S. J. D. Phoenix and P. D. Townsend, Quantum cryptography – how to beat the code breakers using quantum mechanics, *Contemporary Physics* **36**, 165–95 (1995)

[549] C. H. Bennett, G. Brassard, S. Breidbart and S. Wiesner, Quantum cryptography, or unforgeable subway tokens, *Advances in Cryptology. Proceedings of Crypto 82* (Plenum, New York, 1982), pp. 267–75

[550] C. H. Bennett and G. Brassard, Quantum cryptography: public-key distribution and coin tossing, *Proceedings of 1984 IEEE International Conference on Computers, Systems and Signal Processing* (IEEE, New York, 1984), pp. 175–9.

[551] C. A. Fuchs, N. Gisin, R. B. Griffiths, C. S. Niu and A. Peres, Optimal eavesdropping in quantum cryptography: 1. Information bound and optimal strategy, *Physical Review A* **56**, 1163–72 (1997)

[552] R. B. Griffiths and C. S. Niu, Optimal eavesdropping in quantum cryptography: 2. A quantum circuit, *Physical Review A* **56**, 1173–6 (1997)

[553] M. Williamson and V. Vedral, Eavesdropping on practical quantum cryptography, *Journal of Modern Optics* **50**, 1989–2011 (2003)

[554] I. Devetak and A. Winter, Relating quantum privacy and quantum coherence: an operational approach, *Physical Review Letters* **93**, 080501 (2004)

[555] A. K. Ekert, Quantum cryptography based on Bell's theorem, *Physical Review Letters* **67**, 661–3 (1991)

[556] C. H. Bennett and G. Brassard, The dawn of a new era for quantum cryptography: the experimental prototype is working!, *Sigact News* **20**, 78–82 (1989)

[557] A. Muller, H. Zbinden and N. Gisin, Quantum cryptography over 23 km in installed under-lake telecom wire, *Europhysics Letters* **33**, 335–9 (1996)

[558] W. Tittel, J. Brendel, H. Zbinden and N. Gisin, Quantum cryptography using entangled photons in energy–time Bell states, *Physical Review Letters* **84**, 4737–40 (2000)

[559] R. T. Thew, S. Tanzilli, W. Tittel, H. Zbinden and N. Gisin, Experimental investigation of the robustness of partially entangled qubits over 11 km, *Physical Review A* **66**, 062304 (2002)

[560] K. J. Gordon, V. Fernandez, P. D. Townsend and G. S. Buller, A short wavelength gigahertz clocked fiber-optic quantum key distribution system, *IEEE Journal of Quantum Electronics* **40**, 900–8 (2004)

[561] W. T. Butler, R. J. Hughes and seven others, Practical free-space quantum key distribution over 1 km, *Physical Review Letters* **81**, 3283–6 (1998)

[562] R. J. Hughes, J. E. Nordholt, D. Derkacs and C. G. Peterson, Practical free-space quantum key distribution over 10 km in daylight and at night, *New Journal of Physics* **4**, article no. 43 (2002) [on-line journal]

[563] C. H. Bennett, G. Brassard, C. Crepeau, R. Jozsa, A. Peres and W. K. Wootters, Teleporting an unknown quantum state via dual classical and Einstein–Podolsky–Rosen channels, *Physical Review Letters* **70**, 1895–9 (1993)

[564] S. L. Braunstein, A. Mann and M. Revzen, Maximal violation of Bell inequalities for mixed states, *Physical Review Letters* **68**, 3259–61 (1992)

[565] D. Boschi, S. Branca, F. De Martini, L. Hardy and S. Popescu, Experimental realization of teleporting an unknown pure quantum state via dual classical and Einstein–Podolsky–Rosen channels, *Physical Review Letters* **80**, 1121–5 (1998)

[566] A. Furusawa, J. L. Sorensen, S. L. Braustein, C. A. Fuchs, H. J. Kimble and E. S. Polzik, Unconditional quantum teleportation, *Science* **282**, 706–9 (1998)

[567] I. Marcikic, H. de Riedmatten, W. Tittel, H. Zbinden and N. Gisin, Long-distance teleportation of qubits at telecommunication wavelengths, *Nature* **421**, 509–13 (2003)

Index

(Names in references are indexed only if there is no mention in the main text.)

Printed in the United States
By Bookmasters